The Fukushima Daiichi Nuclear Power Station Disaster

When the Nuclear Safety Commission in Japan reviewed safety-design guidelines for nuclear plants in 1990, the regulatory agency explicitly ruled out the need to consider prolonged AC power loss. In other words, nothing like the catastrophe at the Fukushima Daiichi Nuclear Power Station was possible—no tsunami of 45 feet could swamp a nuclear power station and knock out its emergency systems. No blackout could last for days. No triple meltdown could occur. Nothing like this could ever happen. Until it did—over the course of a week in March 2011.

In this volume and in gripping detail, the Independent Investigation Commission on the Fukushima Nuclear Accident, a civilian-led group, presents a thorough and powerful account of what happened within hours and days after this nuclear disaster, the second worst in history. It documents the findings of the Chairman, five commissioners, and a working group of more than thirty people, including natural scientists and engineers, social scientists and researchers, business people, lawyers, and journalists, who researched this crisis involving multiple simultaneous dangers. They conducted over 300 investigative interviews to collect testimony from relevant individuals. The responsibility of this committee was to act as an external ombudsman, summarizing its conclusions in the form of an original report, published in Japanese in February 2012. This has now been substantially rewritten and revised for this English-language edition.

The work reveals the truth behind the tragic saga of the multiple catastrophic accidents at the Fukushima Daiichi Nuclear Power Station. It serves as a valuable and essential historical reference, which will help to inform and guide future nuclear safety and policy in both Japan and internationally.

By the Independent Investigation Commission on the Fukushima Nuclear Accident.

Edited by Mindy Kay Bricker.

Published in association with the *Bulletin of the Atomic Scientists* and the Rebuild Japan Initiative Foundation.

In a subject area beset by controversy and seemingly interminable disagreements, The Bulletin has long served to provide highly reliable analyses of things nuclear. The Fukushima Daiichi disaster is a case in point, with numerous—often conflicting—reports on what went wrong and what went right, who did what, and who did not… and this book does a great job of distilling the principal issues, and providing a clear-headed analysis of what actually transpired. While there is still much to learn about, and from, this nuclear disaster, the interested reader will not find a better current analysis, and I urge all with a stake in learning how to do better in the nuclear power arena to read this book.

Robert Rosner, former director of Argonne National Laboratory, William E. Wrather Distinguished Service Professor, Astronomy and Astrophysics and Physics at University of Chicago, and Co-Director, Energy Policy Institute at Chicago, USA.

The Fukushima Daiichi Nuclear Power Station Disaster

Investigating the Myth and Reality

By the Independent Investigation Commission on the Fukushima Nuclear Accident

Edited by Mindy Kay Bricker

Published in association with the *Bulletin of the Atomic Scientists* and the Rebuild Japan Initiative Foundation

HV
555
.J3
N556213
2014

First published 2014
by Routledge
2 Park Square, Milton Park, Abingdon, Oxon OX14 4RN

And by Routledge
711 Third Avenue, New York, NY 10017

Routledge is an imprint of the Taylor & Francis Group, an informa business

© 2014 Bulletin of the Atomic Scientists

All rights reserved. No part of this book may be reprinted or reproduced or utilised in any form or by any electronic, mechanical, or other means, now known or hereafter invented, including photocopying and recording, or in any information storage or retrieval system, without permission in writing from the publishers.

Trademark notice: Product or corporate names may be trademarks or registered trademarks, and are used only for identification and explanation without intent to infringe.

British Library Cataloguing-in-Publication Data
A catalogue record for this book is available from the British Library

Library of Congress Cataloging-in-Publication Data
The Fukushima Daiichi Nuclear Power Station disaster : investigating the myth and reality / by the Independent Investigation Committee on the Fukushima Daiichi Nuclear Accident ; edited by Mindy Kay Bricker.
  pages cm
"Published in association with the Bulletin of Atomic Scientists and the Rebuild Japan Initiative Foundation."
Includes bibliographical references and index.
 1. Disaster relief--Japan--Fukushima-ken. 2. Fukushima Nuclear Disaster, Japan, 2011. 3. Nuclear power plants--Accidents--Investigation--Japan--Fukushima-ken. I. Bricker, Mindy Kay, editor. II. Nihon Saiken Inishiatibu. Fukushima Genpatsu Jiko Dokuritsu Kensho Iinkai. Fukushima Genpatsu Jiko Dokuritsu Kensho Iinkai chosa kensho hokokusho. Based on (work):
 HV555.J3N556213 2014
 363.17'990952117--dc23
                                    2013043787

ISBN: 978-0-415-71393-1 (hbk)
ISBN: 978-0-415-71396-2 (pbk)
ISBN: 978-1-315-88280-2 (ebk)

Typeset in Bembo
by Saxon Graphics Ltd, Derby

Printed and bound in Great Britain by
TJ International Ltd, Padstow, Cornwall

Contents

Letter from Koichi Kitazawa	vii
Letter from Yoichi Funabashi	xiv
Points of Investigation	xxi
List of Members of the Commission and the Working Group for the Independent Investigation Commission on the Fukushima Nuclear Accident	xxiii
List of Acronyms and Abbreviations	xxvi
Figures	xxviii
Foreword by Kennette Benedict	xli
Prologue	xliv

1	A Fukushima Diary, March 11–16, 2011	1
2	Nuclear Energy Development in Japan	36
3	The Safety Myth	50
4	Actors in Japanese Nuclear Safety Governance	63
5	International Safety	86
6	Accident Preparedness and Operation	95
7	Impact of Radioactive Material Released into the Environment	124
8	Communicating the Fukushima Disaster	145
9	US–Japan Relationship	158

10 Lessons of the Fukushima Daiichi Nuclear Power Station
   Accident and the Quest for Resilience   175

   Epilogue   196
   Expert Reviews   204
   References   229
   Index   239

Letter from Koichi Kitazawa, Chairman of the Independent Investigation Commission on the Fukushima Nuclear Accident

On March 11, 2011, as the accident at the Fukushima Daiichi Nuclear Power Station unfolded, it was only a matter of time before the key issues behind the catastrophe were made clear: namely, the overcrowded site layout and the proliferation of catastrophic dangers. At the Fukushima Daiichi station, six nuclear reactors and seven spent fuel pools were positioned in close proximity to one another. Operators at the site, facing a nightmare scenario of unreliable readings from water-level meters, pressure gauges, and other measurement equipment, were forced to oversee increasingly dangerous situations—playing out at the same time—in multiple reactors and multiple spent fuel pools. The dangers seemed to multiply at every step. While radiation levels increased, debris from explosions in units 1 and 3 flew through the air. Not only were the equipment and facilities damaged, but workers' efforts to address the situation in the reactors and spent fuel storage pools were stymied by the sheer impossibility of the situation.

At the start of our investigation in 2011, we discovered that the prime minister and several top Cabinet officials had feared serious high-risk dangers during the accident—dangers they had not at all disclosed to the public until we published our original Japanese-language report in March 2012; we suspect, however, that much of this information remained largely unreported in the English language. In the later chapters of this book, we describe the threats that kept these government officials terrified for an extended period of time, including the rising pressure in unit 2's reactor containment vessel, which, they knew, could lead to a possible explosion that would instantly release huge quantities of radioactive elements into the atmosphere. Furthermore, because of the hydrogen explosion that occurred in the building of unit 4—a unit that was not even in operation at the time of the earthquake—they further feared that unit 4's spent fuel pool had been exposed directly to the environment. Shunsuke Kondo, chairman of the Japan Atomic Energy Commission, reported to Prime Minister Naoto Kan and other officials what he identified as the "worst-case scenario": If the situation worsened, he said, citizens within a 200-kilometer radius, or greater, would need to evacuate—that is, as many as thirty million people would need to leave their communities, including many in the Tokyo area.

Sharing information at a time of danger: the Cabinet's instructions to site workers and the sense of elite panic

Heeding seismologists' warnings of the possibility of a large earthquake in the Tonankai region, in the southeast, to follow the Tohoku earthquake, in the northeast, the government plunged immediately into accident-response activities. Then-Prime Minister Kan paid an unexpected visit to the Fukushima Daiichi plant on the morning of March 11; he later pronounced, at the Tokyo Electric Power Company (TEPCO) headquarters, that he would not allow the company to withdraw from the nuclear plant, insisting that water be dumped over the nuclear reactor buildings using helicopters from Japan's Self Defense Force. At the government's request, over one and a half months after the accident at Fukushima, Chubu Electric Power shut down its Hamaoka nuclear plant located southwest of Tokyo. All of these incidents can only be understood in the context of the strong fear growing within the Cabinet Office: that "Japan just might not survive" with multiple nuclear reactors crammed into the affected area.

The 2011 accident at the Fukushima Daiichi Nuclear Power Station exposed inadequacies in Japan's information-sharing systems. In particular, when the accident first occurred, information was insufficiently transferred among the Fukushima Daiichi station, TEPCO headquarters, the Nuclear and Industrial Safety Agency (NISA), the Nuclear Safety Commission, and, ultimately, to the Japanese government—creating a climate of fear and doubt. During the course of our investigation, we also learned of instances in which information was concealed due to fits of what we might term "elite panic," when various layers of government, fearful of inciting panic among the general public, refused to pass on critical pieces of information. Examples include the government's failure to disclose to the public data on radioactive contamination distribution predicted by the System for Prediction of Environment Emergency Dose Information (known as SPEEDI), as well as the government's decision not to inform the public of Kondo's worst-case scenario.

In the face of grave danger, what is the best way to disclose information to the public? To whom does information belong? Does the public have a right to know, or is it acceptable for supervisors at various levels to decide independently to keep things hidden? These questions dogged both the Japanese government and TEPCO.

In light of this, it is essential that the government learn from the Fukushima accident—during which, at times, there were debilitating communication problems—and construct systems and technologies to communicate and share large amounts of real-time information and data. Japan must work hard on constructing an organizational infrastructure to allow unfettered information sharing in times of crisis. The nation needs to design systems that allow both information sharing via multiple channels throughout various layers of government, vertically and horizontally, as well as timely information sharing with foreign governments.

After our 2012 report was published, the Japanese government and the Diet, in September of that year, created the Nuclear Regulation Authority, a single agency to regulate the nuclear power industry, and thereby abolished NISA and the Nuclear Safety Commission. The law that established the Nuclear Regulation Authority directs that it exists as an external agency of the Ministry of the Environment and that the Diet assumes responsibility to assign five top professionals to lead the authority. Such affiliation of the new regulation authority is a stark contrast between the one with its predecessor, the Nuclear and Industrial Safety Agency, and the promoter of nuclear power in Japan, the Ministry of Economy, Trade, and Industry.

But still more needs to be done.

Atrophy of Japan's nuclear safety assurance framework

During the course of our investigation, it became clear that Japan's framework for ensuring safety at nuclear plants had suffered a systematic process of atrophy, to the point of resembling nothing so much as a paper tiger. Nothing is more emblematic of this phenomenon than the widespread persistence of the "myth of safety," which is said to have been constructed to win over the local populations near the sites of nuclear plants. But at some point, the nuclear power advocates, themselves, began to believe in, and to be bound by, their own myth of safety, and a culture emerged within which it was taboo to suggest that safety could ever be further improved. Both electric power companies and nuclear reactor manufacturers somehow believed that improving the safety of something that is already safe enough is a logical impossibility, and they became unable to conceptualize the notion of improving safety, either by enhancing existing procedures or by instituting new ones. The phrase "safety improvements" even disappeared from proposal documents distributed to power companies from manufacturers, and the idea of modifying design specifications "for safety purposes" fell into disrepute.

An excellent illustration of the culture pervading the nuclear power community is offered by the Nuclear Safety Commission's policy philosophy: namely, the notion that the total loss of AC power for an extended length of time was a contingency that need not be contemplated, because power lines would be restored or emergency AC generators would come on line in such a scenario. How could the Nuclear Safety Commission, whose very mission is to ensure the highest possible safety levels, have designed policies built atop such fallacious reasoning? Because of this philosophy, power companies grew somewhat lazy when it came to preparing for severe accidents. This phenomenon can only be interpreted as a craven attempt by the safety commission to sacrifice safety in order to lessen the burden on the power companies. "There is no question that TEPCO had a strong influence in the Nuclear Safety Commission at that time," said one high-level government official who had supported nuclear power. "Of course, once policies like this were put in place, it became essentially impossible for anybody to say, 'This is wrong!' or to attempt to make any amendments."

The United States and Europe—in response to safety concerns after the accidents in 1979 at Three Mile Island and in 1986 at Chernobyl, as well as security concerns after the multiple terrorist attacks on September 11, 2001—have, over the past decades, adopted various protection measures against severe accidents, including some minor upgrades in technology such as new types of sensors and improved vent valves with elongated shafts. However, the Japanese government, together with Japanese power companies, ignored most of these new protections and, consequently, were left with inadequate protections against serious accidents. The fact that a nation such as Japan—where the probability of massive earthquakes is dozens of times higher than the world average—had been taking such an attitude toward the incorporation of new safety features can only be viewed by the international community as greatly shameful.

In the course of this investigation, we spoke with former high-ranking government officials experienced in nuclear safety, as well as members of TEPCO's management team, who all used different words to say essentially the same thing: "There was awareness that safety precautions were problematically insufficient. However, there was nothing that I, as an individual person, could do to make a difference; it would be like trying to dam a flood with a toothpick." With the myth of safety gradually building up over time and crowding out all other points of view, all those interviewed described the same general attitude: "Nobody wanted to say anything that would cause trouble, especially to their affiliated organizations. Instead, people just went with the flow." Some would understand this attitude as a reflection of certain unique attributes of Japanese society. But if Japan is a society in which "reading the atmosphere" has become a fundamental survival skill, then it is a society that has proven itself incapable of safely operating the complex, high-risk, large-scale technology, such as nuclear power generation.

Nuclear power community

During one of our interviews with a former high-level official at the Ministry of Economy, Trade, and Industry (METI), he painted a vivid portrait of the relationship between his ministry, which regulated the nuclear power industry, and the businesses it regulated: "You see, TEPCO is a company that flat-out rejects any requests by independent power-generation companies for permission to use TEPCO's transmission lines to transmit their power." He went on to explain that TEPCO uses government regulations as an excuse for this.

> They [TEPCO] say, "Well, we don't have any problems with it ourselves, but unfortunately it can't be done because of government regulations"; and in so doing, they maintain a monopolistic stranglehold over the system. We were basically on our hands and knees trying to regulate and provide guidance to these people. In the mid-'90s, we revised our regulations. There was a perception that this was done in order to improve the situation with the Agency for Natural Resources and Energy, which was controlled by TEPCO ... We were supposed to be regulating them, but in fact they were just using us as a tool. I'm not saying that the Nuclear and Industrial Safety Agency can't stand up to TEPCO, but, as a matter of fact, NISA can neither stand up to TEPCO nor properly act as its regulator.

Safety regulations must be able to exist in substantive counterpoint to the proponents of nuclear power. Without fundamental legislative and organization reforms to destroy the excessively comfortable relationship that has developed in this community, it will be extremely difficult to ensure the safety of nuclear power in Japan.

In addition, our investigation revealed that the "nuclear power village" contains many different types of structural collusion: politicians from both majority and opposition parties receive contributions from electric power company executives and labor unions; mass media companies receive enormous advertising support from electric power companies; nuclear power researchers receive huge research grants from electric power companies; former government officials pass through a revolving door into electric power companies or organizations working on nuclear power issues; electric power companies dispatch employees to government agencies or organizations working on nuclear power issues; the national government provides assistance to cultural institutions and school-teacher groups to support nuclear-power-friendly education for children; regional governments receive subsidies from the national government; power companies donate money to regional governments for cultural facilities and other infrastructure projects. Thus, through these and other means, a full-scale "village"—a giant, interwoven, interdependent community—is created, one that possesses inertia for change.

It is clear that, no matter how much Japan might try to establish regulatory bodies or safety-assurance committees within this village, before long these groups inevitably fall back into their familiar back-scratching, business-as-usual business patterns. Fundamental reforms are needed in legislation, operational systems, and organizational structures, but even these alone will not suffice;

even more fundamental organizational restructuring is needed to ensure a pervasive spirit of "civilian control"—that is, the ability to attract a continuous stream of people from *outside* the village to serve in positions of importance within it.

Building a safer country

Of the approximately 20,000 people who perished in the Tohoku earthquake disaster, 93 percent are estimated to have died from the tsunami. To date, the nuclear accident has not been directly responsible for any deaths, but today—more than three years later—more than 150,000 people continue to live as refugees from radioactive contamination. On top of this, many hundreds of thousands of people live with daily anxiety over the unknown effects, both today and long into the future, of radioactive contamination.

News of the Tohoku earthquake disaster and its aftermath was broadcast around the globe, and so the plight of the disaster evacuees garnered attention not only within Japan, but from the wider world as well. The strong Japanese spirit of community, which binds the nation's people together and motivates the citizenry to care tirelessly for one another without succumbing to panic, gives hope for the nation's future rebirth and recovery. A great many foreign countries sent the Japanese population heartwarming words of encouragement, massive financial assistance, assets to assist in its recovery, as well as emergency rescue teams. With this report, we, the members of this investigation commission, extend our deepest gratitude to the foreign countries that rushed to assist our nation in that time of need. The world's sympathies and encouragement will linger in the hearts of the Japanese people and have provided invaluable spiritual sustenance to stoke our nation's recovery.

Nonetheless, the tragic accident at TEPCO's Fukushima Daiichi Nuclear Power Station released large quantities of radioactive material that contaminated the nation's air and water. The Japanese government, though admittedly challenged by a crisis involving multiple simultaneous dangers, was delinquent in notifying other countries of radiation leaks. For this, on behalf of the entire Japanese population, we extend our apologies to the world.

Our report documents the findings of a working group of some thirty people, both young and mid-career professionals, including natural scientists and engineers, social scientists and researchers, business people, lawyers, and journalists, who researched the following chapters by gathering resources and conducting investigative interviews to collect testimony from relevant individuals. The responsibility of this committee was to act as an external ombudsman, summarizing its conclusions in the form of our original report, published in Japanese in February 2012—and now in the form of this English-language translation—to share with the world.

Needless to say, a civilian accident-investigation committee is entirely devoid of any official authority. We extend our deep gratitude to all sources who voluntarily granted us an interview based on their professional responsibility and conscience as an individual. By elucidating the fuller truths behind the tragic saga of the multiple catastrophic accidents at the Fukushima Daiichi Nuclear Power Station, we hope that our 2012 report in Japanese and our 2014 report in English extract useful lessons to ensure a better future not only for the children of our nation, but for future generations everywhere.

Koichi Kitazawa
Chairman of the Independent Investigation
Commission on the Fukushima Nuclear Accident

Letter from Yoichi Funabashi, Program Director of the Independent Investigation Commission on the Fukushima Nuclear Accident

In late March 2011, news started to spread that the Japanese government was planning to form a commission to investigate the Fukushima accident. Though ostensibly good news, I could not help but feel that a single government-organized commission would be insufficient. If solely the government's investigation committee—composed of members chosen by the government itself—were to produce reports on this nuclear disaster, the citizens of Japan and of the world would naturally find it difficult to read them without suspicion, regardless of the quality. Independent perspectives are also needed. After all, there are many approaches to expose the full truth of a tragedy, and each approach teases out necessary but different information. Undoubtedly, the government's commission would frame its investigation to address the question of how the government performed when faced with crisis. It is an important question, yet it is not the only one. Equally crucial, in the case of the Fukushima accident, is to reflect on past and present government and bureaucratic policies, nuclear safety regulation systems, and the vested interests of politicians—not to mention the social awareness and organizational tendencies in relation to these issues.

For Japanese citizens, the government's handling of this event was a case of déjà vu: That is, historically, Japan has slacked in its duty to establish a formal covenant with its citizens to investigate problems and find workable future solutions. After the Sino-Japanese War and World War II, for example, the government did not convene an investigation commission, nor did the Diet attempt to review and examine the background, causes, or responsibility for the disastrous events that befell Japan. Instead, this work was, and continues to be, bravely carried out by civilian researchers. Similarly, unanswered questions persist on the twenty-year deflation that followed the end of the Cold War and the consequential deterioration of Japan's national strength; the precise root of the problem, either in government administrations or policies, remains unknown even today.

The mere thought of the Fukushima accident meeting the same fate filled me with great dread in the days and weeks after the nuclear disaster. In terms of Fukushima, both the government and the Diet must identify the most serious errors in government policy and in the nation's disaster-response framework—and must clearly present these findings to the Japanese people. Thus, in late March 2011, I set out to form a private commission to investigate and analyze Japan's nuclear and electric power companies, along with its nuclear power industry as a whole. All of these entities belong to a massive system known as the "nuclear power village," the pervasive influence of which seeps into politics, government agencies, universities, legislation, and even the media. Needless to say, within each of these existing organizational frameworks are many institutions doing outstanding research while managing the herculean task of shrugging off the influence of the nuclear power village. But I, personally, wanted an independent commission to investigate this issue without officially collaborating or partnering with existing universities, foundations, or think tanks—therein lies a path littered with far too many unknowns and potential obstacles. To me, the surest and most pristine way to accomplish such an investigation was to start with a clean slate and build a new, independent think tank.

At the outset, I partnered with Masaakira James Kondo, Twitter's managing director for East Asia, and Takeshi Niinami, president of Lawson, Inc. Together, the three of us recruited supporters of our cause and established our foundation, the Rebuild Japan Initiative Foundation, by the end of September 2011. To officially launch the Fukushima investigation—the first initiative to be sponsored by the foundation—we needed first to establish a commission, today known as the Independent Investigation Commission on the Fukushima Nuclear Accident. Koichi Kitazawa, former president of the Japan Science and Technology Agency, generously agreed to serve as our chairman. Shortly thereafter, we established a seven-member executive committee, and our first executive committee meeting convened on October 14, 2011.

The members of our executive committee are among Japan's finest leaders and experts in their respective fields. Moreover, each member is a prominent public intellectual with deep involvement in work undertaken on behalf of Japanese society and the public good, in realms not limited to their own areas of expertise. Though the executive committee was originally established with seven members, Kiyoshi Kurokawa, academic fellow at the National Graduate Institute for Policy Studies, stepped down in the early stages when he was appointed to chair the Fukushima Nuclear Accident Independent Investigation Commission of the National Diet of Japan.

Under the guidance of the executive members, the investigation commission created a working group to execute the actual investigation. This working group was composed of roughly thirty members—scholars, academics, lawyers, and journalists—specializing in various fields such as nuclear engineering, philosophy of science and technology, scientific administration, sociology, political science, regional government, energy policy, risk management, disaster prevention, international politics, nuclear disarmament, and the nonproliferation of nuclear weapons. The group met for its preparatory meeting on August 27 and subsequently gathered for an additional ten meetings, each of which were held on Saturdays for eight hours; during these meetings, members immersed themselves thoroughly in debate and discussion.

Throughout the investigation and in the compilation of this report, the commission and working group adopted one simple motto: *Truth, Independence, Humanity*. And this continues to be the Rebuild Japan Initiative Foundation's motto to this very day.

Throughout the course of the investigation, the commission largely turned to the public for insight on how citizens felt the Tokyo Electric Power Company (TEPCO) and the government behaved during the accident; we not only asked people to provide relevant details, but we asked them to contribute their true, unfiltered experiences of the accident. This inspired us to launch a website called "Have Your Say," by which we could collect such firsthand accounts. The prologue of this book is an example of one such narrative collected through this portal. It is the actual experience of a Fukushima Daiichi Nuclear Power Station worker, who, on the afternoon of March 11—in the immediate aftermath of the earthquake and tsunami, after the station blackout, after the extreme pressure change, and in the midst of the possible threat of a hydrogen explosion—worked at the key anti-seismic building of the power station. Upon discovering this story, the commission sent one of its working group members to interview this worker personally; he graciously provided his consent to use his account in this book.

The executive committee members analyzed these oral histories and extrapolated data to establish a framework to distinguish among immediate causes, contributing causes, and remote causes of the accident. Toward this process, the working group invited the following government officials to participate in two- to three-hour interviews.[1] In 2011: Tomihiro Taniguchi, former deputy director general, International Atomic Energy Agency (August 27); Banri Kaieda, former minister of economy, trade, and industry (October 1); Hiroyuki Fukano, director general, Nuclear and Industrial Safety Agency (October 15); Tetsuro Fukuyama, former deputy chief Cabinet secretary (October 29); Goshi Hosono, minister of environment, minister of state for nuclear disaster, and former special adviser to the prime minister (November 19); Kenkichi Hirose, adviser to the Cabinet Office (November 26); Yukio Edano, minister of economy, trade, and industry; former chief Cabinet secretary (December 10); Haruki Madarame, chairman, Nuclear Safety Commission (December 17). In 2012: Naoto Kan, former prime minister (January 14); and Yoshioka Hitoshi, vice president Kyushu University, and member of the Government Investigation Commission (January 21).

---

[1] Note that the titles below reflect what were current at the time of the actual interviews.

The working group also conducted additional interviews in a roundtable format with government officials and experts. With the exclusion of those who wished to remain anonymous, the following individuals participated in 2011: Toshiso Kosako, former adviser to the Cabinet Office and professor at University of Tokyo (December 20); Kenichi Shimomura, councilor to the Cabinet Secretariat (December 21); Yasutaka Moriguchi, deputy minister of education, culture, sports, science and technology (December 22); Shunsuke Kondo, chairman, Japan Atomic Energy Commission (December 26); Kohei Otsuka, former senior vice minister of health, labor and welfare (December 27). And the following participated in 2012: Nobuyuki Fukushima, member, Japan House of Representatives (January 10); Kazuo Sakai, director, Research Center for Radiation Protection at the National Institute of Radiological Sciences (January 13); Yutaka Kukita, deputy chairman, Nuclear Safety Commission (January 20); Hiroshi Tasaka, former adviser to the Cabinet Office and professor at Tama University (February 2).

In addition to the above-mentioned interviews, the working group members conducted many off-the-record interviews with a range of government officials and experts in order to collect various pieces of additional background details. By the end of the project, the commission interviewed approximately 300 individuals.

Following the commission's report, which was released in February 2012, the Investigation Committee of the Diet and government published their investigations on July 5 and July 23, 2012, respectively. As a follow-up to its interim report published in December 2011, TEPCO released its final analysis on June 20, 2012. All of these reports shed light on important issues surrounding the Fukushima Daiichi Nuclear Power Station accident. However, much has also been discovered through extensive media coverage by NHK, *Asahi Shimbun*, *Tokyo Shimbun*, and *The New York Times*—not to mention the hard work of various individual journalists and their continuous research, interviews, and reports. I am glad that we—the Independent Investigation Commission on the Fukushima Nuclear Accident—were able to contribute to some extent in the context of this multifront, multifaceted effort.

Because high radiation levels continue to be a problem around the affected area in Fukushima Prefecture—making field inspections difficult—lingering technical causality issues remain unresolved even three years after the accident. For example, a full understanding of the compounding effects of earthquake- and tsunami-inflicted damage is yet to be realized; further, the extent of breakage in unit 2's nuclear containment building is still unclear. That said, some accident-response aspects were brought to the foreground as the tragedy unfolded at the nuclear station—namely, TEPCO plant manager Masao Yoshida's choice to disobey orders from TEPCO headquarters and to keep cooling the reactors.

Clearly, TEPCO's management problems as showcased before the world in March 2011 still need to be thoroughly investigated. The events between nightfall on March 14 through daybreak on March 15, 2011—during which time the unit 2 reactor's pressure could not be controlled and a possible explosion was imminent—remain a mystery to this day. Then-Prime Minister Kan and his deputies have claimed that TEPCO's strategic executives were faced with critical conditions and considered evacuating workers at the Fukushima Daiichi Nuclear Power Station—an issue that has yet to be brought to full light.

In March 2013, TEPCO's Nuclear Reform Monitoring Committee published a report in which it admitted—for the first time—that the accident was a "human-generated disaster." This is, in fact, made obvious in the following pages, as our commission illuminates the clear lack of facilitation and preparedness for a severe incident at a nuclear power station in Japan. Yet the lessons to be learned from Fukushima are not only safety issues, but security issues as well. That said, it remains to be seen whether TEPCO will work to deepen security at its plants.

Perhaps the biggest takeaway from the Fukushima Nuclear Power Station disaster concerns the nonexistence of a nuclear regulatory organization in Japan—one that is independent from nuclear enterprises, politics, or academia; in short, independent from any community within the "nuclear power village." To work toward this objective post-Fukushima, the Japanese government established the Nuclear Regulation Authority—an independent outfit headed by Chairman Shunichi Tanaka, former president of the Atomic Energy Society of Japan and former director of the Japan Atomic Energy Agency (previously known as the Tokai Research and Development Center of the Japan Atomic Energy Research Institute)—as well as the Nuclear Regulation Agency as its secretariat, and the two regulatory outfits debuted in September 2012. Since then, however, the Nuclear Regulation Authority has been caught in constant crossfire between pro- and antinuclear groups. And the pressure has not been mild. In August 2013—the time of writing this introduction—a national consensus remained unreached on both the nation's regulation of nuclear power stations and its nuclear energy policy issues. One of the reasons behind this delay can be found in the strong sentiments expressed by Hirohiko Izumida, governor of Niigata Prefecture: "Without the general review of the Fukushima Daiichi Nuclear Power Station, especially on facilities, security cannot be achieved just by setting hard measurements," he has contended. "The new regulation standard would sadly not be able to live up to the citizens' expectations."

This perspective cannot be discarded in light of the ongoing ramifications of the Fukushima Daiichi Nuclear Power Station disaster. Since the disaster in 2011, water has continuously been poured into the nuclear reactor for cooling purposes. Despite such efforts to curtail soil and water contamination, in May

2013 TEPCO detected radioactive substances in the soil near the Fukushima power station; further, the company found that water, contaminated by this very soil, had also seeped into the Pacific Ocean—possibly since May 2011.

The problem with this development was that TEPCO did not readily admit that this possibility even existed. In fact, it was not until two months after its findings—on July 18, 2013—that the company confirmed this information. Though Naomi Hirose, the company's CEO, instructed his staff to inform the Japanese fisheries association of this data within twenty-four hours, TEPCO pushed back its general public announcement to dovetail with the regulatory press conference slated for July 22, the day after elections in the lower house. Though TEPCO denied public and press accusations that the company purposely delayed its announcement to avoid upsetting the elections, the public remained skeptical that either TEPCO or the Nuclear Regulation Agency has changed in the aftermath of the accident.

Doubtful that the electrical companies, the Nuclear Regulation Agency, or the politicians have sincerely learned from past mistakes, the Japanese population continues waiting for the full truth of the Fukushima disaster to be revealed. At the time of this writing, TEPCO still refused to fully disclose its videoconference footage recorded in March 2011, as the crisis developed at Fukushima; live videos show meetings among officials from TEPCO headquarters, the Fukushima Daiichi and Daini Power stations, and the off-site centers. These tapes are the black box of the Fukushima accident—critical in understanding what, exactly, happened in the first days of the crisis. This constant hedging only calls into question the company's organizational culture and management practices—the very elements that, in its own March 2013 report, the company lauded as being the bedrock of a secure environment.

It has been two years now since our commission presented its Fukushima Daiichi Nuclear Accident Report to Japan in 2012; with this book, we now present the commission's findings for the first time in English.

The aim herein is to review and examine the numerous disasters that unfolded in the days and weeks after the Tohoku earthquake shook Japan on March 11, 2011. Focusing mainly on the Fukushima Daiichi Nuclear Power Station operated by TEPCO, our primary objective on the following pages is to review, investigate, and analyze the background of this disaster. We examine and evaluate the countermeasures and crisis management systems in place within TEPCO and at the national and municipal government levels.

While gathering information and synthesizing and scrutinizing our data, we found key points that are necessary to tease out in this report. Namely, our primary aims were as follows: to analyze why Japan is seen as "a secure and safe nation" and extrapolate why this historical misconception is a remote cause of the Fukushima disaster; to investigate Japanese government accident responses and leadership roles, namely crisis management within the Cabinet; to examine the water injection actions taken by first responders—the Self Defense Forces, the police, and the fire department; to explore why the government did not fully utilize SPEEDI to draw up evacuation zones; and to review US responses and US–Japan cooperation, and contextualize how the accident at Fukushima fits into the wider international picture.

This investigation is purely a citizen's report—independent of government agencies or corporations. In all of this, the commission's intent is to share with the world the lessons to be drawn from the experience. As we move forward, and as the Fukushima site is cleaned up over the next forty years, it is important to remember past mistakes, errors in judgment, and oversights to ensure accountability and that history does not repeat itself.

When the Independent Investigation Commission on the Fukushima Nuclear Accident released its Japanese report in 2012, media covered and referenced the content globally. In fact, it was front-page news in *The New York Times* on February 28, 2012. Additionally, Staff Director Kay Kitazawa and I had the opportunity to introduce its contents in an article for the March/April 2012 issue of the *Bulletin of the Atomic Scientists* titled "Fukushima in review: A complex disaster, a disastrous response."

We would like to express our sincere gratitude to Routledge—an internationally renowned publisher, particularly in the natural and social sciences realms—for publishing the English-language version of our investigative report. We are also honored to have received valuable feedback concerning the report's significance and future areas for further investigation from the following four scholars in their respective capacities: Frank N. von Hippel, professor of public and international affairs at Princeton University; Jessica Tuchman Mathews, president of the Carnegie Endowment for International Peace; James M. Acton, senior associate in the Nuclear Policy Program of the Carnegie Endowment for International Peace; and Paul 't Hart, professor of public administration at the Utrecht School of Governance. Finally, a special thanks goes to Mindy Kay Bricker and her team at the *Bulletin of the Atomic Scientists*; without their tireless editing efforts, this English-language version of the report would not have been possible.

Yoichi Funabashi
Program Director of the Independent Investigation
Commission on the Fukushima Nuclear Accident and
Chairman of the Rebuild Japan Initiative Foundation

Points of Investigation

The experts we interviewed for this report positively influenced our investigation. In fact, the following points—which were frequently raised during the interviews—became enshrined in our daily investigative efforts. In an effort to be transparent in the commission's methods, motivations, and considerations, we would like to share these key points of investigation.

- The Independent Investigation Commission on the Fukushima Nuclear Accident's most critical mission is to answer the following questions: How did the Tokyo Electric Power Company (TEPCO) and the Japanese government, while faced with a catastrophic accident, carry out their respective missions to protect people in Japan? In addition, what responsibility must the government accept for having advanced the adoption of nuclear power as national policy in Japan?
- It is important that the commission evaluate the accident responses of the government, particularly the Office of the Prime Minister. In carrying out this investigation, the commission must carefully consider the decision-making process that led to such responses during the crisis.
- The commission may classify the factors that caused the accident and ensuing destruction as immediate causes, contributing causes, and remote causes. It is important to understand all of these dimensions of contributing factors and to properly identify them.
- In addition to the visible causes and consequences of the nuclear accident, the commission must keep in mind the more obscure and perhaps concealed causes and consequences of the accident; that is, the commission must search for the underlying factors that govern what is visible. Structures, power, mechanisms, behavioral styles, and other elements are all interwoven into a complex tapestry, from which the commission must strive to unravel the true multilayered reality. To this end, the most desirable situation is to have the analysis of natural scientists, together with social scientists.
- The question of information and to whom it belongs is a critical one. When government and industry officials, among others, withhold crucial information from the public for fear of inciting panic, public anxiety

actually worsens as trust in the government's willingness to disclose information plummets.
- The commission must not forget the broader, global context of the accident at TEPCO's Fukushima Daiichi Nuclear Power Station. In a globalized world, international rule-making bodies and partnerships exist to control a variety of growing risks—including nuclear safety—and we must not lose sight of the fact that Japan was not sufficiently engaged in this process.

Yoichi Funabashi

Members of the Commission and the Working Group for the Independent Investigation Commission on the Fukushima Nuclear Accident

Chairman

Koichi Kitazawa
Former Chairman of the Japan Science and Technology Agency

Commissioners

Tetsuya Endo
Former Chairman of the Board of Governors of the International Atomic Energy Agency (IAEA)

Keiichi Tadaki
Attorney at Mori Hamada & Matsumoto

Ikujiro Nonaka
Professor Emeritus at Hitotsubashi University

Mariko Fujii
Professor at the Research Center for Advanced Science and Technology, University of Tokyo

Kenji Yamaji
Director-General at the Research Institute of Innovative Technology for the Earth

Working group members

Nobumasa Akiyama
Professor at the School of International and Public Policy, Hitotsubashi University

Hiroyuki Fujishiro
Journalist and associate professor at the Hosei University

Masaharu Fujiyoshi
Journalist, former reporter for *Shukan Bunshun*

Kenta Horio
PhD candidate at the Nuclear Non-proliferation Research Laboratory, Department of Nuclear Engineering and Management, School of Engineering, University of Tokyo

Akira Igata
PhD candidate at the Graduate School of Law, Keio University

Sachitoshi Isago
Associate professor at the Tokiwa University College of Community Development

Hiroshi Kainuma
PhD candidate in sociology at the Graduate School of Interdisciplinary Information Studies, University of Tokyo

Tadahiro Katsuta
Associate professor at Meiji University Law School

Hiromi Kikuchi
Journalist, former economics reporter for *Asahi Shimbun*

Kay Kitazawa: Staff Director
Fellow at the Rebuild Japan Initiative Foundation

Kenta Murakami
Special assistant professor at the Nuclear Professional School, University of Tokyo

Hironobu Nakabayashi
Fellow of Regional Security Policy Division Research Section, Okinawa prefectural Executive Office of the Governor

Takashi Otsuka: Editor of the Japanese Report
Former science-news editor at *Asahi Shimbun*

Kazuyuki Sasaki
Senior assistant professor at the Graduate School of Governance Studies, Meiji University

Tomohito Shinoda
Director and professor at the Research Institute, International University of Japan

Akihisa Shiozaki
Attorney

Shinetsu Sugawara
PhD candidate at Nuclear Non-proliferation Research Laboratory, Department of Nuclear Engineering and Management, School of Engineering, University of Tokyo (as of the publication of the Japanese Report)

Kazuto Suzuki
Professor at the Public Policy School, Hokkaido University

Hirofumi Tosaki
Senior research fellow at the Center for the Promotion of Disarmament and Non-proliferation, Japan Institute of International Affairs

Shinsuke Tomotsugu
Assistant professor of international relations and area studies, Nagoya College

Goro Umeyama
Consultant (risk assessment and crises management)

Kota Yamaguchi
Attorney

(Note: Some members are not included in the list upon their requests.)

Acronyms and Abbreviations Included in this Report

AEC Atomic Energy Commission
ANRE Agency of Natural Resources and Energy
BEIR Committee on the Biological Effects of Ionizing Radiations
DPJ Democratic Party of Japan
ECRR European Committee on Radiation Risk
ERSS Emergency Response Support System
IAEA International Atomic Energy Agency
ICRP International Commission on Radiological Protection
INES International Nuclear Event Scale
INPO Institute of Nuclear Power Operations
IRRS International Regulatory Review Service
IRRT International Regulatory Review Team
ISO International Standards Organization
JAEA Japan Atomic Energy Agency
JAERI Japan Atomic Energy Research Institute
JAPC Japan Atomic Power Company
JEA Japan Electric Association
JNES Japan Nuclear Energy Safety Organization
JSCE Japan Society of Civil Engineers
LDP Liberal Democratic Party
METI Ministry of Economy, Trade, and Industry
MEXT Ministry of Education, Culture, Sports, Science, and Technology
MITI Ministry of International Trade and Industry (now METI)
MOX Mix Oxide
NEA Nuclear Energy Agency
NISA Nuclear and Industrial Safety Agency
NS Net Nuclear Safety Network
NSC Nuclear Safety Commission
NUPEC Nuclear Power Engineering Corporation
NUSS Nuclear Safety Standards, IAEA
OECD Organization of Economic Co-operation and Development

OSART Operational Safety Assessment Review Team
PWR Pressurized Water Reactor
RaSSIA Radiological Safety and the Security and Safety of Radioactive Sources
SDF Self Defense Forces
SPEEDI System for Prediction of Environment Emergency Dose Information
STA Science and Technology Agency
TEPCO Tokyo Electric Power Company
TranSAS Transport Safety Appraisal Services
UK United Kingdom of Great Britain
US United States
WANO World Association of Nuclear Operators
WENRA Western European Nuclear Regulators' Association

Figures

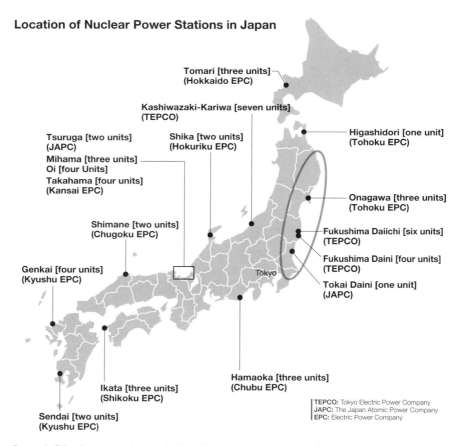

Figure 1 Fifty-four units (thirty Boiling Water Reactor units and twenty-four Pressurized Water Reactor units); a total of 49 gigawatts at seventeen sites. As of May 20, 2011, thirty-five units are now shutdown.

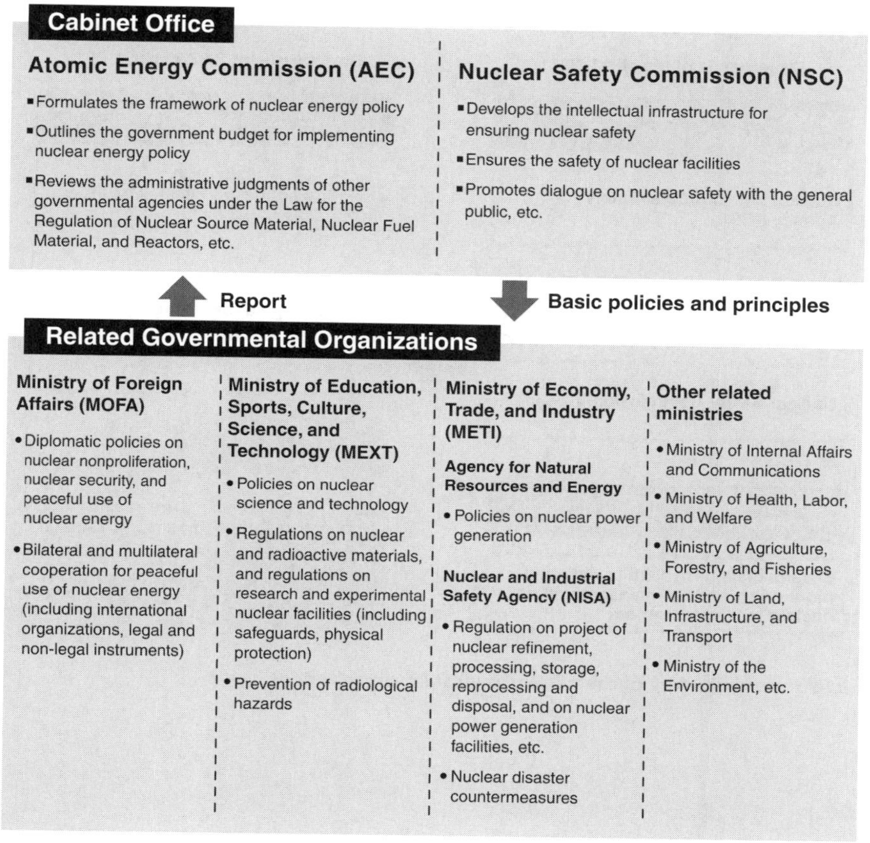

Cabinet Office

Atomic Energy Commission (AEC)
- Formulates the framework of nuclear energy policy
- Outlines the government budget for implementing nuclear energy policy
- Reviews the administrative judgments of other governmental agencies under the Law for the Regulation of Nuclear Source Material, Nuclear Fuel Material, and Reactors, etc.

Nuclear Safety Commission (NSC)
- Develops the intellectual infrastructure for ensuring nuclear safety
- Ensures the safety of nuclear facilities
- Promotes dialogue on nuclear safety with the general public, etc.

 Report   Basic policies and principles

Related Governmental Organizations

Ministry of Foreign Affairs (MOFA)
- Diplomatic policies on nuclear nonproliferation, nuclear security, and peaceful use of nuclear energy
- Bilateral and multilateral cooperation for peaceful use of nuclear energy (including international organizations, legal and non-legal instruments)

Ministry of Education, Sports, Culture, Science, and Technology (MEXT)
- Policies on nuclear science and technology
- Regulations on nuclear and radioactive materials, and regulations on research and experimental nuclear facilities (including safeguards, physical protection)
- Prevention of radiological hazards

Ministry of Economy, Trade, and Industry (METI)

Agency for Natural Resources and Energy
- Policies on nuclear power generation

Nuclear and Industrial Safety Agency (NISA)
- Regulation on project of nuclear refinement, processing, storage, reprocessing and disposal, and on nuclear power generation facilities, etc.
- Nuclear disaster countermeasures

Other related ministries
- Ministry of Internal Affairs and Communications
- Ministry of Health, Labor, and Welfare
- Ministry of Agriculture, Forestry, and Fisheries
- Ministry of Land, Infrastructure, and Transport
- Ministry of the Environment, etc.

Figure 2 The governance structure of nuclear-related policies and activities (as of March 11, 2011).

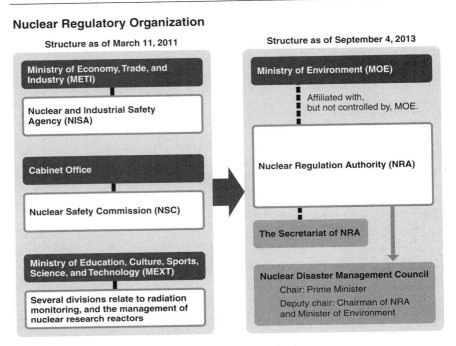

Figure 3 Restructuring of the nuclear regulation organizations.

Hydrogen explosion in the operation floor

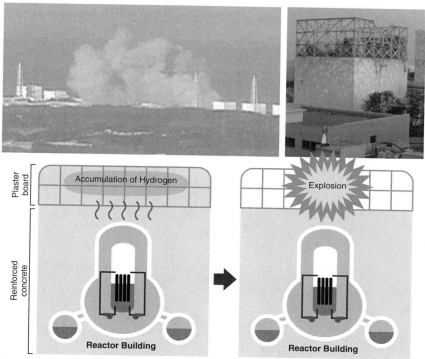

Source: Nuclear Regulation Authority (2011) *The 2011 Off the Pacific Coast of Tohoku Pacific Earthquake and the Seismic Damage to the NPPs.* Available at: http://www.nsr.go.jp/archive/nisa/english/files/en20110406-1-1.pdf (accessed September 5, 2013).

Figure 4 The process of the hydrogen explosion at Unit 1.

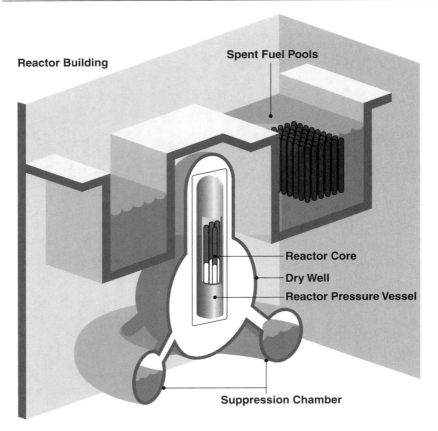

Figure 5 Simplified cross-section sketch of a typical BWR Mark I containment.

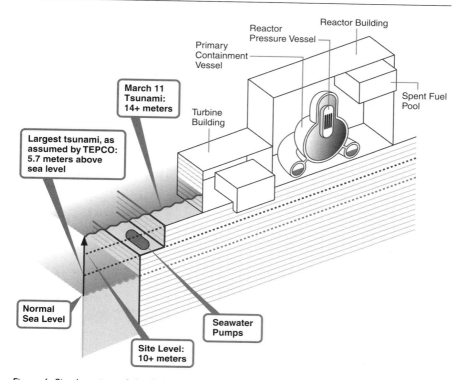

Figure 6 Site location of the Fukushima Daiichi Nuclear Power Station and key facilities.

Effort to Sustain Reactor Water Level

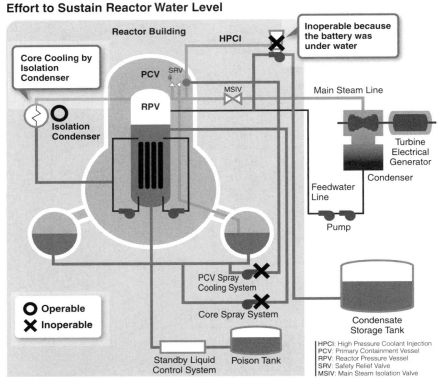

Figure 7 Simplified cross-section sketch of unit 1 and the effort to sustain its reactor water level.

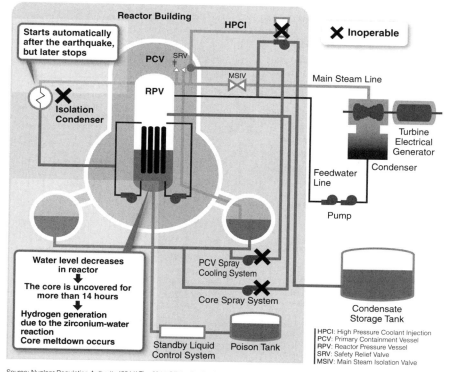

Figure 8 Simplified cross-section sketch of unit 1 and the process of reaching core meltdown.

Seawater Injection Using Fire Pumps Suppression Chamber Venting to Depressurize the Primary Containment Vessel (as of 8 p.m. on March 14, 2011)

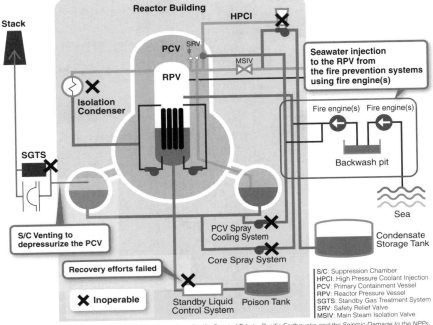

Source: Nuclear Regulation Authority (2011) *The 2011 Off the Pacific Coast of Tohoku Pacific Earthquake and the Seismic Damage to the NPPs*. Available at: http://www.nsr.go.jp/archive/nisa/english/files/en20110406-1-1.pdf (accessed: September 5, 2013).

Figure 9 The diagrammatic representation of seawater injection using fire pump.

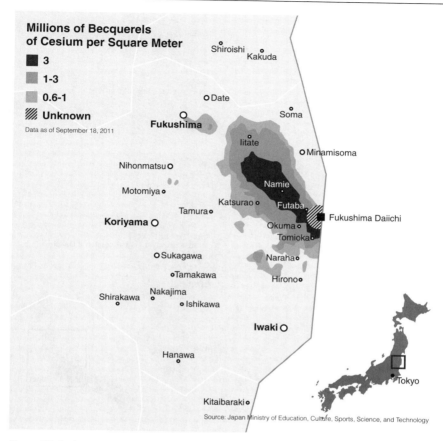

Figure 10 Radiation contamination.

Figure 11 Drinking water advisory.

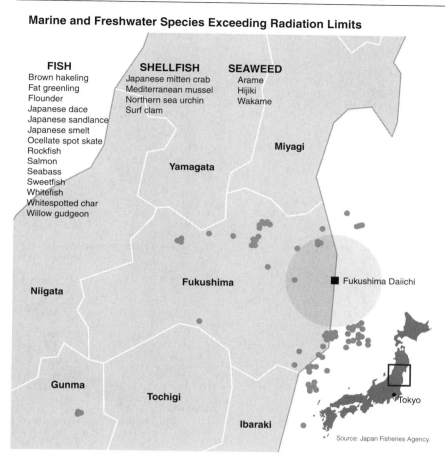

Figure 12 The impact of Fukushima radiation on Japanese fishery products.

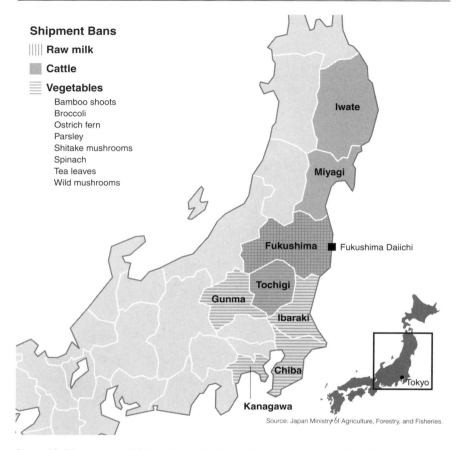

Figure 13 The impact of Fukushima radiation on Japanese agricultural products.

Foreword by Kennette Benedict

The *Bulletin of the Atomic Scientists* is proud to partner with the Independent Investigation Commission on the Fukushima Nuclear Accident to bring their report to an English-speaking audience. The original report, which was published within a year of the disaster, was intended as a citizen-to-citizen report. That is, the commission—as citizens—felt a responsibility to investigate this nuclear disaster and report on its findings to inform Japan and the world about the devastating events that unfolded on March 11, 2011, at the Fukushima Daiichi Nuclear Power Station.

Acting as a civilian accident investigation committee, the commission's expert team of more than thirty scientists, engineers, social scientists, business people, lawyers, and journalists interviewed some 300 individuals in industry, government, regulatory agencies, and academia—from nuclear plant operators to the prime minister. The goal was to gather, as much as possible, the "unfiltered experiences of the accident" free from prevailing assumptions about Japan's technological prowess and purported attention to safety. The result is a set of oral histories that exposes the expectations and pressures that led to unquestioned belief in the safety of nuclear power plants in Japan. The original Japanese-language report was both informative and immediate—and spurred many both inside and outside Japan to action.

Since the publication of the commission's investigation in February 2012, other reports, articles, and books have provided assessments of the disaster, lessons learned, and recommendations about the future of nuclear power in Japan and in other countries. The original report by the commission, with its detailed description of the events and first-hand accounts of the chain of terrible events after the earthquake and tsunami on that day in March 2011, led us to believe that a revised English translation could serve a somewhat different purpose than other reports and books. That is, the *Bulletin* and the commission recognized that an opportunity existed to recast the report's content and to take readers deep into the cultural and historical assumptions surrounding nuclear power in Japan, a context well-known to Japanese citizens, but not available to an English-speaking readership. Working from that realization, the authors and the *Bulletin*'s editor worked to restructure portions of the original report to provide context for the disaster, and to make explicit what was taken for granted by the industry, by policy makers, and by the general public in Japan and elsewhere.

Our goal was to make this English-language version of the commission's report a valuable resource for years to come. The *Bulletin* worked with the Japanese team to unpack the details surrounding the disaster and provide a rich and deep narrative. In keeping with the *Bulletin*'s mission to translate and contextualize technical analysis, the authors have added background that places the March 2011 events into the historical trajectory of the nuclear industry and reveals understandings about nuclear safety at the time of the disaster. Not only do the authors explore what happened in the hours and days after March 11, but they also provide the reader with a history of nuclear energy in Japan—from the atomic bombs that marked their society with death, illness, and trauma, to the US-sponsored Atoms for Peace, to the Japanese nuclear village that created, over the past half century, a myth of absolute safety.

Given the nature of this topic, at times there is more technical information in the book than some readers may be comfortable with, but we sought to make this a reference book for policy leaders as well as an interested public.

The original report was written by a number of authors, and, at times, they differed with one another in their opinions about specific aspects of events. Though some of those differences still appear in this English-language version, we have worked to find consensus where possible among the original authors. Our aim is not to stifle disagreement, but rather to present the most authoritative conclusions possible without needless distractions from the force of the commission's report.

Because Fukushima will be with us for decades, as a technical problem of clean up and restoration, as well as a traumatic event in nuclear history, we intend this book to be a reference and source for years to come. And if—or when—another Fukushima happens somewhere in the world, we will have in this masterful report the lessons that were—or were not—learned.

The commission authors have worked tirelessly to bring this book to fruition. Not only did they spend months investigating and writing the initial Japanese version in 2011 and 2012, but they immediately returned to their findings to translate their work into English. Over the course of yet another year—namely, months of reviewing, editing, and reworking their translated report—they have produced a new manuscript for English-language readers. Their volunteer efforts, through weekends, disrupted vacations, and on the edges of their normal workdays demonstrate their commitment to bringing to the English-speaking world the lessons they have learned from this terrible nuclear disaster.

We are particularly indebted to the lead coordinators of the report, Yoichi Funabashi and Kay Kitazawa, for their unfailing attention to the project and for their patience as they helped us understand the significance of events and details surrounding the disaster. Our thanks also go to the Rebuild Japan Initiative Foundation and the Civil Society Institute for financial support of this project. The *Bulletin of the Atomic Scientists* editor, Mindy Kay Bricker, has devoted untold hours to revising and editing the manuscript for this book. Her editorial instincts about the significance of this collaborative report, and her determination to provide a cultural and technical translation, as well as an English-language translation, have yielded a valuable resource—and a really good read.

Kennette Benedict
Executive Director and Publisher of the
Bulletin of the Atomic Scientists

Prologue

Fukushima 1/50: eyewitness account

In September 2011, six months after the disaster shook the Fukushima Daiichi Nuclear Power Station in Japan, the newly formed Independent Investigation Commission on the Fukushima Nuclear Accident launched a national campaign to collect personal testimonials from those who experienced the accidents at the plant and those who were forced to evacuate the area close to the plant. The commission launched a website to serve as an online meeting point; it published questions, and the public provided in-depth answers. For half a year, the commission received hundreds of responses to its queries. Though the commissioners found all the reactions to be insightful and useful to the investigation, they needed to hear several responses firsthand.

One such story was from a subcontractor of the Tokyo Electric Power Company, known as TEPCO, which owns the nuclear power plant. He was among several hundred workers who were at the Fukushima Daiichi Nuclear Power Station when the great earthquake and subsequent tsunami occurred in eastern Japan. Working in the plant's crisis center, located on the second floor of the earthquake-resistant building, he recounted his story of what happened as the accident unfolded on March 11, 2011.

The TEPCO subcontractor turned to the commission to answer this particular question: What were the decisions facing TEPCO's General Manager Masao Yoshida as he was flooded with information?

Ultimately, Yoshida would make headlines when he famously disobeyed instructions from TEPCO headquarters to stop using seawater to cool the reactors. Though he was later reprimanded, his disregard for corporate instructions was possibly the only reason that the reactor cores did not explode.

> I first felt the earthquake as I walked from the vicinity of units 5 and 6—which are located near the ocean—to the site's entrance gate. Suddenly, the asphalt began to ripple, and I couldn't stay on my feet. In a panic, I looked around and saw a 120-meter exhaust duct shaking violently and looking like it would rupture at any second. Cracks began to appear on the

outside of unit 5's turbine building and on the inside of the entryway to the unit's service building. The air was filled with clouds of dirt.

When the shaking subsided, more than 200 workers, who had been on the ocean side of the plant, came rushing to the gate. To protect the facility, anyone entering or leaving the gate had to pass through a metal detector.

"Let us out of here," we yelled. "A tsunami may be coming!" Screams and shouts filled the air.

"Wait for instructions from the radiation safety group," demanded a security guard.

This response angered the workers. When an earthquake had struck the Kashiwazaki-Kariwa Nuclear Power Plant [in 2007], some workers had jumped over the gate to flee—and they were later charged for having "broken the law."

After keeping us waiting for a few minutes, the guard collected our APDs [alarm-equipped pocket radiation meters] and our ID cards and instructed that we "all seek refuge." I headed for the earthquake-resistant building; however, when I arrived, a ruptured ground pipe was spraying water like a geyser and had caused a mudslide that covered the stairs—which, from top to bottom, spanned some 20 to 30 meters. When I reached the operational headquarters, numerous windows on the second floor had shattered, and the blinds were flapping about in the wind. Three or four cooling towers on the roof had either fallen or were tilted over. Considering that the walls of the newly constructed units 5 and 6 had been damaged, I figured that units 1 through 4, which were older, must have been in even worse shape.

The Crisis Center on the second floor was jam-packed. As we watched the news on TV, we were first worried about the Onagawa Nuclear Plant. NHK News showed aerial helicopter footage of a tsunami hitting fields in Natori City in Miyagi Prefecture [in northeastern Japan, where the Onagawa Nuclear Power Plant is located, more than 200 miles from Tokyo]. But then a section chief came rushing up to Fukushima's plant manager, Masao Yoshida, and reported: "A tank had been washed away and had sunk into the ocean."

We all went pale with shock: The tank that had been lost was a surge tank of suppression pool water at the Fukushima Daiichi Nuclear Power Station.

People continued coming in and out of the Crisis Center, delivering one report after another to Yoshida. Each time, the plant manager's shouts reverberated through a microphone: "That's not the question I asked!" and "Give me the answer to ... this and that!" The workers surrounding Yoshida kept trying to get in touch with people at the reactor buildings at units 1 through 4, but they were unsuccessful, because the dedicated on-site PHS [Personal Handy-phone System] base station had lost electrical power due to the tsunami. Shortly after 4 p.m., we received instructions to "gather whatever you can," including Sunnyhoses, small pumps used for construction work, and emergency light-oil power generators to help drain water from the electric power room that had flooded. Because we had lost all electric power, it was too dark to get to the electric power room.

By this point, around 700 people had taken refuge in the earthquake-resistant building; and, because we had just conducted an emergency-preparedness drill the previous week, things were surprisingly orderly. TEPCO employees handed out water and crackers, and we took turns using a PHS that was able to connect to the outside world to confirm the safety of our families. Meanwhile, some foreigners sat on the floor and chatted, and the female employees screamed during each aftershock.

I suppose it was around the early evening when the following reports arrived: "The water level has begun to fall, and we can no longer see the meters," and "we can't assess the water level." I think the report for unit 1 or unit 2 was: "If the water level continues to fall at this rate, the fuel will be exposed by 10 p.m." Yoshida's only reply to these reports was: "Understood." But he then announced, "All personnel not engaged in work activities, please evacuate." The allocation of vehicles began, but nobody got up to go home. There was a sense that something had to be done, and it was not an atmosphere in which people felt like rushing off on their own.

One report came in that "we are able to see the water level in unit 1." Another report soon followed that said, "Once again, we have lost the ability to assess the water level." Even when the numbers were reported, Yoshida asked: "Are those numbers really correct?" and "Are those the right numbers?" The people delivering the reports could only reply, "Umm, we aren't sure." It had become impossible to trust the numbers.[1]

I don't remember precisely what time it was when we received the report that "the building is in a dire state!" I remember that it was after the water level began to fall, which was after 7 p.m. This report came from an operator at unit 1 or 2; I can't remember which unit it was, exactly, but apparently he took a flashlight and approached the reactor building in the dark. The nuclear reactor building had a double door. Using his flashlight,

---

[1] It is worth noting here that a subcontracted firm's employee, who delivered car batteries and small power-generators to the central control room of units 1 and 2 sometime after 8 p.m., shared this comment on our website: "When we entered the central control room, it was pitch black; a worker had a flashlight in his hand and was trying to read the meter. When I arrived, everybody was ecstatic: 'At last, we have light!' We connected the battery directly to a terminal on the back of the meter's control board and began to read the meter. However, according to reports TEPCO provided to the government's investigation committee, at 9:19 p.m., the water level in unit 1 was 'TAF [top of active fuel] + 200 mm,' a high level. Thus, it is likely that the meter was not functioning properly."

he first opened the outer door and went through it, then approached the inner door and shined the flashlight through the glass window in the door. The operator said he could see billowing white steam filling the space on the other side of the glass window.

"That's raw steam!"

Upon receiving this report, the Crisis Center exploded in chatter: "What are we going to do? ... It's not going to explode, right?"

There could be only two possible explanations for the steam: The first is that it was steam for the heater. However, on top of the fact that the boiler had stopped because of the earthquake, the heater steam pipes were quite narrow. Thus, the consensus emerged that this was probably not the case. This left only the possibility that the nuclear reactor's main steam pipe was sending steam to the turbine building; a rupture in the main steam pipe system would be very dangerous and would mean that no work could be done on that floor. Sure enough, we soon received reports that radiation had been detected as far away as the outer walls of the central control room and the uncontrolled region: This meant that the volume of radiation was extremely high.

"Well, that's the end of this nuclear plant," I thought at the time. "And this is the end of TEPCO." Considering the location of the main steam system, it seemed unlikely that the tsunami had ruptured the pipes.[2]

Yoshida and others had begun conducting meetings, starting in the evening, to determine whether fire trucks could be used to supply water to the reactor. Because the reactor pressure vessel had reached an extremely high pressure, this meant that water could only be supplied at a higher pressure.

A TEPCO-affiliated company, named Nanmei Kousan, performs water-spraying drills and patrols inside the facility on a daily basis; apparently on March 11, one of the two fire trucks being used for training purposes on the ocean side of the plant's campus was damaged by the tsunami. In

---

[2] According to public statements from TEPCO, a statement written on the white board in the central control room of units 1 and 2 said: "hissing sound on the corridor side." However, among several workers who went to the corridor near the unit 1 reactor building on the evening of March 11, no one reported the sound of a pipe rupturing or steam leaking, or the appearance of white billowing steam. According to an intermediate report from the government investigation committee, the observation of "white billowing steam" in the double doors of the unit 1 reactor building was made much later, at 3:45 a.m. on March 12. Moreover, as for the origins of the steam, the report concludes that, contrary to the mention of the "main steam system" in the eyewitness account, "we cannot exclude the possibility that steam escaped outside the dry well through previously developed leaks in pressurized containers, pipes, or ducts in the reactor." Outside these official reports, however, our commission spoke with one eyewitness, an employee of a subcontractor firm, who, at some point after 10 p.m. on March 11, was instructed to "go take a look." He approached the double doors, with flashlight in hand. "I could see white billowing steam through the window in the inner door," he said. "I didn't know what it was, but I had an instinctive sense that going any farther could be very dangerous, so I immediately turned back."

addition, even if fire trucks could have been used to pump water from the truck's water tanks, it remained unclear how to connect to the supply pipes. Employees argued: "We have no hoses! ... We have no plugs! ... We have no fuel!" For disaster-prevention purposes, various tools were readied, but there were no connectors.

Instructions came from TEPCO headquarters: We were to "vent" and "inject water." Yoshida had a direct phone line to TEPCO headquarters, to whom he told: "It doesn't matter *what*: Bring me whatever liquid you can find!"

As we were wondering whether the time had come to vent manually, a dejected TEPCO employee in the room muttered, "Well, this is the end of our company."

I went down to the first floor, where a bunch of people, primarily TEPCO employees and people from affiliated companies, formed into response brigades. They organized themselves into five teams of about twenty people each, and members of the radiation-control group were outfitting them with protective suits. Among them was a woman in her twenties who helped place tape over the seams of the suits.

This impressed me. "She is really dedicated," I thought: She had *asked* to stay behind and help. Because the outside air could enter the first floor of the building, it was extremely cold. Two female employees, who were not involved in radiation-related work, were exposed to radiation in excess of the annual limiting dose of 1 millisievert. The news was that the Nuclear and Industrial Safety Agency had given TEPCO a severe warning.

I'll never forget the expressions on the faces of the employees assembled into those response teams. Their faces, in the face of lethal danger, were white as sheets. They couldn't find words to express how terrified they were. Every single one of them was trembling; they were truly scared. Nobody knew what could happen. Needless to say, there was a chance they could die.

By the time TEPCO had prepared a bus and we had received our evacuation instructions—"Cover your mouth and get on the bus as quickly as possible"—the sun had already risen.

Outside radiation levels continued rising and members of Japan's Self Defense Force began to arrive on-site. We could see them through the windows of the bus. But they were not wearing masks. As the bus drove away, I could not help but think, "I wonder if those guys are going to be all right."

Yoshida directed the evacuation of the TEPCO-affiliated employees, who, beginning on March 12, were gradually evacuated by bus from the Fukushima Daiichi Nuclear Power Station and elsewhere. Ultimately, only some fifty people, primarily TEPCO employees, remained on-site at the Fukushima plant. However, after the explosion in the nuclear power building, it became increasingly difficult to proceed with just fifty people; at that point, TEPCO workers and TEPCO-affiliated personnel who had been previously evacuated were brought back to the facility, thereby beginning the long battle that continues more than three years later.

Chapter 1

A Fukushima Diary, March 11–16, 2011

Japan, more than any other country, understands the inherent dangers of splitting the atom. In 1955, when the Japanese government pledged to adopt nuclear energy for peaceful purposes, the scars of the world's first nuclear war were still fresh in its cities—that same year, the Hiroshima Peace Memorial Museum opened with its exhibits of the devastation wrought by the atomic bomb dropped on Hiroshima.[1]

Yet the Japanese government concluded that the need for cheap and reliable electricity outweighed the risks of nuclear power. Without significant fossil fuel reserves of its own, Japan had been importing oil, gas, and coal to keep its post-war economy growing. To increase the country's energy independence, the 1955 Atomic Energy Basic Act established a program of nuclear research and endorsed the import of nuclear reactor technology. Throughout the 1960s, nuclear power stations were built across Japan.

On a coastal plain in Japan's northeast, hemmed in between a ridge of forested mountains to the west and sandy beaches that edge the Pacific Ocean, sit the towns of Futaba and Okuma. These quiet burgs had relied on fishing and farming for generations, but wanted to reinvent themselves and take part in the post-war boom. In the autumn of 1961, the town councils of Futaba and Okuma both unanimously voted to invite TEPCO, the Tokyo Electric Power Company, to build a nuclear power station on the border of the two towns. To win the townspeople's acceptance, the Japan Atomic Energy Commission and TEPCO reinforced the notion that nuclear power stations were absolutely safe. Citizens need not fear, because nothing would go wrong. Construction of the Fukushima Daiichi Nuclear Power Station's first reactor began in 1967, and it began supplying electricity in 1971.

In a country as prone to earthquakes and tsunamis as Japan, building nuclear reactors on the coast may seem like pure folly. But the government was confident that it had assessed the geological risks, as well as the myriad other risks related to siting a nuclear power facility, and had prepared for them. For

---

[1] Arima, T. (2008) *Nuclear Power Plants, Shoriki, and the CIA—the background history of the Showa Era discovered through confidential documents.* Tokyo: Shincho Shinsho. (In Japanese).

Fukushima Daiichi, engineers had estimated that the largest tsunami that could possibly hit the plant would be about 10 feet[2] high, and when, in 1970, Japan established a regulatory standard for tsunami height, the government found Fukushima to have met this new standard. More than three decades later, the Japan Society of Civil Engineers raised that worst-case estimation for Fukushima Daiichi to about 19 feet, and TEPCO voluntarily elevated some equipment installations near the harbor. The plant was assumed to be well out of harm's way and untouchable by even the angriest ocean.

Over the decades, the nuclear industry flourished throughout Japan. By 2010, a total of fifty-four reactors around the country were keeping on the bright lights of Japan's metropolises and powering its factories. The utilities that owned the nuclear power stations propagated what has come to be called the "safety myth"—the idea that nothing could go wrong in Japan's nuclear sector—and government regulators echoed the message. In 2010, for example, the government of Niigata Prefecture, on Japan's western shores, made plans to conduct a joint earthquake and nuclear disaster drill. This was imminently sensible, since just three years before an offshore earthquake had temporarily shut down a TEPCO nuclear power station on the Niigata coastline. But the Nuclear and Industrial Safety Agency (NISA), the nation's main nuclear regulator, advised the local government that a nuclear accident drill premised on an earthquake would cause "unnecessary anxiety and misunderstanding" among residents. The prefecture instead conducted a joint drill premised on heavy snow.[3]

So when, on March 11, 2011, a massive tsunami rolled in from the Pacific and inundated the Fukushima Daiichi Nuclear Power Station, when the facility's electrical systems went off and stayed off, when the emergency cooling systems faltered and failed, and when the three operating nuclear reactors started to boil away their coolant, the utilities and the government and the people of Japan were unprepared.

In Tokyo, Cabinet members reached for the nuclear emergency-response manual and flipped through it, looking for instructions on what to do in such a dire situation. But those instructions didn't exist. According to the emergency-response manual, this situation could never occur.

The manual assumed that any nuclear crisis would be a local affair centered around an accident at a nuclear power station. It didn't consider—and included no provisions for—a crisis brought on by external forces affecting a broad swath of the country, such as a powerful earthquake and tsunami. The envisioned nuclear response assumed that communication and transport systems

---

[2] 3.2 feet = 1 meter.
[3] Personal communication with Hiroyuki Fukano, NISA director, February 7, 2012. Funabashi, Y. and Kitazawa, K. (2012) Fukushima in review: A complex disaster, a disastrous response. *Bulletin of the Atomic Scientists* 68(2): 9–21. Available at: http://bos.sagepub.com/content/68/2/9.full.

beyond the affected facility would be intact, and that help would rapidly be brought to bear. And while the manual did consider the possibility of a nuclear power station losing all electricity, it assumed that such a circumstance would be temporary. When the Nuclear Safety Commission (NSC) reviewed safety-design guidelines for nuclear plants in 1990, the regulatory agency explicitly ruled out the need to consider prolonged AC power loss. Even if power systems were disrupted in an accident, the commission said, transmission lines would be quickly repaired, or emergency AC power generators would see the plant safely through the crisis.[4]

In other words, nothing like the Fukushima Daiichi catastrophe was possible—no tsunami of 45 feet could swamp a nuclear power station and knock out its emergency systems. No blackout could last for days. No triple meltdown could occur. Nothing like this could ever happen. Until it did—over the course of a week in March 2011.

March 11—at the nuclear power station

For more than 500 years,[5] tension had been building on the seafloor of the Pacific Ocean, where two tectonic plates were stubbornly grinding against each other. The Pacific plate was sliding slowly beneath the Okhotsk plate, which supported the Japanese islands. But the long, slow slide was not going smoothly: Along one fault line between the plates, about 60 miles[6] off the coast of northeast Japan, these stubborn pieces of Earth's surface were stuck. Potential energy gathered and the strain accumulated in the rocky seafloor. Finally, at 2:46 p.m. on March 11, the tension was too much. With a jolt and lurch, the Pacific plate slipped forward and downward, changing the shape of the seafloor along 150 miles of the fault line and shifting the water column above. The earth moved, the ocean moved, and disaster rolled toward the coast of Japan.

Many places were in the path of danger. The rising water would spill over harbor walls, carry fishing boats miles inland, and sweep houses off their foundations. It would surge implacably over landscapes and up hills where townspeople were taking refuge. Entire coastal towns would be destroyed, and more than 15,000 people would be killed. And then there was Fukushima Daiichi, an aging nuclear power station on the coast that would suffer the worst nuclear disaster since Chernobyl.

---

[4] Nuclear Safety Commission (1990) *Inspection guidelines for safety design of light-water type nuclear reactor for power generation*. Report, Ministry of Education, Culture, Sports, Science, and Technology. Tokyo: NSC.
[5] California Institute of Technology (2011) Caltech researchers release first large observational study of 9.0 Tohoku-Oki earthquake. Available at: www.caltech.edu/content/caltech-researchers-release-first-large-observational-study-90-tohoku-oki-earthquake (accessed August 31, 2013).
[6] 1 mile = 1.6 kilometers.

But long before the waves hit, in the seconds after the ground started to shake, Japan's high-tech earthquake alarm system went into action. As the first seismic pulses of the massive 9.0-magnitude quake registered at monitoring stations, automatic systems switched on. In city office buildings, elevators stopped. Recorded warnings blared from speakers. On rail lines all around the country, high-speed bullet trains slammed on their brakes. Updates flashed across television screens. At Fukushima Daiichi, seismographs in the basements of the buildings recorded Earth's violent shaking, and the plant's automatic shutdown systems went into action. Sections of pavement buckled and cratered throughout the power station; and in some secondary buildings, windows shattered and cracks zigzagged up walls.

On the day of the earthquake, only three of Fukushima Daiichi's six nuclear reactors were in operation. Units 1, 2, and 3 were busily splitting atoms. Units 4, 5, and 6 were shut down for routine maintenance, and in unit 4 all the nuclear fuel had been removed to the spent fuel pool so that a part of the reactor core's internal structures could be replaced.

In Fukushima Daiichi's boiling-water reactors, the radioactive element uranium was the main source of the nuclear fission reaction. When one atom of the uranium isotope U-235 breaks down into smaller parts, it produces both energy (in the form of heat) and neutrons, as well as some radioactive byproducts. When a large enough quantity of uranium fuel is gathered together, it creates a self-sustaining chain reaction, in which the released neutrons smack into other uranium atoms and cause them to split in turn. The energy from the fission reaction is used to boil water into steam, which drives huge turbines to produce electricity. Fukushima Daiichi's six reactors could produce 14.2 gigawatts of heat and 4.7 gigawatts[7] of electric power in total, more than double the output of the Hoover Dam's power station.

Inside a nuclear core, pellets of uranium are contained in long, narrow fuel rods made of an alloy of zirconium. There are thousands of these fuel rods inside a reactor's innermost chamber, called the reactor pressure vessel. Water circulating inside the pressure vessel keeps the fuel rods from overheating and also boils to create the steam that drives the turbines. When the earthquake shook the ground beneath Fukushima Daiichi, the three operating reactors were scrammed by an automatic procedure (what is known as passive design): Within seconds, control rods shot upward into the three reactor cores. The control rods are made of substances that absorb neutrons and therefore halt the nuclear fission chain reaction. Between 2:54 and 3:02 p.m., operators in the control rooms confirmed that units 1, 2, and 3 were "noncritical," meaning that the chain reactions had been stopped.

But even when a nuclear reaction is halted, the radioactive byproducts of that reaction continue to generate a dangerous amount of heat. Water must be

---

[7] Institute of Nuclear Power Operations (2011) *Special report on the nuclear accident at the Fukushima Daiichi Nuclear Power Station.* Report. Atlanta: INPO.

continuously circulated through a reactor's core to keep the water from boiling away and exposing the fuel rods, which would allow them to overheat and melt. That kind of core damage is commonly referred to as a meltdown, and it has the potential to cause explosions that can release vast clouds of radioactive elements into the atmosphere. That's why nuclear power stations have multiple emergency cooling mechanisms—a nuclear core *can't* be allowed to overheat.

The March 11 earthquake had brought down transmission lines and wrecked electrical substations, causing Fukushima Daiichi to lose external power to all six of its reactor units. From the plant's control rooms, operators watched and listened as alarms blared and backup systems clicked on. They were ready for this situation. Within seconds of the loss of external power, the plant's emergency diesel generators kicked in. Thirteen of these massive machines were operational at the time, and they were well capable of powering the plant's cooling systems. Operators looked on as a cooling system called an isolation condenser kicked on inside unit 1, the oldest of the station's six reactors. In units 2 and 3, mechanisms called reactor core isolation cooling systems—which provide water to a reactor to help cool it when the main steam lines are closed—went to work. Operators also prepared to activate a cooling system for the suppression chambers—large, doughnut-shaped vessels, which contain tons of water, and where steam from the reactor core is condensed.

All of Japan's nuclear power plants are designed to withstand a certain amount of shaking during an earthquake. According to the seismographic records from Fukushima Daiichi, the vibrations exceeded the plant's design standards at units 2, 3, and 5. But despite this thorough rattling, the gauges on the operational reactors recorded no drop in internal pressure or water level in the half-hour after the quake, suggesting that it caused no significant structural damage to the reactors' essential components. The power station's workers may have been relieved at first, but they soon had fresh reason to worry.

Within minutes of the earthquake, the tsunami alerts flashed on television screens, predicting that a 10-foot-high wave would hit Fukushima Prefecture. The Fukushima Daiichi plant had been originally designed to withstand a tsunami of 10.2 feet; that was the official estimation of the largest tsunami that could conceivably hit the facility, and it was the water level stated in the power station's 1966 permit application. In 2002, the Japan Society of Civil Engineers released a new tsunami assessment for the country's nuclear power stations and raised the estimation for Fukushima Daiichi to 18.7 feet. In response, TEPCO raised some equipment installations on the ocean side of the plant. Overall, the plant was assumed to be safely beyond the reach of the largest ocean swells. The main buildings for units 1 through 4 sat nearly 33 feet above sea level, and units 5 and 6 were even higher at 42 feet above sea level.

At 3:27 p.m., the first tsunami wave rolled over the seawalls protecting Fukushima Daiichi's manmade harbor. The tidal gauge registered a water height of 13 feet above normal. At 3:35 p.m., the next set of waves began to roll in, and they destroyed the tidal gauge. Much later, TEPCO staffers would measure the high-water marks on buildings throughout the site and estimate that the water reached 46 to 49 feet above sea level.

The water swept away equipment and tanks near the harbor and flowed into the turbine buildings, which faced the sea. The waves damaged the seawater intake pumps for systems that cooled reactor water. And the water flooded into the basement rooms of the turbine buildings, where most of the emergency diesel generators were housed. Immediately, ten of these crucial generators were knocked out of service by either the flooding, loss of seawater circulation for their coolant, or both. Three air-cooling generators were above the high-water line, but two of those—in units 2 and 4—were useless because metal-clad switches, which are connected to the generator to distribute high-voltage AC power to the systems and components, were flooded. Of the twelve generators that were supposed to power the plant through any emergency, only one, in unit 6, was still operating when the tsunami waves abated. That generator would keep unit 6 and its sister unit 5 essentially intact and out of danger throughout the crisis.

Between 3:37 and 3:42 p.m., units 1 through 5 lost all AC electrical power. The DC power distribution panels in units 1, 2, and 4 also flooded, causing those units to lose all DC electrical power as well. In the joint control room where operators oversaw units 1 and 2, all the lights, including emergency lights and read-outs on crucial instrument panels, faded out. The room was plunged into total blackness. Operators had no way now to determine the basic conditions of reactors 1 and 2: They could not keep an eye on the water levels inside those reactors, and they could not verify whether any emergency cooling systems were working to inject water into them.

As the water moved around the reactor buildings and poured into basements, it brought immediate tragedy to the site. Following the earthquake, workers had fanned out across the site to assess the damage. In the unit 4 turbine building, two workers were mid-inspection when the water flowed in; they were caught in the onslaught of water and drowned.

At 3:42 p.m., TEPCO declared that the loss of all AC electrical power in five of the reactors qualified as a "special event" under the provisions of Japan's Nuclear Emergency Act. The utility's officials notified the national and local governments.

March 11—at the Cabinet

Back in Tokyo's Kasumigaseki district, where bureaucrats for all of Japan's ministries keep the country running, officials gathered together mere minutes after the earthquake. By 3:05 p.m., Prime Minister Naoto Kan and a core team of top government officials had assembled in the Emergency Control Center, located in the basement of the Cabinet building.

Throughout the first crucial week following the tsunami and the onset of the nuclear accident, when a series of disasters seemed to be bringing the country inexorably toward nationwide catastrophe, the response was split in two. At the Fukushima Daiichi Nuclear Power Station, workers fought desperately to stabilize the plants and to halt the release of radioactive materials that could contaminate a wide swath of northeastern Japan. Meanwhile, in the Cabinet building, the prime minister and his closest advisers struggled to understand the situation and to issue effective orders that would protect the people.

In the heat of the crisis, this small group of politicians and technical experts became directly involved with managing the on-site response at Fukushima Daiichi, ignoring the typical chain of command and the division of responsibilities among ministries. In retrospective interviews, some observers have deemed this an "ad-hoc" approach and declared that it was detrimental to the overall goal of managing the risk to the nation of Japan. One of the politicians belonging to the Cabinet's core response team likened the scene at the Cabinet, in its hectic rush for on-site information and the confusion arising out of the events, to a kid's soccer game where a crowd of children all scrambled for a single ball.

In the minutes following the earthquake, the Cabinet building's basement control center became crowded with anxious officials. Prime Minister Kan and Minister for Disaster Management Ryu Matsumoto took seats around the big round table, along with Chief Cabinet Secretary Yukio Edano and Deputy Chief Cabinet Secretary Tetsuro Fukuyama. On the walls, large screens displayed videos from the Ministry of Defense and emergency TV broadcasts. About twenty top government officials, each representing different ministries and agencies, gathered around the table, which was equipped with special emergency telephones and microphones. Over those emergency phone lines, the reports came in. The Ministry of Land, Infrastructure, Transport, and Tourism called in a report on rail and road status. The Meteorological Agency reviewed the magnitude of the quake's aftershocks, and the national police and fire agencies reported on the number of emergency calls.

With each minister's subordinates standing by for orders, a total of more than 100 people were crowded into the Emergency Control Center. In neighboring rooms, a large contingent of lower-level staff assembled. Altogether, the scene at the Emergency Control Center resembled the clamorous trading floor of an old stock exchange. There was too much to do all at once. Chief Cabinet Secretary Edano was communicating with local governments in earthquake-stricken areas, but was simultaneously grappling with the issue of how to return millions of commuters from Tokyo to their suburban homes—when none of the trains were running. And then the tsunami hit.

At 3:42 p.m., TEPCO informed the central government that a "special event" had occurred at the Fukushima Daiichi Nuclear Power Station, as defined by the Nuclear Emergency Act. The tsunami had struck a mighty blow, and the power station was now in a terrifying situation: full station blackout. The officials in the Cabinet's basement room, most of whom had little knowledge of nuclear power systems, learned that, without electricity from the grid or from emergency generators, the plant's operators had few tools to manage the plant. Tetsuro Fukuyama, deputy chief Cabinet secretary, recalls, "That is when the pervading tension seemed to heighten."

At the time of the earthquake, Banri Kaieda, head of the Ministry of Economy, Trade, and Industry (METI), was attending a committee meeting at the Diet. When the minister returned to METI shortly thereafter, he received reports from the agencies under his control, including NISA, the regulatory body with oversight of all nuclear power stations. Kaieda listened to the first report of the scram of the Fukushima Daiichi reactors, and was told that it was a standard emergency shutdown, and that control rods were safely in place. Relieved, he turned his attention to reports of local power outages and a fire at an industrial complex near Tokyo Bay. On returning to his office after a meeting about the industrial fire, Kaieda was informed for the first time of TEPCO's report of a "special event" at Fukushima Daiichi. On learning of the complete loss of AC power, he immediately recognized the gravity of the situation at the nuclear power station.

In the early evening of March 11, four top politicians and two experts held the first meeting on the response to the nuclear plant accident: Prime Minister Kan, METI Minister Kaieda, Chief Cabinet Secretary Edano, Deputy Chief Cabinet Secretary Fukuyama, and special advisers to the prime minister Manabu Terada and Goshi Hosono, together with NISA Director General Nobuaki Terasaka, and TEPCO fellow Ichiro Takekuro. The gathering did not take place in the Emergency Control Center, itself, but in a small conference room on the mezzanine adjoining the center to avoid the hectic environment within the control center at the time.

At 7:03 p.m., Prime Minister Kan officially declared a state of nuclear emergency on the basis of the Nuclear Emergency Act. At the mezzanine meeting, the politicians discussed Fukushima Daiichi's loss of AC power and decided on their first priority: immediately dispatching power-generator trucks. If a few of these trucks could reach the power station, the officials assumed, the Fukushima Daiichi operators could turn the plant's emergency cooling systems back on in plenty of time to avert disaster. As it was not known which highways and roads were flooded or blocked by rubble, more than forty trucks were dispatched along a variety of routes.

As the evening progressed, Prime Minister Kan took a great interest in the progress of the power-generating trucks. At his office on the Cabinet's fifth floor, he instructed his secretaries and others to keep him informed on where they were and in what number. His staffers carried a whiteboard into his office, and one secretary constantly updated a detailed table; the board listed the time and number of trucks that had left each dispatch point, their locations on their routes throughout the evening, and their expected arrival times at Fukushima Daiichi. In this and other ways, the prime minister personally oversaw details of the emergency response. At various intervals, he asked his advisers whether it would be better to assign police cars to lead the power trucks, and whether the trucks had arrived at the plant yet. When a secretary said they would have the police manage the trucks' progress, the prime minister took no notice and gave instructions that their progress should continue to be reported to him.

In the end, not one of these trucks was any help. The first trucks that the government dispatched from Japan's Self Defense Forces reached the Fukushima Daiichi plant around 9:30 p.m., but couldn't be used immediately because flooding had damaged all of the metal-clad switches; thus, the vehicles couldn't be hooked up to the plant's power systems. The trucks that were ultimately used were TEPCO's high-voltage power-generator trucks, which arrived on-site after midnight.

According to several people who were at the Cabinet building that night, the core Cabinet team's mistrust of TEPCO began on the night of March 11. Top officials pressed TEPCO for a clear answer on why the Fukushima Daiichi operators couldn't restore the power supply despite the arrival of the power-generator trucks. When they got no simple explanation, the core team began to feel misgivings concerning the capabilities of the TEPCO personnel. As later related by Chief Cabinet Secretary Edano: "Frankly, it was at this time that doubts began to grow concerning TEPCO."

March 11—at the nuclear power station

Fukushima Daiichi's crisis control room was located on the second floor of a building specifically designed to withstand earthquakes. As the afternoon of March 11 progressed, hundreds of agitated workers crowded into that building, and people clustered around the television screens that flashed updates and showed footage of the tsunami sweeping through harbors and coastal towns. Powerful aftershocks repeatedly shook the building, causing workers to shout in alarm.

Inside the crisis control room, Masao Yoshida, the manager of the Fukushima Daiichi station, sat at a table and tried to collect information from his staff about the status of each of the reactors. It was not easy: The emergency phone system used for internal communication at the plant had been knocked out when the electricity went down. Workers scrambled to find working mobile phones and sometimes ran across the site to convey messages in person. And with no power at all in the joint control room for reactors 1 and 2, not even batteries to power indicator lights, operators there struggled to determine such basics as the water level inside each reactor pressure vessel.

Since they couldn't confirm that any coolant whatsoever was fed into those reactor cores, operators declared another "special event" under the auspices of the Nuclear Emergency Act. At 4:45 p.m., TEPCO informed the government that reactor core cooling might have failed at units 1 and 2.

In the pitch-black control room for these two units, operators worked with flashlights to determine the status of the last-resort coolant feeding system for the forty-year-old unit 1, the oldest reactor at the site. This cooling system, called an isolation condenser, was designed to operate without power. It simply piped steam from the reactor core through a "heat exchanger" tank filled with cold water; there, the steam condensed back into liquid water, which passively flowed back into the core to cool it. With all power lost to unit 1, this system was the only way to keep the water inside the reactor core from boiling away and exposing the radioactive fuel rods, allowing them to melt.

With the benefit of hindsight, the Fukushima Daiichi crisis appears to have been a chain reaction of failures, in which each breakdown made it harder for the workers to contain the damage and thus led to further breakdowns. The first incident in the chain was the near-total disabling of the isolation condenser system, which led to core damage in unit 1, and the explosion of hydrogen gas inside unit 1's reactor building. That explosion was followed by damage to unit 3's reactor core and then hydrogen explosions inside unit 3's reactor building, as well as at unit 4's reactor building, and damage to unit 2's reactor core. But it all started with a few closed valves on the isolation condenser inside unit 1.

On the evening of March 11, workers tried first to determine if the isolation condenser system was functioning at all. With no indicator lights working, workers couldn't determine if water was flowing through the reactor core. Because pipe ruptures in the system could cause water to leak outside the primary containment vessel, the system is equipped with four valves and with circuitry to detect pipe ruptures. Later investigations revealed that, when power was lost due to the tsunami, a fail-safe mechanism built into this rupture-detection circuitry began to close all four valves, shutting off the flow of water through the isolation condenser. AC power is necessary to ensure that a containment vessel's valves fully close. But, because AC power was cut off in the midst of the closing process, the two valves on the outside of the containment vessel were fully closed, while the two valves inside the vessel were left in a "partially open" state.

During the first hours of the emergency, the operators tried various tactics to determine if the valves were open or closed, and tried repeatedly to manipulate the valves. By 6:18 p.m. on March 11, operators were already concerned that the valves in the isolation condenser system were closed and could not be opened to start the flow of water. However, they were unable to communicate this concern quickly and accurately to the plant's crisis control room.

As a result, plant manager Yoshida and the rest of the staff at the crisis center believed that unit 2 posed a greater danger and proceeded under the incorrect assumption that unit 1's coolant feeding system would continue to function. Interviews conducted for this investigation make clear that this misconception was shared by officials at TEPCO headquarters and passed along to key members of government agencies.

In fact, during those first hours of the accident, the water level inside reactor 1 was already falling fast. Workers finally restored some power to the control room for units 1 and 2 with a small portable generator, and a reading taken at 9:19 p.m. showed that the water level inside unit 1's reactor pressure vessel was only 8 inches[8] above the top of the fuel rods. However, that reading from the

---

[8] 200 millimeters.

water-level instrument is now believed to have been inaccurate—it's likely that the situation was already dire. By 9:19 p.m., no cooling water had been injected into the unit 1 reactor core for nearly six hours; it's likely that the fuel rods were actually exposed to the air at this point, and that the uranium fuel was overheating. As heat pulsed through the rods, their protective zirconium cladding reacted with steam inside the vessel, creating volatile hydrogen gas that began to accumulate inside the vessel. The first meltdown had begun inside reactor 1, and the danger was growing.

Unit 2 contained a different coolant feeding system, called a reactor core isolation cooling system, which functioned by taking steam from inside the pressure vessel and using it to drive a turbine, then using the turbine to power a pump that injected chilled water into the reactor core. Without working indicator lights, workers didn't know if this system was functioning. Later that night, a contingent of workers would wade into a dark and flooded room in unit 2's reactor building in an attempt to check; they couldn't get close enough to confirm its operation, but guessed from mechanical sounds in the room that the system was probably functioning.[9] After this, as subsequent investigations revealed, they could finally confirm that the pressure of the pump was higher than that of the reactor pressure vessel. It meant that water could be pumped into unit 2's reactor core by the system.

The correct urgency of priorities wasn't properly revealed until after 11 p.m. on March 11, when workers in the control room for units 1 and 2 were able to bring the pressure gauges back on line for the two reactors. The readings revealed that the pressure in unit 2's containment vessel was within manageable parameters, but unit 1's vessel had already exceeded the maximum operational pressure.

The message to the on-site operators was clear: Unit 1's containment vessel needed to be vented immediately. Wet venting would release gases, through steam created by the decay heat, into the open air. Even though a small amount of radioactive materials would have been released with the steam, the operators should have brought down the pressure inside the containment vessel. In the crisis control room, all attention shifted to unit 1's heightening danger. A build up of pressure was a recipe for the containment vessel to break and for steam—ferrying a large amount of radioactive materials—to erupt uncontrollably. Unit 1 had to be vented.

March 11—at the Cabinet

Back in Tokyo, on the evening of March 11, the top officials in the Cabinet building were struggling to understand the basics of the situation that they were simultaneously trying to control.

---

[9] Institute of Nuclear Power Operations (2011) *Special Report on the Nuclear Accident at the Fukushima Daiichi Nuclear Power Station.* Atlanta, INPO.

The government's nuclear emergency-response manuals called for a national response headquarters to be established at the Cabinet building and a local response center to be established in the vicinity of the accident. That local control center was meant to be a place to gather regional government officials, police, firefighters, and nuclear operators. The center would gather information and pass it on to the agencies that could take action and was also meant to coordinate the evacuation of nearby towns. That is, it was meant to play the role of a "control tower" for emergency-response measures.

However, with the earthquake and tsunami's devastation spread throughout Fukushima Prefecture, officials had a difficult time establishing an effective local response center. The first location chosen—a safety inspection office about 3 miles from Fukushima Daiichi—had no external power, malfunctioning emergency power systems, and few working communication systems. So officials relocated to a radiation-monitoring center nearby, but there were further difficulties in getting personnel to that location, since many roads and highways were damaged or blocked by debris. Several days into the accident, as radiation levels in the area steadily climbed, the control center would relocate again to the city of Fukushima.

The unanticipated problems in establishing a local response headquarters is just one example of the inadequacy of the government's emergency-response manuals, which simply didn't anticipate such a large-scale disaster that combined a nuclear plant emergency, tsunami, and catastrophic earthquake.

Instead of having a local headquarters to direct the accident response and disseminate information, the actual procedure in the first days of the Fukushima Daiichi accident went something like this: Operators at Fukushima Daiichi phoned in information to TEPCO headquarters, TEPCO staff called officials at NISA, and those officials in turn passed the information along to Cabinet staff and sometimes directly to top Cabinet officials. Those leading politicians, most of whom were completely untrained in nuclear reactor operations, then tried to decide what to do.

Shortly after 9 p.m. on March 11, the top government officials huddled in the Cabinet building's mezzanine room were joined by two nuclear power experts: NSC Chairman Haruki Madarame and TEPCO Fellow Ichiro Takekuro. As soon as he arrived, Madarame recalled later, he had to begin answering rapid-fire questions from the prime minister in "a situation with no handbooks and with no one to confer."

The prime minister's team quickly got a crash course in nuclear power operations: That evening, the administration group plied the experts with elementary questions relating to the plant and its accident response. Madarame has stated that, immediately after he arrived at the Cabinet, he and Takekuro agreed that venting would be necessary to stabilize the overheating reactors. But the other officials asked basic questions such as what a vent was, why venting was necessary, and how much radioactive material would be released by venting.

Chairman Madarame, representing the opinion of the experts, explained that venting wouldn't lead to a widespread diffusion of radioactive elements and that it wouldn't create anything as serious as Chernobyl's famous 19-mile no-entry zone.[10] Madarame initially believed that only a "wet venting" would be necessary at Fukushima Daiichi, in which gas would be vented into a water-filled suppression chamber inside the reactor building, thus releasing very little radioactive material into the atmosphere. A small evacuation zone would serve as adequate precaution, he advised.[11]

The administration group had trouble following the difficult technical explanation, but in this early stage of an unprecedented emergency situation, they still placed considerable trust in the experts.[12] At 9:23 p.m., with a firm consensus on the need for venting, Prime Minister Kan ordered the evacuation of anyone within 2 miles of the Fukushima Daiichi station. A little more than an hour later, TEPCO Fellow Takekuro called the TEPCO head office and urged them to "quickly reach a corporate decision" on venting. Within the Cabinet building, the nervous officials waited.

The venting procedure may have seemed much delayed from the Cabinet building, but at the power station, workers were doing their best to set the process in motion. On March 12 at 12:06 a.m., plant manager Yoshida issued instructions to prepare for venting at unit 1. Venting is usually triggered remotely from the control rooms, but in view of the loss of power, the only possible option was to open the vent lines manually. In the plant's crisis center, workers began reviewing the detailed procedures for working with the vent line valves, but preparations would take some time.

TEPCO's head office had approved the venting, and at 1:30 a.m. on March 12 TEPCO formally proposed the venting procedure to the Cabinet. It was the understanding at the Cabinet that venting would begin immediately after the Cabinet press conference scheduled for 3 a.m., but in fact the operation was delayed further, and many more hours would tick by while the Cabinet officials waited for updates. In interviews conducted for this report, government officials stated that this long delay increased their mistrust and doubts of TEPCO.

March 12—at Okuma

At 5:45 a.m. on March 12, the government issued an evacuation order for everyone living within a 6-mile radius surrounding the Fukushima Daiichi Nuclear Power Station. That included almost the entire towns of Futaba and Okuma.

---

[10] Personal communication with Deputy Chief Cabinet Secretary Fukuyama, October 29, 2011.
[11] Personal communication with related Cabinet personnel, n.d.
[12] Personal communication with Deputy Chief Cabinet Secretary Fukuyama, October 29, 2011.

In Okuma, a town of 11,500 residents, Mayor Toshitsuna Watanabe had been managing his community's response to the tsunami from the town hall. The waters had washed away thirty houses, and eleven of the town's citizens were missing. Town officials were helping their shocked and grieving citizens. Mayor Watanabe was at his office in the town hall, only 3 miles from the Fukushima Daiichi station, when he received the early morning call from the Fukushima prefectural government ordering the evacuation.

The town had never before evacuated due to trouble at the power station. Watanabe had grown up in Okuma, had watched the plant's construction in the late 1960s, and had many friends and neighbors who reported to work at Fukushima Daiichi every day. There were so many layers of protection at the power station, he thought, that even if an accident occurred, it would be swiftly contained.

Even that morning when the evacuation order came, Watanabe still didn't believe that the plant's problems were very serious. He kept watch as fire engines paced through the streets, broadcasting the evacuation order on their loudspeakers. Okuma's citizens began boarding buses at 8 a.m., heading for a gymnasium in the town of Tamura, 28 miles away and across a ridge of mountains. Watanabe watched the buses fill, then climbed into a car to follow them down the highway. He expected to be gone from his home for five days at the most. Instead, two years later, he still lived in exile. Gradually, the 66-year-old mayor came to accept that he would probably never live in his hometown again.

March 12—at the Cabinet

In the late hours of March 11 at the Cabinet building, officials were still hearing alarming reports about unit 2 at the Fukushima Daiichi Nuclear Power Station. That night, at 10 p.m., NISA predicted that unit 2's fuel would begin to melt in the next few hours and stressed the need for venting. METI Minister Kaieda soon received information that the unit's cooling systems were operational, but he didn't effectively share this information with his government colleagues.[13]

Officials from METI, NISA, and TEPCO conducted a joint press conference in the dark hours of the morning on March 12. At 3:06 a.m., the officials announced that the venting was scheduled to begin immediately—although they gave contradictory explanations about which of the power station's reactors were to be vented. Chief Cabinet Secretary Edano held a press

---

[13] Cabinet Office, Government of Japan (2011) *Interim report of the Government Investigation Committee on the Accident at TEPCO's Fukushima Nuclear Power Stations.* Report, Government Investigation Committee on the Accident at the Fukushima Nuclear Power Stations, December 26. Tokyo: Cabinet. Available at: http://jolisfukyu.tokai-sc.jaea.go.jp/ird/english/sanko/hokokusyo-jp-en.html.

conference at 3:12 a.m. to announce the Cabinet's approval of immediate venting for unit 1.

Yet the venting did not happen. When Minister Kaieda asked TEPCO Fellow Takekuro, "Why hasn't the venting begun?" his response was: "I don't know." Minister Kaieda's doubts concerning the power company deepened.[14]

Following the press conference, at about 3:45 a.m., the TEPCO head office performed an estimate of radiation dose exposure at the time of venting in the surrounding region, and shared the results with the Fukushima Daiichi operators. The estimate predicted that people 2.6 miles to the south of the plant would get a dose of 28 millisieverts if the plant conducted a "dry venting," in which gas is released directly from the containment vessel to the atmosphere, without passing it through the water in the suppression chamber. The Fukushima Daiichi Nuclear Power Station reported this to NISA at 4:01 a.m., but it did not communicate this to the top echelons of the Cabinet. NSC Chairman Madarame later remarked that officials "forgot" the possibility of dry venting. "In that case," he said, "a 3-kilometer [2-mile] evacuation is not enough. A 10-kilometer [6-mile] evacuation is essential."

Around 5 a.m., Prime Minister Kan also learned that no venting had yet been performed. When he sought an explanation from TEPCO Fellow Takekuro, his response was that electrically driven venting could not be performed because of the power outage.[15] At 5:44 a.m., the evacuation order was expanded to include everyone within a 6-mile radius of the Fukushima Daiichi station. This measure was taken due to fears of an eruption if unit 1 couldn't be vented and pressure continued to build.

Early on the morning of March 12, Kan decided that he must visit the Fukushima Daiichi Nuclear Power Station. A number of political leaders initially opposed this plan, including Chief Cabinet Secretary Edano, METI Minister Kaieda, and Deputy Chief Cabinet Secretary Fukuyama. Edano pointed out that it would inevitably result in political criticism, but Kan replied by asking whether that possibility was more important than bringing the nuclear plant under control. Edano then responded, "Go ahead if you are aware of the consequences."

Meanwhile, at the Fukushima Daiichi station, there was little enthusiasm for the prime minister's visit. According to witnesses, plant manager Yoshida expressed reluctance concerning the sudden visit by Kan. A heated discussion ensued during a videoconference with the TEPCO head office, in which Yoshida asked what could possibly be gained by having him meet the prime minister.[16]

At 6:14 a.m., Kan boarded a Self Defense Forces helicopter from the roof of the Cabinet building. NSC Chairman Madarame was on board the helicopter

---

[14] Personal communication with METI Minister Kaieda, December 10, 2011.
[15] Personal communication with Deputy Chief Cabinet Secretary Fukuyama, October 29, 2011.
[16] Personal communication with NISA personnel, February 3, 2012.

as well, and he wanted to convey various concerns to Kan during his flight to the site. But Kan shouted at him that he already knew the basics and told him to just answer his questions. Madarame has stated that he had wanted to convey more information, but that it would have taken a very brave heart to speak loudly and strongly when confronted by Kan.

A series of questions and answers then followed during the flight. At one point, the prime minister asked whether a hydrogen explosion would occur. Madarame answered that no explosion would occur inside the primary containment vessel, because the oxygen in the vessel had been completely replaced by nitrogen, an inert gas.[17] His answer was based on the assumption that the containment vessel was still functional, and that it would prevent gases from leaking into the larger building. When a hydrogen explosion actually did occur in unit 1 later that same day, Madarame immediately lost credibility with the prime minister.

While Kan was in flight toward Fukushima Daiichi, government ministers in Tokyo were still trying to push through the venting. At 6:50 a.m. Kaieda of METI stated, "Do it under my authority." The order to perform venting was finally issued orally at 6:50 a.m., under provisions of the Nuclear Regulation Act. The reactor number was not specified at that time, but internal government documents indicate that NISA viewed the order as applying to both units 1 and 2.

Kan arrived at Fukushima Daiichi at 7:11 a.m. Immediately after boarding the bus awaiting him there, he began cross-examining TEPCO Vice President Sakae Muto, who was seated next to him, and pressing him to explain why venting had not yet begun. When Muto explained that the electric valves could not be opened due to a lack of electric power, the prime minister shouted at him that he had not come to listen to excuses.[18] He repeatedly demanded that the venting be conducted as soon as possible.

The discussion continued in the Fukushima Daiichi crisis control center. Kan finally relented when Yoshida stated to him that the venting would be accomplished even if it meant the formation of a suicide squad. At 8:03 a.m., Yoshida directed that the venting operation be performed with a target starting time of 9 a.m., and the prime minister left Fukushima Daiichi at 8:04 a.m. After returning to the Cabinet, he conveyed his view of the on-site situation to Edano: "Yoshida is competent. We have to use him as the key person."[19]

At 8:27 a.m., TEPCO learned that the evacuation was not finished in one nearby town, and TEPCO and Fukushima Prefecture mutually confirmed that venting would begin after resident evacuation from the 6-mile zone was completed. The Cabinet was not informed that workers were waiting for the evacuation to be completed, and when Edano learned of this several months

---

[17] Personal communication with NSC Chairman Madarame, December 17, 2011.
[18] Personal communication with NSC Chairman Madarame, December 17, 2011.
[19] Personal communication with Chief Cabinet Secretary Edano, December 10, 2011.

later, he expressed complete surprise. Completion of the evacuation was reported at 9:02 a.m.; at 9:04 a.m., workers entered unit 1's reactor building to perform the venting operation.

According to the report of the government's investigation committee, every attempt to begin the venting was already in progress at Fukushima Daiichi that morning. The decisions by the Cabinet, the order issued by METI Minister Kaieda, and Prime Minister Kan's visit were not found to have contributed in any respect to its early implementation.

March 12—at the nuclear power station

Throughout the morning of March 12, the response teams at the Fukushima Daiichi Nuclear Power Station were active on three distinct efforts. One team had prepared for the TEPCO power supply trucks that had arrived in the night, and in the early morning hours, the workers were trying to connect the trucks' mobile generators to the plant's electricity system. It wasn't easy. In unit 2, they found one power center that had survived the flood and was able to distribute low-voltage AC power; they began laying long, heavy cables from the truck to the power center. If they could supply electricity to the power panel, they knew, they could turn on a standby liquid control pump that would inject cooling water into unit 2.

At the tsunami-smashed facility, the crew began work in the dark, stumbling over debris and wading through standing water. The 200-meter-long power cables weighed up to 1 ton each. Usually trucks are able to handle such cables, but that night and throughout the next morning, forty workers did the job by hand. The power-delivery preparations were finally completed around 3:30 p.m. on March 12.

While the power-cable team did their work, a separate team had been scrambling through the night to implement a quicker way to inject cooling water into the overheating reactor cores.

At 1:48 a.m. on March 12, workers realized that the diesel-powered fire-prevention pump was not operating, which meant that pump could not be used to send water to the core. Workers brought fresh diesel fuel to the pump and fresh batteries to its engine, but were unable to get it working. But operators knew that, after the 2007 Chuetsu-oki earthquake, TEPCO had installed a connection allowing the fire-prevention system to be linked to mobile fire engines outside the building. Therefore, they decided to use fire engines to inject coolant into the reactor core. But they needed to locate a fire engine and a water source for this operation.

At 2:45 a.m. on March 12, workers realized that pressure had dropped in unit 1's reactor pressure vessel, which would allow them to inject water using the fire-prevention line without depressurizing the reactor pressure vessel.

There were three fire engines that should have been available to connect to those lines, but one had been wrecked by the tsunami, and a second was stuck near units 5 and 6, unable to reach the site of the main crisis because of earthquake-shattered roads. Crew members slowly drove the only useable fire engine through the power station as other workers cleared debris from its path and eventually smashed a lock on an electronic security gate to allow the truck access to unit 1.

Even once the fire engine reached unit 1, obstacles remained. Crew workers had to use the truck to collect water from a freshwater storage tank, and then drive the truck slowly and carefully under a damaged building to unit 1, where the vehicle was finally connected to the water lines so it could empty its load. Ultimately, the truck had to be driven back to the storage tank to do it all again. After this laborious process, workers finally found the available water reservoir to establish a direct connection to the intake lines on unit 1's reactor building, allowing for a continuous injection of freshwater. At 5:46 a.m., workers constantly injected water into the overheated reactor core of unit 1 for the first time since the tsunami knocked out Fukushima Daiichi's power—more than fourteen hours before.

Between 6 and 7 a.m. on March 12, two Self Defense Forces fire engines arrived at Fukushima Daiichi; later, at 10:52 a.m., one fire engine arrived from a nearby nuclear power station. Workers injected a total of 80 tons of water before the water storage tank ran dry at 2:53 p.m. Plant manager Yoshida then ordered workers to use the trucks to take seawater from a flooded pit. Saltwater is never allowed inside the core of a nuclear reactor, because it would corrode the steel walls of the vessel and leave behind mineral residue on fuel rods, which would interfere with the nuclear reaction. Yoshida's order to use seawater was as good as an admission that unit 1 could not be saved. The workers could only hope to avert a broader disaster.

The third front of the response was the team preparing to vent the containment vessel in unit 1. By 8:37 a.m. on March 12, the prime minister had come and gone, and workers in the Fukushima Daiichi crisis center notified the prefectural government that they were preparing to vent unit 1 at around 9 a.m. They then waited for word that all nearby residents had been evacuated before beginning to vent.

At 9:02 a.m., the crisis center notified the team leaders at unit 1 that the evacuation of citizens was complete and instructed them to begin the venting procedure. (In fact, some citizens were still within the 10-kilometer radius at that time.) At 9:04 a.m., the operation began. For the procedure, it was necessary to set two valves on the vent lines to the "open" position. At 9:15 a.m., the first team succeeded in manually setting the first valve to "open" on the second floor of the unit 1 reactor building.

At 9:24 a.m., the second team proceeded to the first basement floor of the reactor building and began to move through the dark rooms in order to set the second valve to "open." As they followed their flashlight beams deeper into the basement, however, their handheld dosimeters began to flash troubling numbers. Radiation levels were extremely high; if the workers carried on, they would likely receive doses in excess of the 100-millisievert limit that, in emergency situations, a nuclear operator is permitted to receive in a single year. At 9:30 a.m., the workers had no choice but to abort the valve-opening operation and turn back.

In the crisis control center, the operators realized they had to open the valves by restoring remote-operation capability, despite the greater uncertainty of this procedure. The workers scrambled to locate a portable air compressor that could be connected to the pipe system in order to blast the valve open with air and struggled all morning and afternoon to put it into place. Finally, at 2:50 p.m., the containment vessel's pressure fell significantly, suggesting that the valve had indeed opened, letting pressurized gas rush through the pipes to the nearby ventilation stack. Thus, more than fourteen hours after Yoshida issued the venting order, and more than five hours since workers began their mighty effort to carry out the operation, unit 1 finally began venting successfully. It seemed like a huge step toward stability and safety.

By 3:30 p.m. on March 12, the workers at the Fukushima Daiichi plant had been laboring for twenty-four hours without rest, fighting to restore control of the station. By that time, the fire engines had pumped more than 21,000 gallons[20] of freshwater into unit 1, and they were preparing to start injecting seawater. Power was about to be turned back on via the generator trucks and the power center in unit 2, which would restore the unit's motor-driven pump. And the venting was underway at unit 1's primary containment vessel, which workers hoped would relieve pressure and prevent the containment vessel from breaking.

But it was simply too late: Too much had already gone wrong. Inside the reactor core of unit 1, the fuel rods had melted and slumped to the bottom of the reactor vessel, creating a molten sludge of uranium, zirconium cladding, and other metals. The zirconium had reacted with steam inside the reactor pressure vessel to create volatile hydrogen gas—which is highly combustible when combined with oxygen and will explode with the smallest spark.

At 3:36 p.m., that spark appeared. An explosion shattered the unit 1 reactor building, blowing away thick concrete walls and, on the upper floors of the reactor building, leaving behind only a skeleton of steel reinforcement beams.

The flying debris and falling rubble injured five people (three TEPCO staffers and two contract workers). Uninjured workers rushed to the emergency control center to regroup. The blast spread highly radioactive materials throughout the site, making the grounds of Fukushima Daiichi even more dangerous for its workers. The explosion pummeled the fire engine with shrapnel and tore through the fire hoses. And it bombarded the generator trucks and the heavy cables that had been so carefully stretched through the site to the power distribution centers. All operations, including the effort to inject seawater into the unit 1 reactor core, came to an immediate halt.

---

[20] 80,000 liters, or 80 tons.

And the task of containing the crisis within the Fukushima Daiichi Nuclear Power Station became exponentially more difficult.

March 12—at the Cabinet

Even before the explosion at unit 1, tension mounted at the Cabinet building. Top administration officials were becoming increasingly mistrustful of both TEPCO and the Nuclear and Industrial Safety Agency. While NISA should have had eight safety inspectors on-site at Fukushima Daiichi to serve as its "eyes and ears," those inspectors were evacuated early on the morning of March 12. NISA officials were reduced to calling TEPCO headquarters and asking for updates, and the information they received was often inadequate and dated. That only increased top officials' distrust of the agency and left them with a feeling of, "then it's up to us" to directly control the escalating crisis.[21]

On the first day of the accident, key officials conducted their meetings in the Crisis Management Center in the basement of the Cabinet building. However, that location had a serious drawback: As a security measure, cell phones were forbidden inside the crisis center, and only two telephone lines were available in the conference room. The officials felt they couldn't make well-informed decisions in such isolation, and in the early afternoon of March 12, they moved to the Prime Minister's Office and the adjoining reception room on the building's fifth floor.

Prime Minister Kan had long distrusted Japan's bureaucracy, and this attitude colored his accident response. "Kasumigaseki [home to Japan's national government] only provides information convenient to their own purposes," he said during a later interview with our team. According to a Cabinet staff member, Kan repeatedly expressed frustration at the lack of any information from the bureaucracy during the accident; each time, the related ministries would then rush to produce descriptions and explanations, only to be rebuffed when they presented these to the prime minister. Kan had had enough of their long administrative presentations, he told them.[22]

Instead, Kan relied on trusted advisers, such as his special advisers Hosono and Terada, even though they had no role in the organizational command line. The prime minister also used his own cell phone to call technical experts outside the government for advice and information, which he did not always share; government officials seated with him were often reduced to inferring the content of the conversation from the fragments they heard.

Kan took personal responsibility for small particulars of the accident response. For example, when he heard that replacement batteries were needed at Fukushima Daiichi, he took out his cell phone and called the plant to ask for details of the batteries needed: their size, height, and weight, and whether they

---

[21] *Yomiuri Shimbun* (2011) Prime Minister Kan interview. (In Japanese.)
[22] Personal communication with related Cabinet personnel, January 23, 2012.

could be transported by helicopter. As his ministers stood watching, Kan diligently took notes on the exchange. One observer recalled shuddering and wondering what would become of the nation with the prime minister spending his time managing such details.

When the explosion rocked Fukushima Daiichi's unit 1 at 3:36 p.m. on March 12, Kan was in a conference with opposition party leaders. On returning to his office, he was informed that white smoke was rising from unit 1. When Deputy Chief Cabinet Secretary Fukuyama asked NSC Chairman Madarame about the source of the smoke, Madarame suggested that volatile materials inside the building might be burning.

Staffers moved a television and a white board into the reception room adjoining the Prime Minister's Office. About an hour after the initial report of white smoke, one of the prime minister's special advisers burst into the room and switched television channels. The screen now showed repeated video images of unit 1's reactor building being shattered by a large explosion and emitting large volumes of white smoke. Kan registered his surprise at the great disparity of this scene from the explanation previously provided by Madarame and asked him: "Isn't that white smoke? Isn't that an explosion? Didn't you tell me that no explosion would occur?" Madarame only responded by groaning and cradling his head in his hands.

When Kan demanded an immediate report from NISA and TEPCO, the utility responded only that staff members were proceeding on foot from the Fukushima Daiichi control center to perform observations at unit 1. Fukuyama asked Madarame whether they were actually viewing a disaster on the scale of Three Mile Island or Chernobyl, but Madarame gave no clear answer. Madarame later testified that, in viewing the televised images, he immediately realized that unit 1 had experienced a hydrogen explosion, but that he was simply too stunned to tell anyone. He was also concerned because he had previously told the prime minister that such an explosion couldn't occur.

Top government officials still didn't have a clear understanding of what had happened hours later, at 5:45 p.m., when it came time for Chief Cabinet Secretary Edano's scheduled press conference. Based solely on information that no increase in radioactive dose had been found at the main entrance to the Fukushima Daiichi site following the explosion, Edano began calling on the public to remain calm while recognizing the occurrence of "an explosion event of some nature." Reporters pressed him for more information on unit 1 and whether the 10-kilometer evacuation zone was sufficient. Edano's responses remained noncommittal, as he himself lacked answers to those questions. He later stated that it was the hardest press conference he had ever conducted.

Eventually TEPCO and NISA provided a report to the Prime Minister's Office that confirmed the hydrogen explosion. It was not until another press conference at 8:41 p.m.—about five hours after the explosion—that the government gave a specific explanation of the event to the panicked and confused Japanese public. That evening, Kan also extended the evacuation order to all residents living within 12 miles of the Fukushima Daiichi Nuclear Power Station.

While Edano conducted the initial press conference about the explosion, top officials on the fifth floor continued to talk about a topic that had been raised earlier that day: the importance of continuing to feed water into unit 1 and the possibility of using seawater if freshwater injection proved difficult. Saltwater, however, is corrosive and would damage the reactor, ensuring that it never again would produce electricity for Japan. Nonetheless, at 5:55 p.m., METI Minister Kaieda issued an administrative order, under the Nuclear Regulation Act, to have the unit 1 reactor filled with seawater.

Prime Minister Kan entered the room at 6 p.m. When informed by Kaieda of the plan to inject seawater, he raised several questions. "Do you understand the problems?" he asked. "It contains salt. Have you considered its effects?" Madarame explained risks such as channel blocking and corrosion during the course of the discussion, while Prime Minister Kan, in very strong terms, asked all present whether there was risk that recriticality could occur in the damaged fuel inside unit 1 if seawater were injected.

Yet Madarame replied: "The possibility of recriticality is not zero."[23] Prime Minister Kan responded: "And isn't that a serious problem?" In the ensuing discussion, TEPCO Fellow Takekuro noted that it would take an hour and a half to prepare for the injection of seawater because of damage to the hoses. At the end of the meeting, the prime minister instructed the members in the room to evaluate the risks of using seawater and to consider injecting boric acid to prevent recriticality.

Some participants were surprised by the comments of Madarame and the reaction of the prime minister at the meeting and worried that they might prevent the seawater injection. Immediately after the meeting, a few officials assembled in a small room near the Prime Minister's Office; those present included Madarame, Takekuro, METI General Affairs Division Director Tadao Yanase, and NSC Deputy Chairman Yutaka Kukita.

Madarame stated, "When asked in that way, as an engineer, that was my only possible answer,"[24] but the participants all immediately agreed that seawater injection was urgently needed. Yanase noted, "If we fail in this, the consequences will be serious," and the participants then confirmed the content of the explanations to be given to the prime minister in order to allay his doubts and rehearsed these explanations. Madarame commented that he would like to ask Kukita to undertake the explanation concerning recriticality, and Kukita assented.

---

[23] This is the phrasing as accepted by Chairman Madarame and internally adopted by the government relating to its response to the Diet and other matters; however, related personnel who attended the meeting recalled that Chairman Madarame actually said "There is a possibility of recriticality."

[24] Personal communication with related Cabinet personnel, January 24, 2012.

The meeting with Kan resumed at 7:40 p.m. As planned, Takekuro explained TEPCO's desire to carry out the seawater injection. Kukita then explained, in reference to the earlier response by Madarame, that "the possibility of recriticality is extremely small, and the need for seawater injection is extremely high." Special Advisor Hosono also reported the confirmation that the water injection hoses could be used. Kan accepted this series of explanations and, at 7:55 p.m., instructed Kaieda to have seawater injection performed. Madarame did not participate in the discussion.

Ironically, while the officials on the fifth floor discussed the risks and benefits of injecting seawater into unit 1, that procedure had already begun. At the Fukushima Daiichi Nuclear Power Station, workers had begun the operation at 7:04 p.m. This information had reached the crisis center in the basement of Prime Minister Kan's office, but it hadn't made it upstairs.[25]

March 12–14—at the nuclear power station

On the afternoon of March 12, workers assessed the damage from the explosion at unit 1. While the building had shattered, workers believed that its pressure vessel was still largely intact and resumed their preparations to inject seawater into the core. They hoped to keep the fuel rods covered with water to keep the melted uranium from eating through the steel of the pressure vessel, which would allow a much larger release of radioactive materials into the environment. Workers stretched long hoses from a pit near unit 3's turbine building and connected them to the fire-prevention lines in unit 1's turbine building.

At TEPCO, the initial order from METI Minister Kaieda to inject seawater into unit 1 was shared at 6:05 p.m. in an internal meeting. Even before the order came, however, preparations had been underway, and at 7:04 p.m., seawater began flowing into the reactor core of unit 1. But shortly thereafter, plant manager Yoshida received a telephone call directly from TEPCO Fellow Takekuro at the Cabinet, who essentially told him that the prime minister's approval had not yet been obtained and that the seawater injection operation should therefore be suspended.

According to the report from the government's investigation committee, Yoshida then contacted TEPCO Vice President Muto and others at the company's head office to discuss response options. They told Yoshida that the head office considered the suspension unavoidable. Nevertheless, Yoshida was deeply concerned that a delay in water injections would cause more damage inside unit 1, and he decided to take it upon himself to continue the seawater injection. He directly instructed the water injection manager to continue.

---

[25] Cabinet Office, Government of Japan (2011) *Chronology report on Fukushima accident*. Report. Tokyo: Cabinet.

While the TEPCO head office reported that the injections would be suspended, in fact, they never were.[26]

At around 8:45 p.m. on March 12, workers began mixing boric acid with the seawater in the pit. Boric acid absorbs neutrons and would therefore prevent a nuclear fission chain reaction from starting up again in the sludgy, melted nuclear fuel inside unit 1.

While workers contended with the water injections at unit 1, conditions in unit 3 had been rapidly deteriorating. Since the earthquake and the scram of the active Fukushima reactors, a coolant feeding system had been continuously injecting water into the reactor core of unit 3, keeping it relatively stable. However, at 11:36 a.m. on March 12, that coolant feeding system shut down unexpectedly. The water level in the reactor began to fall, and soon after 12 p.m. the high-pressure coolant injection (HPCI) system, another method of cooling the reactor core that does not require an AC power supply, automatically started up. But that worked for less than a day; in the evening, the injection of water from the system gradually diminished.[27] Adding to the workers' difficulties, at 8:36 p.m. the batteries for unit 3 started to dry out, unbeknownst to the operators, and the water-level meter for unit 3 clicked off. That failure meant that workers could no longer monitor the water level in the troubled unit 3's reactor core.

Before dawn on March 13, workers manually shut down the failing HPCI system and decided to switch over to injecting water into the reactor via fire-prevention pipes, powered by a diesel-powered pump. But first the pressure had to be reduced inside the pressure vessel, otherwise the pump wouldn't be powerful enough to force water inside. At 2:45 and 2:55 a.m., workers attempted to bring down the pressure by manipulating safety release valves from the control room, but their efforts failed. With the loss of battery power in unit 3 the valves couldn't be powered. With a growing sense of desperation, workers attempted to restart both the reactor core isolation cooling system and the HPCI system, but failed in both cases.

Top management at the Fukushima Daiichi crisis center, at around 3:55 a.m., recognized that the HPCI had shut down, and managers realized that unit 3's reactor core was now in serious danger of overheating. They had to open those safety release valves. Around 6 a.m., operators ordered a search for batteries to power the valves. Workers fanned out across the facility and raided the car batteries from employees' personal automobiles that had been parked in high places and escaped tsunami damage. Workers lugged these batteries to the

---

[26] In its outcome, the discussion of seawater injections at the Cabinet had no effect. Yet it is possible that, if the Cabinet's request for suspension had been accepted, it might have led to a dangerous situation. It must also be noted that, although it did not result in a problem in the present case, Yoshida's contradiction of the stated intention of the Cabinet and the TEPCO head office constitutes a problem in terms of crisis management.

[27] The main purpose of the HPCI system is to inject a large amount of water in a short period of time. In other words, these systems were not designed to be used for days at a time.

control room for units 3 and 4, and used them to power the safety release valve control board. This ingenious solution allowed workers to bring down the pressure inside unit 3's reactor pressure vessel.

Workers had been preparing simultaneously for the water injection via the fire-prevention lines and, on Yoshida's instructions, had been stretching the hoses to a pit filled with seawater. By about 7 a.m. the connection was complete. However, around that time the Cabinet issued its own instructions, stating that workers should use freshwater rather than seawater if possible, and a TEPCO general manager passed on the directive to Yoshida. He decided to change over to freshwater injection, the fire hoses were re-laid, and freshwater began to gush into unit 3's reactor core at 9:25 a.m. on March 13.

However, for about seven hours, no drop of water had entered the core. During that interval, the water level in the reactor core decreased, the coolant exposed the fuel rods, and the rods began to melt. The zirconium cladding began to react with the steam, creating another dangerous build up of hydrogen gas.

About three hours after the freshwater injections began, the fire-protection tank that had been supplying the water ran dry. It was time to turn back to seawater. Workers extended a fire hose from a pit near unit 3's turbine building to inject seawater into the fire-prevention lines; that injection began at 1:12 p.m. In effect, the changeover to freshwater contributed little or nothing to an improvement of the situation and may have resulted in unnecessary exposure of the workers to radiation during their rerouting of the water lines. The Cabinet's instruction also delayed the progress of the on-site work.

During the morning of March 13, pressure built up in unit 3's containment vessel, the concrete and steel structure that serves as a secondary line of defense surrounding the pressure vessel. Throughout the day, workers tried a variety of difficult maneuvers to open a vent line for the vessel. To open two valves along the vent line, they had to move portable generators and air compressors into position around the reactor building, often working in pitch-black rooms. It was an ongoing battle. On the morning of March 13, steam billowed from the vent stacks, indicating that the lines were open. But a few hours later, the valves along the line were found closed, and workers had to force them open again. The struggle continued all day and night.

At around 1 a.m. on March 14, gauges showed that the pressure inside unit 3's containment vessel was on the rise again. Between 5:20 and 6:10 a.m., the valves of a different vent line were remotely set to the "open" position from the control room and a second vent line was established. Despite this operation, the reading on unit 3's pressure gauge continued to rise, and around 6:30 a.m. on March 14, plant manager Yoshida ordered workers to leave the vicinity of unit 3 and to take temporary shelter in the crisis control center. The evacuation order was withdrawn after the pressure stopped increasing.

However, workers were in harm's way at 11:01 a.m. on March 14, when another hydrogen explosion rocked the Fukushima Daiichi Nuclear Power Station. The unit 3 reactor building shattered, and the shrapnel injured four

TEPCO employees and three employees of collaborating companies. In addition, just prior to the explosion, a Self Defense Forces team had arrived with seven water tankers; at the time of the explosion, this team was in the process of filling a pit near the unit 3 turbine building with freshwater, and four team members were injured.

Except for a few operators in the control rooms for units 3 and 4, workers fled from the area around unit 3. The new explosion forced workers to stop injecting water into the reactor cores of units 1 and 3. The pit in front of unit 3's turbine building, which had been the source of water for units 1 and 3, was now littered with highly radioactive debris and rendered unusable. The explosion also damaged fire engines, pumps, and hoses that were being used to feed water into the reactor cores. Each disaster made the Fukushima Daiichi workers' tasks much harder, and increased the ominous feeling that this catastrophe could not be contained.

March 13–15—at the nuclear power station

After the reactor scram of unit 2 on March 11, water injection into the reactor began immediately via a coolant feeding system[28] that didn't require AC power. Fortunately, the system functioned for several days while workers focused on the crisis first at unit 1 and then at unit 3.

On March 13, plant manager Yoshida directed his staff to prepare for seawater injections into unit 2's reactor core, and workers stretched hoses from a seawater-filled pit near unit 3's turbine building to unit 2's fire-prevention system. But the preparedness was offset by the explosion at unit 3 on the morning of March 14. The explosion damaged the provisional venting system in unit 2, which also had been prepared on March 13. It was the worst time for unit 2's reactor core cooling system to lose its capacity.

Around noon on March 14, the water level in unit 2 began to decline precipitously, and at 1:25 p.m. workers determined that the reactor core isolation cooling system had shut down. Workers scrambled to connect new hoses to a shallow section of the harbor and lined up a fire engine to pump the seawater through to unit 2. However, as had been true at unit 3, workers couldn't inject the seawater until pressure decreased inside the reactor.

Workers brought car batteries to the control room for units 1 and 2 to power the safety release valves and opened the valves to bring down the pressure inside the reactor vessel. Around 7:03 p.m., it finally became possible to use fire engines to drive the injection of water into unit 2's reactor core, but the system was barely in place before workers encountered a new problem. At 7:20 p.m. that night, a worker realized that the fire engine had run out of fuel. The exhausted workers refilled the truck's gas tank, and at 7:57 p.m., the seawater again gushed into unit 2's reactor core. However, it was too late to

---

[28] The reactor core isolation cooling system.

avoid damage to the core. The control room's indicators had shown that the fuel rods inside reactor 2 had been completely uncovered at 6:22 p.m. The damage was well underway.

Efforts to vent unit 2's containment vessel had been proceeding as well, but workers encountered difficulties at every turn. Operators first tried to use a vent line that led from the suppression chamber, the doughnut-shaped structure that encircles the bottom of the reactor vessel. But workers struggled to open valves, and a "rupture disc" that was supposed to burst open at high pressure and allow for venting probably did not rupture—what is certain is that the pressure inside the containment vessel did not go down. Workers then tried to establish a different vent line and worked on valves and connections. To this day, it remains uncertain whether unit 2 was ever successfully vented.

Workers were still laboring to start the venting operation at 6:10 a.m. on March 15, when the pressure gauge for unit 2's suppression chamber, which had been acting strangely, suddenly went down to zero. Workers were confused and shocked, and some reported hearing a loud noise inside unit 2. There was no obvious damage to the unit 2 reactor building, but the roof of the neighboring waste treatment building was damaged. Based on this, it was first thought that an explosion had occurred near the unit 2 suppression chamber, and the incident was widely reported as another explosion. But later analysis of seismographic data showed that nothing with the full force of an explosion occurred at unit 2.

The same couldn't be said of unit 4. That reactor building unexpectedly shattered with an explosion at nearly the same moment in the early morning of March 15. The roof blew off, and workers ran for shelter.

Unit 4 hadn't been active at the time of the earthquake and tsunami. It was undergoing regularly scheduled maintenance, so the reactor was disassembled and all of its fuel rods had been removed from the core and placed in a spent fuel pool. Operators had assumed that it didn't pose an immediate threat. The blast was caused by hydrogen gas that flowed into unit 4's reactor building from the adjacent unit 3. But when the roof blew off, operators had no idea of the cause of the explosion. Many experts believed at first that the unit 4 spent fuel pool must have run dry, leaving the fuel rods exposed and creating clouds of hydrogen gas.

At around 7 a.m. on March 15, approximately 650 workers were temporarily ordered to leave the site for their own safety. Nearly the entire plant work force left Fukushima Daiichi, leaving behind about seventy supervisors and a skeleton crew of workers needed to conduct critical operations.

March 14–15—at the Cabinet

In the dark predawn of March 15, TEPCO President Shimizu called the prime minister's reception room and spoke with METI Minister Kaieda. There was a growing possibility of an explosion at unit 2, said Shimizu, and he was therefore seeking approval for evacuation of on-site personnel from the Fukushima Daiichi plant.

The idea of an evacuation had already been raised. On the night of March 14, plant manager Yoshida had instructed that preparations be made for evacuation of all but the minimum number of operators because of his concern about the possible development of a "China Syndrome" state of nuclear plant meltdown, in which the melted nuclear fuel would eat through every layer of protection, and the release of radioactive materials would be impossible to stop.

Kaieda understood the request from Shimizu to be a proposal for a "total evacuation" of all operators, and responded, "No, control [of the nuclear power station] would become impossible." On hanging up, Kaieda had the spine-chilling feeling that the situation could become extreme.

Chief Cabinet Secretary Edano was called to the reception room. When Kaieda described the TEPCO request to him, Edano was in full agreement that it must be denied. Edano then received a similar call from Shimizu; Edano took the call there and expressed his reservations concerning the proposal.

Shimizu persisted, asking Edano, "Can't something be done? The site simply can't hold out any longer." Edano replied, "This is not a matter in which I can say yes," and continued his refusal. However, he said that he would discuss the matter internally and ended the telephone call.[29]

Special Advisor Hosono then called Yoshida at the Fukushima Daiichi station and asked him about the situation there. Yoshida responded: "We can hold on here, but we don't have enough tools. We need pumps capable of high-pressure injection."[30]

Opinions were sought from Chairman Madarame; Masaya Yasui, the director of the Agency for Natural Resources and Energy; and other NISA officials. They advised against withdrawal, saying that, with no personnel present, it would be difficult to continue any reactor cooling or water injection to spent fuel pools. As for what actions could be taken if personnel did remain on-site, Madarame could only comment, "There must still be something that can be done."[31]

None of the officials gathered in the prime minister's reception room said that the TEPCO evacuation should be permitted, but a feeling of pessimism grew.[32] As the officials conferred there in the dark of night, they wondered if TEPCO would give up on stabilizing the plant and worried that the catastrophe would spread throughout Japan. The situation seemed to be beyond human control. This, officials said later, was their worst hour.

---

[29] Personal communication with Chief Cabinet Secretary Edano, December 10, 2011.
[30] Cabinet Office, Government of Japan (2011) *Chronology report on Fukushima accident*. Report. Tokyo: Cabinet.
[31] Personal communication with NSC Deputy Chairman Kukita, January 20, 2012.
[32] Personal communication with related Cabinet personnel, January 21, 2012.

With lives hanging in the balance, officials awoke Prime Minister Kan at 3:20 a.m. on March 15 to inform him of TEPCO's request. Kan gathered five other administrative members in his office: Edano, Kaieda, Fukuyama, Hosono, and Terada. When the assembled officials reported TEPCO's proposal to Kan, he firmly rejected it. The TEPCO evacuation proposal was "out of the question," the prime minister said.[33]

Kan and the others then moved the meeting to the reception room. Because of the importance of the decision, they summoned Deputy Chief Cabinet Secretary Hirohisa Fujii and Minister for Disaster Management Matsumoto. Also joining them were Madarame and NISA members who had been standing by. Together, they confirmed anew the intention to oppose TEPCO withdrawal.[34]

According to Fukuyama, the prime minister challenged what would happen to units 1, 2, and 3 if the operators stopped working and if the site were just left alone. And was the intention, he wondered, just to leave the spent fuel pool at unit 4 alone? He also raised the ultimate concern: If workers stopped adding water at the plant and left it alone, he said, the continuing release of radioactive materials would contaminate all of eastern Japan.

The communication problems between the Cabinet and TEPCO also became a subject of discussion. As a result, Kan called on Hosono to accept a standing assignment to TEPCO, and he instructed a secretary to determine whether it was legally possible to station Hosono at the TEPCO head office. He issued instructions to call TEPCO and inform them immediately, and a message was conveyed around 4 a.m. for TEPCO President Shimizu to come to the Cabinet.[35]

In response to the summons from the Cabinet, Shimizu arrived at 4:17 a.m. on March 15. Kan said to him, "What about this? There will be no withdrawal. Right?" Shimizu replied, "Of course there will be no withdrawal."[36] Kan

---

[33] Personal communication with Deputy Chief Cabinet Secretary Fukuyama, October 29, 2011.
[34] It must be noted that, in regard to the above course of events, TEPCO has stated that it did not at any time propose a total evacuation and that it was assumed from the beginning that the necessary number of personnel would remain on-site. In regard to the telephone calls from President Shimizu on the night of March 14, NISA Director General Nobuaki Terasaka stated that Shimizu "referred to the gravity of the situation at unit 2, but I have no recollection of being told of a complete withdrawal." TEPCO officials have claimed in investigations and press conferences that they planned to leave certain personnel on-site. However, they have not explained the number, positions, or other particulars regarding any "necessary personnel" stated to have been scheduled to remain at the site. In addition, all of the many Cabinet personnel interviewed in this investigation agreed that the TEPCO proposal was understood to be for a complete withdrawal. It is therefore difficult to conclude that sufficient grounds exist to support TEPCO's statements.
[35] Personal communication with Deputy Chief Cabinet Secretary Fukuyama, October 29, 2011.
[36] Diet, Government of Japan (2011) Diet deliberations. House of Councilors Budget Committee. April 18.

repeated, "There won't be any such thing. Right?" In a faint voice, Shimizu replied, "No."[37]

Kan told him, "We are making a standing assignment of Hosono to TEPCO. Prepare a desk for him." Shimizu looked rather surprised but replied, "Yes, I understand." Kan further stated that he himself would go to the TEPCO head office and asked how soon they could be received there. Shimizu replied, "In one or two hours." Kan said, "We don't have that much time. We are going in thirty minutes. Make the preparations."[38]

At 5:26 a.m., the government formally announced the establishment of a Joint Emergency Response Headquarters at the TEPCO main office. According to top officials interviewed later, this was the turning point that finally allowed the government to begin to gain control of the situation. Kan, Kaieda, Hosono, and Terada then drove directly to the TEPCO building, where they were met with the glare of the camera lights and a large contingent of reporters. They proceeded to the large operations room on the second floor, which was crowded with TEPCO personnel, all clearly prepared for action. Multiple large-screen monitors in the room were connected by videophone lines to the Fukushima Daiichi plant and other nuclear power stations.

In front of the packed hall, Kan took a microphone in hand and appealed to the TEPCO employees. "There will be no withdrawal," he said. "A withdrawal would mean the collapse of TEPCO."[39]

At approximately 6 a.m., however, a booming sound was heard in the vicinity of the suppression chamber in unit 2. Operators at Fukushima Daiichi told TEPCO's main office that they believed an explosion had occurred inside unit 2's reactor building. Approximately 650 personnel were temporarily evacuated from the Fukushima Daiichi Nuclear Power Station, with approximately seventy personnel essential for continuing operations remaining there.

March 15 and beyond—at TEPCO headquarters

After Prime Minister Kan spoke to the TEPCO employees on the morning of March 15, he and his group were escorted into a small room in the TEPCO building. There, TEPCO Chairman Tsunehisa Katsumata and President Shimizu presented a simulation on the subsequent course of the accident at the Fukushima Daiichi Nuclear Power Station.

In one part of the simulation, they noted that the current evacuation zone, a 12-mile radius surrounding the plant, "will be sufficient." In response, Kan

---

[37] Personal communication with Deputy Chief Cabinet Secretary Fukuyama, October 29, 2011. Personal communication with NSC Chairman Madarame, December 17, 2011.
[38] Personal communication with Deputy Chief Cabinet Secretary Fukuyama, October 29, 2011.
[39] Tokyo Electric Power Company, Inc. (2011) *Fukushima nuclear accidents analysis report (interim report) supplement*. Report, December 2. Tokyo: TEPCO. Available at: www.tepco.co.jp/en/press/corp-com/release/betu11_e/images/111202e14.pdf.

noted the large quantities of spent fuel in the pools at Fukushima Daiichi and asked, "Is that really enough?" Over the previous few days, officials had been growing more concerned that water might be leaking or evaporating from the pools containing spent fuel rods. If the rods were exposed, they could heat up and emit radioactive particles into the air, which would drift over farmland and towns.

Shimizu replied, "In that case, perhaps it should be 30 kilometers [19 miles]," an immediate revision of his prior statement. One Cabinet staff member who was present later remarked that Shimizu's quick revision of TEPCO's stated view on the size of the evacuation zone was surprising and perplexing. Kan then said, "If it is going to be necessary, then it must be contemplated." He ordered an immediate study on expanding the evacuation zone, and at 11 a.m., the government instructed residents in the region between 12 and 19 miles from Fukushima Daiichi to take shelter indoors.

Kan returned to the Cabinet building, but Hosono remained at the Joint Emergency Response Headquarters, where he could coordinate the government's actions with TEPCO. The most vital task immediately confronting officials centered on the water injection operation for the spent fuel pools, particularly for units 3 and 4. The water temperature was rising in the pools, and TEPCO worried that evaporation from the pools would leave the fuel rods uncovered. The pools are in unshielded rooms outside the containment vessel, and under normal conditions the water in the pools prevents radioactive emissions. However, if the pools' water evaporated away and the fuel was hot enough, the fuel rods could release a large amount of radioactive material into the environment.[40] Officials drew up plans for water drops by helicopters and injections via fire engines to replenish the water in the pools.

Coordinating the Self Defense Forces, police, and firefighter contingents for the water injection task proved to be challenging. Under the prime minister's instructions, the Self Defense Forces were assigned overall management of the water injection project and took the lead in operations, while the police and firefighters were subordinate. Together, they prepared for an all-out assault to stabilize the spent fuel pools.

March 15 and beyond—at the nuclear power station

Workers at the Fukushima Daiichi site had begun to worry about the spent fuel pools as early as March 13, when they observed white smoke rising from the wreckage of the unit 1 reactor building. The workers believed that the

---

[40] The temperature of spent fuel does not become dangerously hot if there is circulating air to keep the temperature down. At units 1, 2, and 4, the reactor building was damaged; thus, there was a possibility that radioactive materials could release into the atmosphere, but that possibility was low.

emissions were water vapors rising from the spent fuel pool in the building, which suggested that the fuel rods in the pool might be in danger of exposure. In the on-site crisis center and at TEPCO headquarters, operators began debating how to manage the spent fuel pools throughout the plant. Without any source of power in units 1 through 4, the pumps that circulated cooling water through the pools in those reactor buildings were out of commission. Workers had no way to add water to the pools.

In the subsequent days, the explosions that tore the roofs first off unit 1, then unit 3, and finally unit 4 removed the only barriers between the pools and the atmosphere. The explosions also sprayed debris throughout the buildings like shrapnel, and workers worried that the flying chunks of metal and cement may have damaged the pools structurally, allowing water to leak out.

Workers worried particularly about unit 4. When the earthquake and tsunami hit, that unit had been in the midst of a routine refueling operation, and the internal structures of the reactor core were scheduled to be replaced. All the fuel from the reactor core had been moved to the spent fuel pool for the duration of these procedures, and all those fuel rods were capable of generating a massive quantity of heat.

The zirconium alloy used to coat fuel rods rapidly oxidizes when heated to temperatures above 1,700 degrees Fahrenheit.[41] When many fuel rods are stored in a small space, heat from the oxidation of one rod's cladding can be transmitted to other rods, damaging all the rods and threatening to cause a larger zirconium fire that would send plumes of radioactive smoke into the air. That's why it's essential to prevent the temperature of fuel (even spent fuel, which creates less heat) from rising, by methods such as storing the fuel in water or air circulation.

Many experts expressed concerns over the state of unit 4's spent fuel pool in the days of the accident. For example, Gregory Jaczko, then-chairman of the US Nuclear Regulatory Commission, testified to the US Congress on March 16 that, in his opinion, unit 4's spent fuel pool had run dry.[42] Japanese officials denied Jaczko's statements at the time, and later investigations revealed that the fuel rods in the unit 4 pool—and in all the other pools—were never uncovered.

However, in the heat of the crisis, workers believed that the unit 4 pool posed a grave threat—not just to the Fukushima Daiichi plant, but also to the surrounding area of Japan. The response coordinated by the government called for Self Defense Forces helicopters to fly over units 3 and 4 and drop water from their tanks to refill the pool.

On the afternoon of March 16, a TEPCO employee and a Self Defense Forces officer flew a helicopter over unit 4's reactor building to assess both the

---

[41] 900 degrees Celsius.
[42] Bingham, A. (2011) NRC chair: "No water in the spent fuel pool" at unit 4. In: ABC News, *The Note*, March 16. Available at: http://abcnews.go.com/blogs/politics/2011/03/nrc-chair-no-water-in-the-spent-fuel-pool-at-unit-4/ (accessed August 31, 2013).

state of the site and to take radiation measurements needed to prepare for water-dropping operations. The two were able to see into unit 4's pool, and they confirmed that there was water in the pool and that the fuel was not exposed. They then turned their attention to the unit 3 pool. They made four trips in the helicopter, starting at around 9:48 a.m. on March 17, to drop seawater on the upper portion of unit 3; in total, 30 tons of seawater was dropped. But the helicopter had to fly at a high altitude because of the radiation levels, and the strong wind scattered the stream of water. Little water made it to unit 3's pool, and the operation was abandoned after the fourth water drop.

Workers and responders then began to shift their focus to ground-based efforts to inject water into the pools. Starting at 7:05 p.m. on March 17, a team from the Metropolitan Police Department used a high-pressure water-spray vehicle to send a jet of water toward unit 3's spent fuel pool. These water cannons were only marginally successful at propelling water all the way into the pool on the top floor, but efforts continued for days to come. Workers—along with Tokyo Fire Department firefighters, Self Defense Forces officers, and TEPCO's contract workers trained by the US Army—attempted to use high-pressure fire engines to inject water on March 20 and 21, which proved to be more successful.

Finally, the government secured several concrete-pumping vehicles and sent them to Fukushima Daiichi. Known as *kirin*, or "giraffes," these vehicles are typically used to deliver a compressed stream of concrete at construction sites and are equipped with 190-foot-long arms. The first concrete pump vehicle was brought to the site on March 22 and was used to jet seawater into unit 4's spent fuel pool. On March 27 and March 31, concrete pump vehicles began spraying water into the unit 3 and unit 1 pools, respectively.[43]

In unit 2, where the reactor building had not been shattered by an explosion, fire hoses were connected to the pipes of the fuel pool cooling and cleaning system in order to inject seawater, with the operation beginning on March 20.

Despite the pervading sense that one disaster would continue to follow another at the Fukushima Daiichi station, after the morning of March 15 there were no more explosions. Workers were able to resume their efforts to circulate water through the damaged nuclear cores and to reconnect power lines to restore electricity and restart standard emergency systems. Gradually, they began to gain control of the plant.

On March 20, workers connected a power line to the unit 2 power center, and in the next several days cables were extended from there to unit 1. In the control room for units 1 and 2, the lights finally flickered on again on March 23. The lighting had been restored to the control room for units 3 and 4 the previous day. As for unit 5, electric power was recovered on the morning of March 12, by sharing the power of one emergency diesel generator that

---

[43] Because the unit 1 pool contained fewer fuel rods and therefore generated less heat, spraying water into that pool had been a lesser priority.

survived the tsunami at unit 6. The power was delivered through a cable, which was pre-installed for a station blackout event.

The Fukushima Daiichi Nuclear Power Station was still a place of great danger even after electricity was restored: The damaged reactor vessels leaked radioactive water, and over the subsequent months TEPCO struggled to build systems to decontaminate the water and recycle it to create closed cooling loops for the reactors. Workers began to clean up radioactive debris throughout the site, and to build shielding structures around the damaged vessels. TEPCO has estimated that it will take thirty years to fully decommission the plant and decontaminate the site.

Before the end of March 2011, the immediate crisis was at an end. The clean up and reckoning had just begun.

March 15 and beyond—at the Cabinet

From March 14 to 15, as the situation at unit 2 worsened dramatically and the state of the spent fuel pool at unit 4 became a grave concern, Prime Minister Kan increasingly felt the need for the formulation of a worst-case scenario. He and his staff wanted to understand what else could conceivably go wrong and how bad it would be.

Chief Cabinet Secretary Edano later described his own fears of a chain reaction of accidents, in which repeated hydrogen explosions would make the Fukushima Daiichi site so radioactive that workers would have to abandon the site, leading to further loss of control at the reactors and spent fuel pools. Vast clouds of radioactive materials would rise from the plant and float across Japan, Edano feared. The ensuing widespread contamination could lead to the abandonment of other nearby nuclear power stations, causing more problems, and the catastrophe could reach Tokyo itself.

These fears, Edano said later, peaked on the night between March 14 and 15. He described the scenario he dreaded: "If [Fukushima] Daiichi goes bad, then so does [Fukushima] Daini; if Daini goes bad, then so does the Tokai Daini Nuclear Power Station. This was a diabolical concatenation."

"To make sure it didn't happen," Edano continued, "we had to keep everything under control so that, whatever happened, we didn't get into a situation where nobody could get close enough to do anything. ... That was the devil's scenario that was on my mind. Common sense dictated that, if that came to pass, then it was the end of Tokyo."

The prime minister had lost faith in NSC Chairman Madarame, who had denied the possibility of hydrogen explosions at Fukushima Daiichi. At the suggestion of Hosono, he therefore entrusted Japan Atomic Energy Commission Chairman Shunsuke Kondo to formulate the worst-case scenario. Led by Kondo, a group of experts worked on the report from March 22 to 25, with the experts performing the computer analysis operations through the night.

The report concluded that, even if another hydrogen explosion rocked the power station, the amount of radioactive materials released wouldn't require an expansion of the evacuation zone. However, if the unit 4 spent fuel pool ruptured and the fuel rods were exposed, the radioactive emissions that would drift over the landscape would pose a threat to residents living within 31 miles of the plant, and those people would have to flee to safety. Beyond that zone, contaminated soil could require residents within 68 miles to relocate, and voluntary evacuation zones could extend much farther. It would take several decades, the report concluded, for the radiation levels in the soil to come down.

If the contamination extended beyond 155 miles, the report stated, the zone of danger would envelop the greater Tokyo area, and some thirty million residents might have to consider evacuation. One Cabinet adviser stated he was left speechless by the grave vision portrayed by Kondo's worst-case scenario.

Within the government and at TEPCO, officials believed that the unit 4 spent fuel pool might easily become the trigger for this worst-case scenario. NSC Deputy Chairman Yutaka Kukita later described a scenario in which "fuel melts, fires start, and fuel rods fall through the shattered pool floor—that would be the worst." Kondo was also concerned about the strength of the unit 4 pool floor. He particularly feared that, "if aftershocks occurred, they might collapse the pool floor, resulting in water leakage and the cessation of water injection." After leaving the office, Kan stated that the greatest fears during the crisis had centered on that pool.

In the end, the crisis stopped short of a chain reaction of disasters throughout Japan and a large-scale diffusion of radioactive materials. But during the first week following the earthquake and tsunami, a number of dangerous situations developed at the Fukushima Daiichi Nuclear Power Station in which a single mistake could have led to a far larger disaster.

Chapter 2

Nuclear Energy Development in Japan

After World War II, Japan was faced with the need to secure electricity in order to restart industry and rebuild its economy. In October 1953, Prime Minister Shigeru Yoshida, following more than seven years of negotiations, obtained a total of $42 million[1] in loans from the Export-Import Bank of the United States to develop power, mainly from hydro generation; however, Japan still needed to diversify its electricity sources. At the time, Japan's electric power industry and business sector were keenly interested in introducing nuclear power generation, which promised an abundant supply of electricity at low cost. Japan viewed nuclear power as a golden opportunity if the United States could provide similar support, including loans for power plant construction and technical know-how.

Around 1951, nuclear energy drew the attention of Yasuhiro Nakasone, then a member of Japan's House of Representatives. He seized the occasion of the US policy shift—or "Atoms for Peace," as it was called—to push for the development of nuclear energy in Japan. A mere two months after the United States announced the program, Nakasone led a bipartisan group of legislators in getting a budget proposal passed in the House earmarked for nuclear energy. The $830,000 in funds promoting science and technology included $653,000 for nuclear reactor construction, $42,000 for the survey of uranium resources, and $28,000 for purchasing books, reports, and other print materials on nuclear energy.[2]

And so began the determined push for the development of Japan's civilian nuclear program.

The 1950s

After the nuclear energy budget went into effect, the government set about creating the organizational framework for nuclear energy development and use. But peddling nuclear power to citizens was far from an easy sell in Japan,

---

[1] Around $308 million in 2013 prices.
[2] These values are in currency values of the time. In today's values, they would approximately be $6 million, $6 million, $310,000, and $210,000 respectively.

where antinuclear and anti-American roots ran deep since the years following the war. Not only had the country suffered as the only one to be devastated by two nuclear bombs in Hiroshima and Nagasaki, but it was again reminded of the dark depths of the nuclear world on March 1, 1954, when Aikichi Kuboyama and other fishermen set out in his tuna fishing boat, the *Daigo Fukuryu Maru* (*Lucky Dragon 5*). On that day, the United States was conducting Operation Castle Bravo, a series of nuclear tests on Bikini Atoll. As Kuboyama's trawler steered near, the unassuming chief, who had dreamed of quitting fishing to become a florist, was exposed and contaminated by radioactive fallout. He died less than seven months later from acute radiation syndrome and popularly became the first victim of the hydrogen bomb. This incident sparked a massive national protest against atomic and hydrogen bombs that ultimately mobilized some thirty million people; it also fueled and perpetuated anti-American sentiment that had been percolating since the end of World War II.

Nevertheless, Japan moved swiftly forward on the nuclear power front. On May 1, just two months after the incident on the *Lucky Dragon*, Japan established the Preparatory Council on Peaceful Uses of Atomic Energy, the highest decision-making body for Japan's administration of nuclear energy. Chaired by the deputy prime minister and with Cabinet ministers as members, the council's most important decision was to conclude an agreement between Japan and the United States for nuclear energy and enriched uranium for Japan's nuclear program.[3] A year later, Japan would make its nuclear ambitions unarguably clear: In October 1955, Japan adopted the Plan for Nuclear Energy Research and Development, which set a goal of achieving practical nuclear power generation within ten years.

On June 1, 1954, the Ministry of International Trade and Industry (MITI; today known as the Ministry of Economy, Trade, and Industry [METI]) created a commission to deliberate how to implement the funds budgeted for nuclear energy. The commission decided to send a nuclear energy fact-finding delegation overseas, which traveled abroad in December 1954; and in July 1955, it proposed an interim plan for research reactor construction.

Recognizing the growing worldwide sentiment toward civilian use of nuclear energy, the United Nations, in August 1955, held the first International Conference on the Peaceful Uses of Atomic Energy in Geneva. A cross-party delegation led by Nakasone, as well as three other legislators and people from academia, represented Japan. Upon returning home, they launched a joint committee on nuclear energy with members from upper and lower houses, and by November 1955, they had drafted most of the proposals for various laws concerning nuclear energy. With the joint committee including members of the Liberal Party, Democratic Party, and Japan Socialist Party, it was apparent

---

[3] Known as the "Agreement between the Government of Japan and the Government of the United States of America Concerning Civil Uses of Atomic Energy."

that the country was largely united in its pursuit of nuclear energy. On December 16 of that year, the national legislature, or Diet, passed three laws for nuclear energy—the Atomic Energy Basic Act, the Act for Establishment of the Atomic Energy Commission, and the Partial Amendment of the Establishment Law of the Prime Minister's Office[4]—which all went into effect on January 1, 1956.

According to the new laws, the prime minister would issue the ultimate approval for the development of nuclear energy, while the Science and Technology Agency, established in the Prime Minister's Office, would be responsible for the administrative work. The new laws also required the prime minister to respect the views of the Atomic Energy Commission (AEC), set up in the Prime Minister's Office, on matters of safety regulations. Established with the objective of implementing nuclear energy policies systematically and democratically in accordance with the Atomic Energy Basic Act, the commission's purpose was to secure future energy resources and advance science and technology in industry through research, development, and utilization of nuclear energy, thereby contributing to the nation's social welfare and standard of living. But both the Atomic Energy Basic Act and the commission offered little guidance on *how* to pursue safety, save that, indeed, the commission was to foster and support the advancement of nuclear energy policy, establish nuclear energy safeguard regulations, and protect society from nuclear material hazards. In terms of the safety, security, and safeguard of nuclear technology and fissile material, safeguarding was considered to be the AEC's most important responsibility; that is, the commission's objective was to promote nuclear fuel cycle policy, which would improve Japanese energy autonomy.

The director general of the Science and Technology Agency also served as chair of the Atomic Energy Commission. Thus, the Science and Technology Agency, established on May 19, 1956, took on a central role in Japan's administration of nuclear energy. Formulation of additional laws relating to nuclear energy proceeded,[5] and government agencies and national research institutes involved in nuclear energy were set up one after another.

By the time the three laws for nuclear energy were enacted in 1956, the government's only real concrete plan was to construct a research reactor and power demonstration reactor.

The business sector, following the adoption of the nuclear energy budget, had been carrying out its own research on commercial use of nuclear energy. The Electric Technology Research Institute, created by contributions from nine electric power companies, had formed a study group on nuclear energy in

---

[4] This law included an amendment that established the Atomic Energy Bureau.
[5] These included the Japan Atomic Energy Research Institute Act and the Nuclear Fuel Corporation Act, both penned in April 1956.

1951.⁶ Then in December 1954, major corporations with an interest in nuclear energy formed the Nuclear Power Research Council; members began collecting research materials on nuclear energy and sharing this with the business sector. *Keidanren*, Japan's biggest corporate interest organization, established the Study Group on Peaceful Uses of Atomic Energy in April 1955. These three groups gave birth to the Japan Atomic Industrial Forum, Inc. (JAIF) in 1956. Thereafter, corporate groups focused on Japan's heavy electrical-equipment manufacturers: Mitsubishi, Hitachi, and Toshiba. Relationships based on respective technical cooperation with Westinghouse, General Electric, and other overseas large-scale electrical-equipment manufacturers also began laying plans for introducing nuclear energy technology from abroad.

Matsutaro Shoriki, a member of the Diet and former owner of the *Yomiuri Shimbun*, launched the Commission on Peaceful Uses of Atomic Energy. With no particular nuclear science background to speak of, Shoriki brought together the business world and electric power utilities that had sights set on developing nuclear energy into a business. Taking advantage of his connections to the United States forged through the media business, he brought the American Atoms for Peace delegation to Japan in May 1955. Through such activities, Shoriki firmed up his political standing as a promoter of nuclear energy. When the Atomic Energy Commission launched in January 1956, Shoriki became its first chairman and announced his intention to build a nuclear power plant that would become profitable within five years.

Responding to demands from the business world, Shoriki came up with the idea of purchasing a nuclear reactor from abroad and constructing a power reactor. Plans proceeded on purchasing a British-made reactor and building a nuclear power plant in Tokai-mura. The backers were at odds, however, as to whether the plant should be administered by the private sector or the government. Shoriki, as AEC chairman and director general of the Science and Technology Agency, was in favor of government-led nuclear development in association with the nine electric power companies. Ichiro Kono, then-director general of the Economic Planning Agency, believed that the government-led project would not be profitable and therefore would burden the utilities. The compromise resulting from this political impasse was the Japan Atomic Power Company, a joint public–private venture set up as a special corporation with approximately 20 percent government and 80 percent private ownership.

A framework was created whereby nuclear power plants, built with overseas technology, were planned and authorized by the national government, but operated by private electric power companies. The features of this "state-planned, privately operated" approach are seen in the Act on Compensation for Nuclear Damage enacted in 1957, which stipulates that, while operators

---

⁶ The institute later became the Central Research Institute of Electric Power Industry in July 1952.

must carry insurance and are liable for a maximum of 5 billion yen in damages, the government effectively guarantees damages exceeding that amount.

The nuclear community was also split between those calling for Japan to develop its own nuclear energy technology and advocates of imported technology use. The Science Council of Japan, acting as academia's mediator, and some others in academic circles pushed for self-development of technology, running counter to the business sector, whose policy was to use established technology from abroad in order to catch up quickly and cheaply. At the time, overseas technology was neither established nor inexpensive, but eventually the government adopted the approach favored by the business community. As the plants introduced early on were based on turnkey agreements—by which the overseas manufacturer undertook all operations from plant design to materials procurement, construction, and trial operation, delivering to the customer a plant in a fully operational state—problems arose. For one, the Japanese user was excluded from the reactor-design process, thereby creating a clear obstacle to obtain technical skills. (Unit 1 at the Fukushima Daiichi station, for example, was designed by General Electric and constructed as a turnkey reactor. In fact, it was designed to withstand a tornado; thus, the emergency diesel generators were located in the basement. In 2011, when the tsunami hit the Fukushima Daiichi Nuclear Power Station, the generators were submerged under seawater and the plant eventually lost its emergency power supply.) The conflict between the "self-development" and "technology import" approaches marked the beginning of the bifurcation in nuclear energy development frameworks that continued for a long time thereafter.

The framework for regulating nuclear energy safety also split into multiple directions. The system envisaged by the three laws for nuclear energy—that is, the prime minister and the Science and Technology Agency administering nuclear energy, while being advised by the Atomic Energy Commission's team of highly independent outside experts—did not, in fact, function.[7]

The result was that administrative authority remained divided among different agencies, and safety regulations were carried out by multiple government agencies depending on the matter. With government agencies competing over the expansion of jurisdictions, the conflict deepened between the Science and Technology Agency, in charge of research reactors, and MITI, responsible for commercial reactors. Problems arose from this situation: namely, that the industry did not always make use of safety studies or reactor-technology

---

[7] There were three reasons that this was not successful: (1) the involvement of the electric power industry in the commercial nuclear power business meant an increase in duties falling under the jurisdiction of MITI; (2) the legal standing of the Atomic Energy Commission was limited to that of a deliberative body, rather than becoming a decision-making body; (3) the relevant pre-war regulations to regulate utilities were not brought inside the framework of the new laws. Because of this, safety regulations concerning the development of power generation and ships remained within the jurisdiction of other ministries (MITI and the Ministry of Transport [today known as the Ministry of Land, Infrastructure, Transport, and Tourism]).

research, and the academic community did not carry out incisive research on commercial-reactor safety. At the same time, nuclear energy promotion and regulations increasingly merged in each ministry and program.

The outcome with safety regulations was that the regulatory organs ultimately allowed the electric power companies to confirm only a superficial level of compliance, and it became difficult for the government to gain a unified view of the overall state of safety. That is to say, nuclear safety was not considered to be an issue on a national level, and disaster preparedness and response came to be seen as the responsibilities of the operator. As such, neither a government agency nor a power company assessed accident-preparedness plans for one nuclear power plant in Japan.[8]

In October 1957, after the core of unit 1 at the Windscale plant caught fire in the United Kingdom, the UK government requested that disclaimers be included in plant purchasing agreements, which revealed the need to develop a framework for potential damage compensation. Realizing the impossibility for a commercial company to assume the enormous liability that comes from a nuclear accident, the Japanese government began to think of compensation strategies, which the government would pass in the next decade.

The 1960s

Though the government identified the promotion of nuclear power as a national policy, problems nevertheless remained: Specifically, the heavy costs associated with plant construction, waste processing, and various other essentials were often too great to ensure the overall operational efficiency of private-sector companies. Moreover, if an accident like that at Windscale were to occur in Japan, the cost of compensation would become too large to be borne by the companies.

In 1961, the country's Act on Compensation for Nuclear Damage[9] ostensibly solved this problem to exempt electric companies from damage liability over $14 million,[10] beyond which the Japanese government would provide unlimited restitution. The Act stipulated, however, that the Diet must ultimately decide on the provision of such aid—but it did not explicitly state that the national government *must* provide aid to the nuclear operator. In respect to victims, the Act provided that "necessary measures" shall be taken for the purpose of aiding and rescuing victims and preventing the escalation of a disaster; similarly, this did not place an obligation on the national government to compensate victims. In the aftermath of the Fukushima accident, this vague outline of the

---

[8] Nuclear Energy Legal System Research Group (2009) *Science and legal structure study group report*. Report, Department of Nuclear Engineering and Management. Tokyo: School of Engineering, University of Tokyo.
[9] Japan Act on Compensation for Nuclear Damage, Article 16.
[10] $1.25 billion in 2013 prices.

government's responsibility made it difficult for evacuees to claim their compensation.

It was here, in this vaguely stated Act, that the "myth of safety" began to be woven into national policy and public understanding. Since no case of major damage had ever occurred, it was naively understood that commercial insurance would be sufficient to cover any damage. In fact, in 1966, Finance Minister Mikio Mizuta made a statement at the Diet that it was appropriate to assign the central compensatory role to the existing commercial insurance system: At that point, he said, nuclear operators around the world had faced relatively minor—not major—damage. The nuclear industry had not caused an undue burden on insurance companies, he noted, and rationalized that this was an example that could be followed.[11]

But this legal ambiguity led the electric power companies to believe that, so long as nuclear energy was part of national policy, the government would assist in compensation payments in the event of a major accident. At the same time, the ambiguity enabled the government—and in particular, the finance minister—to interpret the provision as meaning solely that the obligation for compensation rested with the electric power companies.

The 1970s

As a nuclear newcomer, Japan's power companies strongly favored early adoption of commercial reactors. Industry called for the import of commercial reactors from abroad,[12] holding an expectation that nuclear energy would provide a stable source and supply of power. Meanwhile, the domestic nuclear industry developed its own technological capabilities to construct nuclear reactors by purchasing nuclear technologies and components from foreign companies; this motivation was encouraged and supported by the Science and Technology Agency, which designed and ushered forward the country's research and development program.

With society's trust in science and technology at an all-time high, the prevalent view was that nuclear energy safety could be assured by maintaining the safety standards established in other countries, from which the nuclear energy technology was being imported. It was then that Eizo Tajima, a professor and physicist who served as a member of the Nuclear Energy Safety Experts Working Group of the Japan Atomic Energy Commission, observed that this de facto structure of delivering judgments on safety was constructed in a way that ensured no clear domestic standard of criteria for approval and

---

[11] Takemori, S. (2011) *The National Policy-Private Operation Trap: The Hidden Nuclear Policy Struggle*. Tokyo: Nikkei Publishing. (In Japanese.)

[12] Japan Atomic Energy Commission (2003) *All about nuclear energy: Wisdom of coexisting with the Earth*. Report, All About Nuclear Energy Editorial Committee Edition. Tokyo: JAEC. (In Japanese.)

disapproval.[13] He argued that the mechanisms of safety regulation governance were not only underdeveloped, but provided no real context for an emergency response to a nuclear power plant accident. But when, between 1973 and 1974, he expressed his concerns to Kinji Moriyama, then the Japan Atomic Energy Commission's chairman and president of the Science and Technology Agency, he was told that it was only necessary for the commission to consider the *construction* of nuclear power plants, with no need for input on safety from academics. The commission, Moriyama said, did not need to question or doubt that nuclear reactors were completely safe.

Still, in earnest, local governments, together with the national governmental organs responsible for nuclear energy advancement, scouted Japan's countryside to site locations for the nation's next nuclear power plant installations. "Safety" was their platform to these communities, emphasizing that "nuclear accidents themselves, let alone disasters, must never occur." At the time, the United States had established Emergency Planning Zones, which required homes to be built at least 5 miles away from a nuclear power plant. Japan, however, ran into problems with considering the implementation of a similar plan: In many of the places where they had sited future nuclear power plants, people lived in homes within such hypothetical zones. Assuming that they could not convince residents to move, while inspiring trust that the plants were indeed safe, the Japan Atomic Energy Commission and the electric power companies, instead, did nothing to even attempt to educate communities that nuclear power plants could have accidents.

Throughout the world in the 1970s, public opinion—and in some cases, public protests—became more apparent and widespread in relation to nuclear energy. This was no different in Japan. In 1972, the historically pro-nuclear Japan Socialist Party—which, for decades had touted the economic advantages of nuclear energy—shifted its position to oppose nuclear power plants. If the antinuclear movement hadn't been vocal before, it would be heard loud and clear by 1974, when radiation leaked from the shielding ring of the *Mutsu*, a nuclear powered ship, in a testing area off the coast of the Aomori Prefecture in northern Tohoku. This not only localized the nuclear consequences—the effects to the fishing industry proved to be a serious livelihood concern—but it also illuminated the fallibility of nuclear technologies. The reactor shield had been designed in Japan, and though the Westinghouse Electric Company peer reviewed the design and warned of its errors, the Japanese designer ignored the advice.[14]

Nonetheless, as public doubt of nuclear safety heightened and questions emerged concerning safety regulation at the power plants, a visible antinuclear community began to form. Inspired by the rise of environmental movements

---

[13] Tajima served in this capacity from 1961 to 1974.
[14] *Global Security* (2013) Nuclear ship *Mutsu*—1974 incident. Available at: www.globalsecurity.org/military/world/japan/ns-mutsu-1974.htm (accessed August 31, 2013).

in both Japan and other countries, the antinuclear movement grew of its own accord. In 1975, Jinzaburo Takagi and nine others co-founded the Citizens' Nuclear Information Center, Japan's first nuclear activist group.[15] The center was established to be the community watchdog of the nuclear industry, reporting on the safety and economic issues as they related to society. The center developed into a respected, trusted, and at times prescient cog in the national nuclear wheel. In fact, in 1995—twenty years after the center's establishment and fifteen years before the accident at Fukushima—Takagi emphasized "the dangers posed by the Fukushima No. 1 Nuclear Power Station and other old atomic plants," warning "the government and utilities about their policy of not assessing the safety risks for nuclear power stations beyond their assumed scenarios."

With the emergence of an active and prolific antinuclear movement, cracks began to appear in the vaguely held nuclear safety myth edifice. Consequently, a growing number of antinuclear groups began to sue utilities and the government for their liabilities of safety. For some time, it seemed that the organized efforts might turn the tide against nuclear energy, but the rising cost of petroleum and the energy shortages only underscored its importance. Still, by the mid-1970s, the pace of nuclear power plant construction slowed significantly—during the 1960s and early 1970s, seventeen units existed in Japan. However, the pace of construction accelerated after the oil shock: Within five years—from 1975 to 1980—eight units were constructed; twenty-one more units were constructed during the 1980s. But Japan faced great difficulty in finding candidate sites after the surge in their construction during the 1960s, as well as the mounting public opposition to potential sites. The government responded to this in two ways: By offering financial incentives to local governments and by forming agencies, responsible for nuclear safety, to speak directly to citizens.

In 1974, Prime Minister Kakuei Tanaka led the passage of three power-siting laws[16] that provided both grants to local governments, which cooperated in siting nuclear power plant construction, and incentives to accept their construction.[17,18] Four years later, in 1978, in an attempt to respond to the public trust that had been upset after the *Mutsu* accident, new Prime Minister

---

[15] The co-founders included: Yuichi Kaido, lawyer, antinuclear activist, former chairman of Japan Green Peace; Kunio Shiba, journalist and antinuclear activist; Hideyuki Ban, staff of COOP; Kimiko Fukutake, lawyer and antinuclear activist; Michiaki Furukawa, professor at Yokkaichi University; Yukio Yamaguchi, professor at Tokyo University; Akiko Wada, translator and antinuclear activist; and Hiroyuki Kawai, lawyer and antinuclear activist.

[16] The Act on Tax for Promotion of Power-Resources Development, the former Special Budget Law for the Development of Electric Power, and the Law for the Adjustment of Areas Adjacent to Power Generating Facilities.

[17] These three laws were designed to accelerate the construction of new power plants.

[18] The economic benefits to local areas that accepted the plant construction became quite meaningful even where considerable community opposition existed, and the grants under the power-siting laws effectively motivated local governments expressing doubts about nuclear safety.

Takeo Fukuda, Kakuei Tanaka's rival, revised the 1957 Nuclear Reactor Regulation Act, partially amended the Atomic Energy Basic Act, and revised the Act for Establishment of the Atomic Energy Commission to become the Act for Establishment of the Atomic Energy Commission and the Nuclear Safety Commission. Designed to assume the safety assurance functions of the Atomic Energy Commission and cross-check the safety reviews by each of the government agencies in charge of regulating nuclear energy, the Nuclear Safety Commission—as with the Atomic Energy Commission—was limited, by law, to that of a deliberative body, having the role of adviser to government agencies. Moreover, no independent secretariat was established; instead, the Science and Technology Agency acted as the secretariat of both the Atomic Energy Commission and the Nuclear Safety Commission.

The basis for giving the Nuclear Safety Commission the legal standing of a deliberative body rather than a decision-making body was the belief that, if it were to be made a decision-making body with administrative authority, it would become part of the government and lose its neutrality, thereby weakening its government-monitoring functions; additionally, explicitly defining its administrative powers would have the effect of hindering its ability to act.

But holes remained. The Science and Technology Agency was supposed to check the health of the individual parts and components of power plants, and the Nuclear Safety Commission was set to double-check the agency's reviews. Though the health of the hardware was micromanaged, no one was assigned to oversee the entire operation of the plant.

This overall administrative complexity resulted not only in overlapping jurisdictions, but also in further blurring the lines of responsibility. There were not enough human resources to allocate expert administrative officials to the multiple government agencies. Since there were personnel rotations within every administrative organ, each encountered constant difficulty in finding and allocating officials possessing the required expertise. In the governmental departments, the workers assigned to safety regulation posts did not have enough time to gain sufficient knowledge and experience in nuclear energy technology before being reassigned a different post every two or three years, which was the case under standard personnel rules. Before long, relative amateurs were tasked with creating proposals and planning safety regulation policy.

But the government's administrative measures to reorganize existing departments and establish even more nuclear energy safety regulation did not appease the community that had formed in opposition to nuclear power. In the midst of the government shuffling, a lawsuit, which had been filed in 1973, concerning the safety of the Ikata Nuclear Power Plant, was brought to trial in 1978. The case raised a number of points contesting the validity of the government's safety audits of the plant. During the trial, the government produced various types of written documents to support its argument of the plant's safety.[19] In the end, the court not only approved nuclear reactor installations to lie within the scope of government discretion—but the court viewed mandatory inspections and the documents attesting to the safety of reactor components as concrete evidence that a plant was safe. Ultimately, nuclear regulators substantially increased the number of items covered in inspections, lengthened the time needed to complete inspections, and increased the time and labor to produce and process the review documents.

The 1970s shaped an atmosphere in which it was difficult to disprove the activists' assertions that nuclear plants were unsafe. Not long after the Three Mile Island accident occurred in the United States in 1979, the Nuclear Safety Commission became the subject of some derision by the public, which labeled it as the "Nuclear Energy Safety Advertising Commission" when its chairman expressed a "safety declaration" to the effect that no similar accident could possibly occur in Japan.[20]

The 1980s

With the approach to regulatory safety governance trending toward even more rigorous inspections, the tendency grew to overemphasize administrative work surrounding safety regulations. The Japanese government targeted the safety of nuclear technology, but the inspections were ultimately performed in a bureaucratic manner.

These inspections piqued the interest of the media, which watched the nuclear industry with a newfound vested interest and published damning criticisms. The industry, however, responded by concealing more of its inspections and its findings. This emerging trend, it seems, was likely behind the case that would be uncovered by a TEPCO whistle-blower in 2002,[21] in which the company was charged with having falsified its inspection reports to the government in the 1980s and 1990s. The falsification may have been attributed, in considerable part, to the company's fear that public disclosure of problems, even if small, would lead to society's mistrust in nuclear plant safety and a public opinion irrevocably marked by suspicion.

The 1990s

In terms of nuclear safety, much came to a head in the 1990s with a series of high-profile accidents. In 1995, the Monju Nuclear Plant experienced a sodium

---

[19] Legal Defense Counsel (1979) *Atomic energy and social dispute on safety: criticism of Ikata nuclear plant court decision.* Report, Nuclear Energy Technology Research Edition. Tokyo: Technology and Humanity. (In Japanese.)

[20] Yoshioka, H. (1999) *A Social History of Nuclear Power: Its Development in Japan.* Tokyo: Asahi Shimbun. (In Japanese.)

[21] *Yomiuri Shimbun* (2002) TEPCO: Conceals nuclear plant damage; 29 incidents, 11 document falsifications still uncorrected. August 30. (In Japanese.)

coolant leak, which produced fumes so hot that they literally melted steel in the secondary coolant pipe distribution room. In fact, the agency went as far as falsifying documents, editing video evidence, and issuing a gag order to employees who knew of the doctored videotapes. Two years later, in 1997, a fire at the Tokai-mura reprocessing plant exposed 337 people to radioactive elements. And in 1999, another accident at the same Tokai-mura plant occurred when untrained workers prepared fuel for the fast breeder reactor and criticality was reached; two people died, thirty-nine households were evacuated, and more than 650 workers, residents, and emergency responders were exposed to radiation. Media and the public charged that the Science and Technology Agency's approach, its concealment of information, and other related factors delayed the emergency response to the 1999 accident—this only exacerbated public distrust of the agency. In response to this disaster, the Nuclear and Industrial Safety Subcommittee—an advisory committee for the Agency of Natural Resources and Energy[22]—conducted an investigation and, in 2001, issued a report in which it concluded that the cause of the accident was that Japan's nuclear safety regulations had a narrow, isolated focus on technology to verify nuclear energy safety.[23] This approach made it possible for deliberate falsification of reports, and, as such, the subcommittee advocated a "culture of safety" and a strengthening of management. It urged that safety management must not only include technology, but also suitable training and education of workers to ensure proper organization, management, and operation of the facilities.[24] The findings and suggestions, however, did not seem to have any ultimate, lasting effect on safety management in Japan.

The 2000s

The Tokai-mura accident spurred efforts to further strengthen nuclear safety regulations. In April 2000, the government transferred the Science and Technology Agency's secretariat functions related to nuclear safety regulations to the Prime Minister's Office; it did this while adding staff and assigning outside experts as technical advisers. This was considered punishment for the Science and Technology Agency—and a chance to up its competence and budget on nuclear development and safety. The Cabinet afforded both commissions their own dedicated secretariat, which pro forma increased their independence from related government ministries; in reality, however, critics pointed to the large number of staff coming from those agencies, and there was

---

[22] This subcommittee, organized by MITI (now METI), proposes long-term energy policy and provides expert comments.
[23] Advisory Committee for Natural Resources and Energy (2001) *On securing the nuclear energy safety base*. Report, Nuclear and Industrial Safety Subcommittee. Tokyo: METI.
[24] Advisory Committee for Natural Resources and Energy (2001) *On securing the nuclear energy safety base*. Report, Nuclear and Industrial Safety Subcommittee. Tokyo: METI.

a continuation of the Science and Technology Agency's mind-set—the tendency to conceal bad news.

Yet administrative authority for nuclear energy remained split among multiple agencies—with regulation of commercial-use nuclear power plant reactors and nuclear fuel facilities under METI (formerly MITI); those of commercial-use ship reactors under the Ministry of Land, Infrastructure, Transport, and Tourism (formerly the Ministry of Transport); and those for research and test reactors under the Ministry of Education, Culture, Sports, Science, and Technology (formerly the Science and Technology Agency). The government transferred the Nuclear Safety Bureau's duties and responsibilities to the Agency for Natural Resources and Energy; within that agency, the government also established the Nuclear and Industrial Safety Agency (NISA), a so-called "special organization" responsible for ensuring nuclear safety. Thus, a safety regulating agency was born in METI, a government body promoting the development of nuclear energy.

Even with these organizational changes, safety regulations remained spread across multiple government agencies; and the establishment of NISA only served to blur the verification functions even more. An example typifying the gutting of the cross-check functions and the blurring of the safety-review responsibility is that the Nuclear Safety Commission's guidelines, intended as internal regulations, were in effect used as NISA's examination standards.

In August 2002, NISA found that TEPCO had deliberately falsified inspection reports for its nuclear power plants to conceal possible safety problems. After this incident, NISA responded to the demands of local communities by taking on the role of explaining to residents near plants the criteria on which nuclear plant safety regulations were based. Such activities were outside the original scope of NISA, causing former Director General Hiroyuki Fukano to point out the lack of consensus among staff on the goal of such explanatory sessions. Further, the public raised doubts about NISA's technical expertise since, at these community meetings, university professors or other outside experts—not NISA personnel—broke down technical details for citizens.

Clearly, long before the Fukushima accident, the public pointed out what it saw as the collusion between the regulator and the regulated—both from an organizational standpoint and with respect to NISA's activities in local communities.

While the Nuclear Safety Standards of the International Atomic Energy Agency require that regulation must be effectively independent from promotion, Japan's regulatory system did not meet this requirement at the time of the Fukushima accident. For one, personnel frequently transferred between agencies that were charged to promote nuclear power and the agencies that regulated the nuclear industry. Further, its regulators also lacked expertise and specialized knowledge, thereby leaving technical decisions up to the power companies they were supposed to be regulating. Promotion and regulation had become melded together in projects spread across multiple ministries; moreover, the scope of responsibility of the Nuclear Safety Commission, which was to double-check safety regulations, had become somewhat ambiguous because the commission's functions overlapped with those at NISA. The Nuclear Safety Commission, in responding to the Fukushima accident, was limited to that of an adviser, while NISA was not able to fulfill its emergency-response role. The agency failed for three reasons: First, the agency's director and his deputy were not well-versed in nuclear technologies and they lost Prime Minister Kan's trust; second, the agency was not prepared to respond to a complex disaster involving both an earthquake and tsunami; and third, NISA's role and responsibility was not made clear by the country's complex safety regulations. Since the myth of safety dominated the paradigm of the nuclear safety administration, the administrative system was not designed to meet the emergency situation. In addition, because the framework for regulating nuclear energy safety was carved into multiple pieces, collusion between the regulator and the regulated proceeded, and appropriate safety measures—including the creation and implementation of proper safety standards—were not taken. As a result, problems arose, such as the failure to consider the possibility of a lengthy station blackout.[25]

Following the Fukushima accident, the government began discussions on establishing a new nuclear safety regulation organization—and in September 2012, more than a year after the accident, the government formed the Nuclear Regulation Authority, which is an external organ of the Ministry of the Environment. The nuclear-related functions held by NISA and the Nuclear Safety Commission transferred to the Nuclear Regulation Authority. In addition, the Japan Nuclear Energy Safety Organization, an incorporated administrative agency in charge of carrying out actual nuclear safety reviews under METI, was brought into the jurisdiction of the Nuclear Regulation Authority. Although this reorganization seems to simplify and identify responsibilities, the record of Japanese nuclear safety administration shows that superficial institutional change has not improved nuclear safety regulation—rather, it has only added more confusion by unclearly defining responsibilities. Only time will tell whether this new institution will be successful in permanently changing the mind-sets and paradigm of the so-called nuclear village.

---

[25] The possible choices for creating a safety regulation organization with sufficient independence were along the following lines: Make the organization an Article 3 commission as defined in the National Government Organization Act, with clear administrative authority; make it an external organ of the Cabinet Office (like the country's Fair Trade Commission, for example); set it as a bureau or agency under the Cabinet Office or a ministry directed by a minister other than one responsible for nuclear energy development and promotion (like Japan's Financial Services Agency or Consumer Affairs Agency).

Chapter 3

The Safety Myth

"*Anzen shinwa*," or the safety myth, is one of the most important notions that has existed since the beginning of Japanese nuclear history and is considered one of the causes of the Fukushima accident. This construct is a vague, abstract understanding that nuclear power, unquestionably, is a safe technology. It is a combination of nuclear promoters' thinking that communicating the risk of a nuclear accident is taboo and their overall interest in marketing nuclear power as a "safe" energy source. Historically in Japan, this notion has played a significant role in affecting attitudes and societal acceptance of nuclear power technology, as well as in determining the administrative framework of safety regulations. Of course, this safety myth is only a general, metaphorical term. However, complex governance systems—with no clearly defined accountability among multiple agencies—were established on top of an overly optimistic understanding of nuclear power safety; not only did power companies and the regulatory bodies responsible for securing nuclear plant safety eventually accept this myth, but so, too, did local governments and citizens in regions near nuclear plants. And ultimately the entire Japanese population accepted it. With the business of nuclear power built on such a foundation, it became increasingly more difficult in Japan, if not impossible, to discuss even the possibility of a nuclear accident or the fact that accident defenses had atrophied to a dangerous state of inadequacy.

When the country first introduced nuclear power in the 1950s, it was marketed as "the energy source of our dreams," and, as the technology was rolled out, its safety and technological superiority were emphasized, while its risks were kept under wraps. Touting the safety of nuclear power to cultivate an environment in which the nation would assuredly accept nuclear plant construction was the so-called "nuclear village," which comprised academics specializing in nuclear technology, a business community thirsty for cheap electricity, government agencies overseeing nuclear power—namely, the Ministry of International Trade and Industry, or MITI (now the Ministry of Economy, Trade, and Industry [METI]), and the Science and Technology Agency, or STA (now part of the Ministry of Education, Culture, Sports,

Science, and Technology)—and regional governments in districts eager to host the nuclear plants and related facilities.

Over the following decades, even after the accidents at Three Mile Island in 1979 and Chernobyl in 1986, the myth persisted in Japan that the nation's culture valued safety above everything—that nuclear power in Japan was somehow safer than in other countries and that its nuclear reactor technologies were "different"; rhetoric and ideology were grand, but in fact, this perception prevented Japan from having science-based discussions and from pursuing thorough safety reform.

In fact, Japan made only incomplete efforts to institute safety enhancements based on the lessons learned from these incidents. Moreover, when trouble did strike Japanese nuclear plants, such as the 1999 Tokai-mura accident in Ibaraki Prefecture, the nuclear community often dismissed it. The societal and political culture that subsequently evolved was one in which the nuclear village felt the need to insist upon the absolute safety of the energy source and to forbid speculations over even the slightest possibility of an accident in order to deny legitimacy to the antinuclear power faction and their safety concerns.

Electric power companies and regulatory bodies worked to create a climate in which they altogether avoided being completely cognizant of safety problems. Namely, Japan's state-planned, privately operated system (*kokusaku minei*)—in which the government codified nuclear power policy, but power companies commercially operated the actual nuclear plants—was an important backdrop to this development.[1]

Fission power was initially introduced as government policy, however, due to the lack of mandatory power; private-sector corporations not only shouldered the responsibility of commercial operation, but also increasingly came to bear the primary burden of improving safety provisions. For this reason, even as regulatory bodies monitored the activities of power companies, in practice, private-sector companies made the investment decisions on safety enhancements. The extent of the government's mandate for ensuring safety was thus rather ill defined; in the event of an actual accident, it was unclear precisely who bore the ultimate responsibility, and accident responses were inevitably confused and chaotic.

Japan's system of governance over nuclear safety regulations reflected these factors. The safety myth was perpetuated by the establishment of detailed regulations based on periodic inspections for massive hardware installations, which gave the impression that safety at nuclear plants was being properly monitored.

The causes of the accident at the Fukushima Daiichi Nuclear Power Station not only include technical failures and systematic and administrative failures, but also a set of factors that can be termed societal failures.

---

[1] Thus, unclear policies meant that there was little governance over the utility companies, thereby holding no one clearly accountable.

Unlike technological causes, societal causes, in general, can defy easy explanations, complicating efforts to pinpoint precisely what background factors lay beneath an accident. Nonetheless, any honest attempt to analyze the causes of the Fukushima accident—to understand why Japan's nuclear plants were so inadequately defended against accidents, and how Japan failed to envision the outbreak of a severe accident—must take into account the safety myth. Thus, it is necessary to review the growth of nuclear power in Japan and revisit the very relationship between the energy source and Japanese society to analyze how it progressed to the status it held on the eve of the Fukushima accident. By identifying the primary force behind the creation of the safety myth—namely, the nuclear village—and in analyzing how this confederation of nuclear power advocates evolved throughout recent Japanese history, we can identify the background factors that ultimately led to the Fukushima disaster.

Japanese society, the safety myth, and the nuclear village

The story of the nuclear village[2] is ultimately a tale of the intersecting interests of a cast of three characters: the nuclear power administrators and the promoters of the nuclear power industry, which may be considered to be the "central branch" of the village; the local governments in the regions surrounding nuclear plant sites and related facilities, which may be considered as the "regional branch"; and, finally, Japan's actual citizens, who are outside the realm of political, administrative, financial, academic, and regional-government involvement in nuclear power—and who, before the disaster at the Fukushima Daiichi Nuclear Power Station, were detached participants in the nuclear village, but ultimately were lulled by the safety myth into a false sense of security.

Through the nuclear village's branches and centralized mechanisms—that is, the actions taken by Japan's political, governmental, industrial, academic, and media entities—it is easier to understand how all of these actors came together to promote Japanese nuclear power production in a way that ultimately led to the Fukushima Daiichi Nuclear Power Station accident. Note that we intentionally use the word "ultimately" in this context: It is important to understand that the various actors in this process did not necessarily always set out to promote nuclear power generation, at least not directly. Among the segments of Japanese society who took a position on nuclear power, there were some who were critical of the technological safety, practical administration, or other aspects of nuclear power, while mainstream Japanese culture largely adopted a supportive posture toward nuclear power.

---

[2] Needless to say, the term "nuclear village" is not an official term, but a convenient shorthand frequently used to evoke the fundamentally closed, insular, conservative, and interconnected nature of the nuclear players.

Two citizens of the nuclear village and their roles in Japanese society

In the immediate aftermath of the Fukushima accident, nuclear technology experts frequently appeared in the mass media to emphasize that various nuclear components, or the overall technology, was "safe" or "secure." However, as time progressed and the deteriorating situation at the plant became increasingly obvious—particularly after the hydrogen explosions in the reactor buildings—the public, who followed the news through traditional print, television, or online sources, increasingly came to doubt expert claims that there were no immediate threats to health. Journalists sought out scientists to explain the situation; but, far from achieving what was no doubt the media's goal of explaining the issue and calming society, their efforts served to sow chaos and confusion among the Japanese people.

The safety myth forbids all doubt regarding the safety of nuclear power, replacing it with a certain logic leading inexorably to a predetermined and extreme conclusion: that nuclear power is simply safe.[3] Though brought to the foreground after the Fukushima accident, the premise of such safety and security had in fact existed in one form or another since the enactment of the Atomic Energy Basic Act in 1955. The measures in various safety guidelines had focused heavily on aspects of nuclear technology; quite simply, these guidelines were based on the assumption that an accident would not happen as long as technological systems were thoroughly in check.

The key citizens of the nuclear village—the nuclear power administrators and the nuclear power industry, as well as the local governments in the regions that hosted nuclear plants and related facilities—had spoken with one voice when it came to promoting nuclear power in Japan, and they worked in concert to erect a framework for nuclear power promotion.

The central branch of the nuclear village

Above all else, the central branch illustrates the unique characteristics of the Japanese nuclear power industry and the national system of nuclear power administration. More specifically, Japan exhibits what Hitoshi Yoshioka, vice president of Kyushu University, describes as a "two-tiered sub-government model": One tier is defined by an alliance between the former Ministry of International Trade and Industry (MITI) and the electric power companies, while the second tier comprises the former Japan Science and Technology

---

[3] Personal communication with former Prime Minister Naoto Kan, January 14, 2012. In the interview, he described the "myth of safety" as the failure to envision that anything might go wrong "beyond a certain level"—thus, the insistence that everything was safe. The prime minister went on to note the severity of the problems created by such a mind-set.

Agency (STA).[4] The two tiers pursue distinct strategies, and competition between the two drives the progress of nuclear power in Japan. Meanwhile, the institutions charged with regulating and overseeing nuclear activities were initially placed under the second tier, the Science and Technology Agency, but after the Tokai-mura incident in 1999, the regulating authority shifted to the first tier, under MITI. While STA was in charge of technological development and the promotion of nuclear technology, MITI, with its association to the power industry, promoted nuclear power for industrial purposes.

This framework, in which the national government took up the cause of promoting nuclear power, enabled Japan to join the elite club of the world's advanced nuclear powered nations; however, it created an exclusive environment for those outside the nuclear village—from the Diet down to the citizens—to be involved in the policy decision process. Historically, the electric power industry, in particular, has enjoyed an unusually outsized influence on Japan's economy. The Japan Business Federation (*Keidanren*), the biggest lobby group that represents Japanese industry, is closely connected with the Federation of Electric Power Companies and, thus, has consistently supported nuclear power. The government's promotional efforts protected the electric power companies and, at the same time, boosted overwhelming technological and political strength to influence the nation; even if external actors outside this industrial alliance had somehow mustered the wherewithal to participate in the policy-making process, they likely would have been hard-pressed to achieve limited regulatory impact. At the same time, it would be inaccurate to suggest that the relationship between the industrial world and the regulatory agencies was without tension. Our interviews with staffers in the Prime Minister's Office suggest that, whereas the industrial community insisted on maintaining the regional monopolies enjoyed by power companies in order to secure a stable supply of electric power, the regulatory community attempted—through deregulation of the electric power market and in other ways—to intervene in the workings of the industry and to break up the vested interests of the existing players. However, this attempt at countering the interests of the electric power companies was ultimately unsuccessful, and the profits for those companies remained as lucrative as ever.

Meanwhile, the central branch of the nuclear village did not consist of industry players and bureaucrats alone. For one, the power companies gained significant influence over the Diet and regional legislatures through political contributions and other means. Indeed, members of parliament from the Liberal Democratic Party worked with the Federation of Electric Power Related Industry Worker's Unions of Japan (*Denryoku Soren*), an alliance of labor unions representing electric power companies and other industries, to promote nuclear energy for securing workers' jobs, despite the fact that the labor union supported opposition

---

[4] Yoshioka, H. (1999) *A Social History of Nuclear Power: Its Development in Japan*. Tokyo: Asahi Shimbun. (In Japanese.)

parties, particularly the Socialist Party. But when the Democratic Party of Japan came into power in 2009, the *Denryoku Soren* supported it. In this way, the power companies established a structure under which the governing party would accept and support nuclear power under all circumstances. According to a *Kyodo News* investigation conducted in July 2011, a political fund-raising expense report from the headquarters of the People's Political Association (*Kokumin Seiji Kyoukai*), a political fund-raising organization for the Liberal Democratic Party, revealed that current and former executives of nine electric power companies, including TEPCO, contributed 72.5 percent of the total yen raised from individual donations in 2009. In addition, in 2010, *Denryoku Soren* and the TEPCO labor union together funneled at least 120 million yen (about $1.4 million) to parliament members and regional legislators from the Democratic Party via contributions, purchases of party tickets, and other means.

With such a powerful, centralized nuclear village using all the tools in its arsenal to perpetuate the safety myth in every corner of Japanese society, one might suspect the history of nuclear power in Japan to have been devoid of opportunities to state an opposing argument. This, however, turns out not necessarily to be the case. Societal acceptance of nuclear power faced a turning point in the 1970s,[5] when an antinuclear movement and safety ideology began to spread around the world, as well as in Japan.[6] Before then, however, the sort of clear-cut opposition to the very idea of nuclear power that is seen in today's society did not exist in Japan. For example, nuclear-powered robots positively appeared in many futuristic cartoons and other aspects of pop culture in the 1960s and 1970s.

The media-driven notion of nuclear power as the ultimate energy source traces all the way back to the period of the earliest introduction of domestic Japanese nuclear power production in the mid-1950s.[7] A campaign in the *Yomiuri Shimbun* newspaper designed to engender support for the introduction

---

[5] Beginning in the mid-1960s, a series of revelations of minor accidents and malfunctions in England, the United States, and other advanced nuclear powered nations gradually raised awareness to nuclear power risks.

[6] These movements in Japan never grew to a scale large enough, for example, to bring about the decommissioning of a reactor that had already commenced commercial operation. We can offer several explanations as to why: For one thing, although voices opposing nuclear power would grow shrill whenever any significant trouble arose, after a while passions invariably cooled, media coverage subsided, and the antinuclear movement receded. Meanwhile, the issue never commanded high priority as a political problem, and the oil shock of the 1970s only served to solidify the status of nuclear power as a key domestic energy source. With the 1955 framework in place, and in an era of sustained economic growth, any legitimate questions that might have arisen were forced to contend with a powerful nuclear power promotion system supported by a nuclear power village populated by politicians, bureaucrats, and industry leaders.

[7] Nanbara, S. (1947) Nuclear power and the second Industrial Revolution. *The Mainchi*, October 27. (In Japanese.) The editorial quotes then-Tokyo University President Shigeru Nanbara's commencement address that the "development of nuclear power will enable a second industrial revolution," and goes on to suggest that nuclear power "paves a road for Japan, which has been reduced to just another tiny, powerless nation, to contribute to the future history of the world."

of nuclear power is symbolic. The newspaper's edition for New Year's Day in 1953 carried a feature titled "A Conversation at the *Yomiuri Shimbun*: Let us put hydrogen to peaceful use!" and stated:

> As a country baptized by the searing shock of nuclear explosion, Japan cannot help but take a strong interest in these matters—and we find our hearts strongly compelling us to make peaceful use of nuclear power. The notion that something boasting such fearsome military could be redirected for peaceful purposes would undoubtedly bring to us—indeed, to people the world over—the most sincere and extravagant joy imaginable.[8]

With nuclear power having not then garnered a certain level of affirmation from Japanese society, it is safe to say that the media were attempting to present an image of a happy society strolling off into a rose-colored future. Here, the participation of experts served to capture the imaginations of politicians and the industrial world, and it was precisely this environment in which the foundation was being laid for the nuclear village.[9]

In the 1980s, spurred by the nuclear accidents at Three Mile Island and Chernobyl, antinuclear activism, which had developed the decade prior, began to capture some attention from the Japanese people, including a movement to establish various tactics to ban nuclear power. Borne out of this movement, the Socialist Party, *Sohyo* (the General Council of Trade Unions of Japan), and antinuclear citizen-activist groups spearheaded a petition to submit to the Diet; however, they were unable to offset the powerful pro-nuclear push of the nuclear village. Even the 1999 Tokai-mura accident failed to change the course of Japanese nuclear power in any significant way.

In the 2000s, the Democratic Party (within which labor unions constitute a significant base of support and have strong influence) ascended to control the government, but the antinuclear movement still showed no signs of becoming a realistic force. Among the reasons for this is the fact that, within labor unions, and particularly in the *Denryoku Soren*, there were pockets of strong support for nuclear power, with other unions tending to proceed in lockstep. When he took office in June 2010, even former Prime Minister Naoto Kan, who had been critical of nuclear power throughout his political career, proposed exporting nuclear power infrastructure overseas as part of his "new growth

---

[8] *Yomiuri Shimbun* (1953) Let us put hydrogen to peaceful use. January 1. (In Japanese.)
[9] Needless to say, the negative repercussions of nuclear power were far from unknown at this time. Memories of the nuclear explosions at Hiroshima and Nagasaki remained fresh, and the *Daigo Fukuryu Maru* incident of March 1954, in which a Japanese fishing boat was contaminated by fallout from a US nuclear weapons test, sent shock waves rippling throughout Japan, a phenomenon captured by the subsequent creation of the film *Godzilla*. However, against this societal backdrop, it was precisely the attractive notion of putting this devastating instrument of war to productive, peaceful use that served to motivate the introduction of nuclear power in post-war Japan, culminating in the 1955 passage of the Atomic Energy Basic Act.

strategy"; phrases such as "nuclear renaissance" and "environmentally friendly energy source" became ubiquitous throughout Japanese society.

On July 13, 2011, at a meeting of the Japan House of Representatives' Special Subcommittee on Disaster Recovery, Makoto Yagi, chairman of the Federation of Electric Power Companies and CEO of Kansai Electric Power Company, stated that total spending on television, newspaper, and other mass media marketing—not only to advertise nuclear power, but also to further public understanding of the electric power industry as a whole—annually averaged some $24 million over five years. Of course, individual power companies also purchased their own advertising, and the total mass media spending was several times this figure: Between 1970 and 2011, electric companies and nuclear companies spent $14.65 billion and $2.93 billion, respectively, for public relations and advertising efforts.[10]

The regional branch of the nuclear village

An observer uninitiated in the realities of the nuclear plant site-selection process might wonder: How could any region possibly want to receive and maintain a nuclear plant within its borders? Surely the local residents would bristle at having a nuclear plant near them and do everything in their power to chase it away. But this is not how things work in practice. In fact, the inescapable reality is that the regional branch of the nuclear village has intimate ties to the nuclear power industry—ties that will be difficult to sever anytime soon.

Since April 2001, the candidates espousing the standard-issue line of support and acceptance for nuclear power have won mayoral elections in cities near nuclear plants in various regions of Japan, including Niigata, Hokuriku, and Hokkaido. Even in the town of Kaminoseki in Yamaguchi Prefecture, which is in the early stages of being selected as a nuclear plant site, the pro-nuclear candidate defeated the antinuclear candidate in September 2011 by a margin of 963 votes. Similarly, in regions where nuclear plants have not been restarted after shutdowns for routine inspections, local governments have been deluged with requests to restart the plants—a situation that has prevailed even in Fukushima Prefecture, the epicenter of the recent nuclear disaster.[11]

---

[10] Hokkaido Electric Power, $93 million; Tohoku Electric Power, $192 million; Hokuriku Electric Power, $87 million; Tokyo Electric Power, $472 million; Chubu Electric Power, $187 million; Kansai Electric Power, $354 million; Chugoku Electric Power, $127 million; Shikoku Electric Power, $68 million; Kyushu, $192 million. The amount of annual spending more than doubled after the accidents at Three Mile Island and Chernobyl, from $91 million in 1979 to $235 million in 1986. In 2010, TEPCO, alone, spent $295 million ($77 million devoted to television and radio advertisement campaigns; $50 million for newspaper and magazines; $47 million on public facilities and events).

[11] On November 20, 2011, when Fukushima Prefecture Governor Yuhei Sato announced plans to decommission reactors, including TEPCO's Fukushima Daini Nuclear Power Station, the towns of Tomioka and Naraha, where the plant is located, were confused by his unilateral decision since the plant, indeed, withstood both the earthquake and tsunami.

What could possibly explain such unwavering support? One possible culprit is the problematic financial structure. Acceptance of a nuclear plant rewards a region with a significant cash influx, divided between citizen-level expenses and regional-government-level expenses.

At the citizen level, nuclear plants bring massive sources of stable employment to regions that previously had relied on primary industries; such a development not only puts money in people's pockets, but also stanches the flow of employment-seeking out-migration. But these are not the only benefits: The periodic plant inspections that take place once every thirteen months bring about a temporary surge in employment in the hospitality industry and other areas of the economy. In Fukushima, nuclear power is a massive industry that directly employs more than 10,000 people continuously and has extremely consequential residual impacts.

At the local level, the fixed asset tax and power-siting laws ensure that revenues begin to pour into regional government coffers as soon as construction begins on a power plant and continue to flow after the plant has gone online.[12] While the details vary from region to region, other benefits include public investment in facilities and infrastructure and relatively favorable treatment of medical expenses.

However, this situation did not continue forever. In particular, regional governments saw taxes decline due to the annual depreciation of fixed assets, while the cost of maintaining public facilities, installed when plants were first constructed, held steady or even grew. With revenue decreasing as costs held firm or grew, regional governments began to take on operating debt.

Consequently, regional governments called for more nuclear reactors or supporting facilities in an effort to secure increased subsidies and other new sources of revenue. Of course, regional governments also attempt to lure industries other than nuclear power, but the nuclear industry is attractive in terms of the number of job opportunities and stable wages it can provide. In the region surrounding the Fukushima nuclear power station, to take just one example, considerable effort had been expended after the site was selected to establish alternative industries, such as manufacturing factories and chemical plants; however, none of these efforts were successful.

In particular, a series of factors present since the 1980s—such as manufacturing offshoring, government failure to develop regional regeneration plans, as well as reduced subsidies paid to regional governments—have narrowed the range of options available to localities for maintaining a solid financial picture. Consequently, regions hosting nuclear plants have tended to call for *more* nuclear plants. Some observers have compared the unique financial structure of nuclear power to a powerful narcotic, capable of ensnaring regional governments

---

[12] These laws include: the Act on Tax for Promotion of Power-Resources Development, the Special Budget Law for Development of Electric Power, and the Law for the Adjustment of Areas Adjacent to Power Generating Facilities.

in a helpless state of addiction. For example, in 2003, only four years after the Tokai-mura accident, former Futaba Town Mayor Tadao Iwamoto—a former Socialist Party member of the Fukushima prefectural legislature who had actually opposed nuclear power in the years before his political career started in 1985—made the following comments in an interview:[13]

> Over the course of my long relationship with TEPCO, I have come to believe sincerely that power plants might experience a minor mishap here and there, but never anything that would seriously impact safety. ... I think it is important to look forward. If you sit around all day hammering idly away at various problems, you wind up becoming the kind of person who looks backward. I always do everything in my power to remain focused on the future. ... Within the structures of today's nuclear reactors, the worst possible accident is a leakage of radiation, but I am confident that Japan's nuclear power plants are equipped with adequate capacity to keep radiation completely confined. I believe that there is no possibility of an accident like the Three Mile Island accident in the United States or the Chernobyl accident in the former Soviet Union occurring at a Japanese nuclear plant. ... Indeed, if I didn't believe these things, it would be difficult for me to remain personally involved in the administration of nuclear power. No matter what happens, I hope always to believe in the advancement of nuclear power. This is the one thing that we must never allow to collapse. I take this as my own personal point of pride.

Sentiments like this (*we have no choice but to trust the national government, trust TEPCO, and trust in nuclear power*) crop up frequently in the statements of mayors and other local leaders in, or surrounding, communities hosting nuclear plant sites. Moreover, the citizens themselves recognize the essential role of the nuclear power industry as an enabler of their lifestyles; with as much as one-third of the entire local population employed at jobs related to nuclear power in one way or another, the industry served to counter the problems of depopulation and employment-seeking out-migration faced by Japan's regional communities.

The depth of dependence on the nuclear power industry runs deep in Japanese regions and, historically, has proceeded—and proceeds to this day—essentially unscathed by major accidents and systematic or policy-related failures.

Outside the nuclear village

For those players on the outskirts of the nuclear village—namely, the general public existing outside the political, administrative, industrial, academic, and

---

[13] Council for Nuclear Fuel Cycle (2003) Nuclear power plants represent a fateful partnership: An interview with Futaba Town Mayor Tadao Iwamoto. *Plutonium 42*. (In Japanese.)

local-government entities that promote nuclear power in Japan—what was their relationship with the nuclear power enterprise?

At the community level, the safety myth that prevailed among Japanese citizens is perfectly illustrated by comments like this one: "The likelihood of an accident at a nuclear power plant is less than the chance of getting hit by a car as you walk down the street. There's no sense worrying about something like that."[14] The citizens of Japanese nuclear plant zones were not merely ignorant, nor was it simply that they could not break their cycle of economic dependence; instead, they independently constructed and chose to immerse themselves in a myth of safety that silenced all attempts to question what the actual risks might be.

For example, the Ministry of Education, Culture, Sports, Science, and Technology, an institution that promoted nuclear power, took pains to stress the safety and advantages of nuclear power in all areas of education that it supervised. This included booklets distributed and used as sub-textbooks in schools and science museums across Japan, thus ensuring that Japanese citizens were indoctrinated in the myth of safety from their earliest childhood years. In order to impede the spread of antinuclear activism through Japanese society, the nuclear village mobilized its full financial and political clout, as well as its technical expertise, to ensure that such movements never spread beyond a small subset of the population, countering opposition movements by emphasizing the safety of nuclear power and bolstering the myth of safety. Consequently, the vast majority of the Japanese population saw nuclear power as a complex and inscrutable technical issue; many Japanese considered the question of whether to support or oppose nuclear power as impossibly intricate and simply did not know whom to believe.

Compared with the technological, systematic, and administrative factors that lay behind the Fukushima Daiichi Nuclear Power Station accident, the societal influences might be considered less direct causes of the disaster. But these very factors were only possible because the Japanese public had adopted an aloof and uncritical attitude toward nuclear power—an attitude that underpinned a consistent acceptance of the energy source.

After the Fukushima accident, the central branch of the nuclear village, despite having just lived through an accident of the gravest possible severity, was quick to release statements declaring the accident resolved, and shortly thereafter began clamoring to restart nuclear reactors that had been shut down for routine inspections. To this end, the Democratic Party of Japan-led government, under the initiative of Prime Minister Naoto Kan, proposed so-called "stress tests" to reassure the public of nuclear plant safety by demonstrating that each plant was deemed safe enough to go back into operation. Such an initiative appears to be an attempt to return immediately to the old system of

---

[14] Kainuma, H. (2011) *A Theory of Fukushima: How Did the Nuclear Power Village Arise?* Tokyo: Seidosha. (In Japanese.)

nuclear power policy, basing safety evaluations solely on the new stress tests, and thus short-circuiting efforts to obtain a comprehensive understanding of the causes of the Fukushima disaster, to reform safety regulations in fundamental ways, and to regain the trust of the Japanese people in the safety of nuclear power. Meanwhile, as for the regional branch of the nuclear village, local elections in regions near nuclear plants continue to champion pro-nuclear candidates, while the antinuclear faction is almost totally rejected.

After the March 2011 disaster, many residents of Fukushima and surrounding regions were forced to evacuate—and many of these residents were not allowed to return home until nearly three years later, experiencing severe challenges in their daily life;[15] this reality, however, did not prevent citizens near nuclear plant sites from choosing continued coexistence with the nuclear power enterprise.

The formidable power of the central and regional branches of the nuclear village has clearly influenced societal acceptance of nuclear power. But the sway of the nuclear village and its foundation of rigid societal trust underpinned the eventual failures of the nuclear power community, the shortsightedness of technological imagination, and the failure of implementing precautionary defenses. Further, such impudence contributed to establishing a complex system of safety regulatory governance in which the locus of responsibility was always ambiguous.

Simply stated, the safety myth constructed by the nuclear village was merely the first domino to fall before the Fukushima Daiichi Nuclear Power Station accident, laying the groundwork for the sequence of tragic mishaps that befell the plant when the accident struck.

In light of our analysis, it is clear that, as Japan moves forward with nuclear power, it must construct an environment in which ordinary citizens have sufficient perspective to assess critically the powerful bonds that exist among the political and industrial communities, the media, and the various academic societies comprising the central nuclear village. Only in doing so can transparency be ensured in the political and economic decisions that drive the progress of nuclear power and achieve a system of information sharing in which the nation's citizens are empowered to make their own independent decisions about nuclear power policy. The newly established Nuclear Regulatory Authority (NRA) was designed to be independent from political and economic pressure from the nuclear village and to provide information in transparent ways. In fact, the NRA established new safety measures in July 2013, which regulate electric power companies more than before March 2011. However, there are many defects in these new safety regulations due to pressure to issue them as quickly as possible. Although it seemed that the myth of safety had

---

[15] Challenges included separation from family members, stress related to frequent moves, stress related to living in temporary shelters, workplace and job loss, compensation fights with TEPCO, and the loss of family members and property.

been shattered by the Fukushima accident, there is a strong force to reintroduce, yet again, this myth.

But the ramifications of the regional branch's policies—including those concerning both the labor environment and regional economic dependence on nuclear power—must be considered. For example, by forcing mandatory disclosures of contributions, advertising campaigns, and other funds related to nuclear power promotion, the nation can establish an environment in which the regional branch of the nuclear village is freed from dependence on that revenue stream for the basic sustenance of daily life. One possibility is to move away from the system of subsidies created by the power-siting laws, a system which creates incentives to cling to old, risky facilities and equipment, and instead to adopt a system like that used for reactor decommissioning, in which subsidies incentivize the scrapping of old facilities with new installations rebuilt in their place. But the process of demythologizing that will be needed to dislodge the golden idea of safety from Japanese consciousness will not be easy. Nonetheless, it is essential: As long as the societal factors that allowed the entrenchment of this myth persist, as long as the central nuclear village is allowed to restart nuclear plants without addressing the inadequacies of existing safety provisions, as long as the regional nuclear village remains economically dependent on nuclear power, and as long as the ordinary Japanese citizen remains apathetic, the possibility of another severe nuclear accident will remain with us forever.

Chapter 4

Actors in Japanese Nuclear Safety Governance

Ineffective governance of Japan's nuclear safety regulatory structure was an underlying cause of the accident at the Fukushima Daiichi Nuclear Power Station—that is, had such governance been functioning effectively, it would have led to some degree of preparedness against the earthquake, tsunami, and the total loss of power at the plant. Thus, the government and nuclear industry characterized the accident as an "unforeseeable" occurrence, begging the question of the precise nature and role of nuclear administration and nuclear safety regulation in Japan, the organizations involved, and the obligations of the responsible parties. As we laid out in Chapter 2, the mechanisms of nuclear safety regulation governance became pluralistic and complex, and responsibilities blurred as these mechanisms developed over time. But what, exactly, were the reasons for this configuration of nuclear safety regulatory governance in the first place? It may be argued, in fact, that it is impossible to understand the true nature of the Fukushima Daiichi Nuclear Power Station accident without an understanding of the very structure of nuclear safety regulatory governance particular to Japan.

The International Atomic Energy Agency (IAEA) makes it abundantly clear who, precisely, is assigned responsibility for a safe nuclear power plant: Safety resides with the license holders. Thus, within Japan's licensing system for nuclear plant operators, the primary responsibility for ensuring nuclear power safety lies with electric power companies. That, at least, is the content of the agency's first safety principle; its second principle, however, is that the government is to establish an effective legal and administrative framework for safety—including independent regulatory institutions—and to monitor that safety precautions are properly observed. In other words, although power companies are responsible for safety, the question of precisely what defines safety is not one that can be answered by corporations alone, but, instead, evolves to factor in societal conditions and must ultimately be decided by governments with the mandate of the people.

The relationship between the power companies and the regulatory bodies varies widely from nation to nation. The primary influences are the relative scale of the power companies and the regulatory bodies and the consequential technological power balance. When the IAEA defined its safety standards in the 1970s, it did so with full cognizance of the increasing number of nations that were then just beginning to use nuclear power technology; for these nuclear power nations, the standards require that both plant operators and regulatory agencies gather technical knowledge and expertise.[1] Thus, the agency's standards for plant operation and maintenance mandate periodic testing and inspections and mechanisms for learning from the feedback they generate. The IAEA's fundamental safety principles, however, require that regulatory bodies exist in a state of technological independence. Specifically, regulators must be independently capable of defining safety mechanisms without input from power companies.

The focus of nuclear safety regulation governance, as it emerged in the years before the accident at Fukushima, was to ensure safety in all phases of nuclear power generation. Specifically, the government set the standards, established the relevant legal system, and sought compliance by the related companies. In form, the Japanese regulatory governance, which has harbored a great variety of interests and expectations of many different parties, was in accordance with the international safety regulation standards (see Chapter 5). Theoretically, a clear division of labor was built into the overall structure of nuclear safety regulation, with electric power companies holding the primary responsibility for safety, the Nuclear and Industrial Safety Agency (NISA) overseeing the electric power companies, and the Nuclear Safety Commission producing relevant guidelines. In reality, since its inception, though, governance has been based on an assumption of safety regulation that is applicable to ordinary conditions, rather than to those that could arise during a major emergency. The basic problem lies with the Nuclear and Industrial Safety Agency and the Nuclear Safety Commission: NISA, although formally independent, had neither sufficient capabilities for practical and effective safety regulation nor sufficient technical resources compared with that of the electric power companies; the commission had an experts council, which created relevant safety standards, but lacked sufficient legal authority and investigatory analysis capabilities. And the electric power companies, it is worth noting, simply did not try sufficiently to fulfill their own responsibilities to strengthen safety regulations.

The Japanese government report submitted to the IAEA in June 2011 on the Fukushima Daiichi Nuclear Power Station accident concluded in its summary that, as severe-accident measures did not carry the force of law, the content of their establishment and maintenance was lacking in rigor. Kenkichi Hirose—

---

[1] In Japan, many electric power companies, including TEPCO, are large-scale enterprises, and it is easy for such companies to accumulate specialized technical knowledge and expertise in-house. In contrast, the regulatory institutions exhibit a personnel system that is typical of government agencies in that they tend to cultivate generalists rather than specialists; consequently, these institutions lack the skills needed to serve as repositories of expert technical knowledge.

the Nuclear and Industrial Safety Agency's director general from 2005 to 2007—who directed the compilation of that report, has in retrospect expressed the opinion that, if such safety measures had been a legal requirement, then both the side issuing the orders and the side receiving them would have considered them much more carefully. He further observed that the report exemplified the "administrative guidance" unique to Japan—that is, an administrative organ, in order to achieve a particular mission and objective, advises and recommends certain behavior to private and corporate citizens without establishing a law. The administrative guidance is not exactly a legally binding decision, but it is regarded as a quasi-legal instrument to enforce administrative decisions.[2]

Though Japan's governance system is in formal accordance with international guidelines and rules, its particular—and perhaps peculiar—role of actors in the governance system is worth exploring to understand which parties actually held responsibility for safety regulation, the degree of resources that could be allocated to safety regulation, the degree of technical expertise and skill held on the part of the regulators, and the stance and actions of the electric power companies and whether they are open to criticism.[3]

Our investigation found that a lack of resources encouraged a formalistic approach to nuclear safety regulation and that a dependence on paper-based inspections had the contrary effect in that electric power companies, ultimately, loosened their control over safety practices altogether.

Nuclear Safety Commission

In response to the rise of the antinuclear movement following the *Mutsu* nuclear ship accident, the Nuclear Safety Commission separated from the Atomic Energy Commission in 1978 with the objective of improving safety regulation. As a neutral organization with the authority to issue recommendations to related organs through the prime minister, it consisted of five specialized

---

[2] Personal communication with former NISA Director General Hirose, November 26, 2011.

[3] As noted in Chapter 2, the dual configuration of Japan's nuclear energy administration—the Ministry of Education, Culture, Sports, and Technology (MEXT) and the Ministry of Economy, Trade, and Industry (METI)—blurred the line between nuclear advancement and nuclear regulation. This dual configuration of the Nuclear and Industrial Safety Agency was created in the 2001 reorganization of government ministries and agencies. In Japan, the government's administration of nuclear energy has always been divided between the Science and Technology Agency (merged into MEXT in 2001) and the Ministry of International Trade and Industry (merged into the Ministry of Economy, Trade, and Industry in 2001). In fact, safety regulation became something of a blind spot during the central government reorganization of 2001, which, ultimately, impeded appropriate management of SPEEDI and any growth of technical expertise on safety regulations; consequently, this impediment engendered deficiencies in regulatory governance of safety that contributed to the accident at the Fukushima Daiichi Nuclear Power Station.

experts who studied and formulated guidelines for not only reactor facilities, but for nuclear fuel processing and reprocessing facilities, as well.

The commission issues safety guidelines in a broad range of matters related to nuclear power plant operation—including plant siting, design, safety assessments, radiation dose target values, and severe-accident measures—and determines safety standards for nuclear operations and performs related examinations. Commissioners indeed hold specific expertise; however, they do not personally confront nuclear operators, detect problems in their operations, or set safety standards for necessary repairs. Accordingly, the commission tends to consider nuclear plant safety from a rational and theoretical basis, regardless of on-site circumstances.

The guidelines formulated by the Nuclear Safety Commission, moreover, are just that—*guides*—and therefore are not legally binding. On the one hand, the commission's guidelines extend to minute details, with extremely fine-grained standards and examination; on the other hand, the commission does not have the authority to directly supervise nuclear operators. The actual performance of safety maintenance and supervision of nuclear operators is the province of the Nuclear and Industrial Safety Agency.

Detailed crisis response

The Nuclear Safety Commission's finely specialized expertise becomes strongly apparent in its formation of the various special review boards and committees serving under it—such as the Nuclear Reactor Safety and Nuclear Fuel Safety Special Review Boards. Beyond these, however, there are six expert committees,[4] yet another tier of numerous subcommittees, specific survey and conference bodies,[5] and advisory bodies.[6] Their constituent members include large numbers of university professors and others from academia, as well as experts from the Japan Atomic Energy Agency and several public-service corporations,[7] many of whom had been employed by nuclear-technology manufacturers and electric utilities. It must be noted, however, that the large

---

[4] The expert committees are: the Nuclear Energy Safety Standards and Guidelines; Radioactive Waste and Waste Treatment; Radiation Protection; Radioactive Materials Transport; Nuclear Energy Accident and Malfunction Analysis and Assessment; Nuclear Energy Safety Research; and Nuclear Energy Facilities Disaster Prevention.

[5] These bodies include the Anti-Seismic Safety Assessment Special Committee; the Testing and Research Reactor Anti-Seismic Conference; the Reprocessing Facilities Safety Investigation Project Team; and the Designated Radioactive Waste Treatment Safety Investigating Committee, among others.

[6] These bodies include the Emergency Technology Advisory Committee, the Nuclear Powered Vessel Damage Response Emergency Technology Advisory Committee, and the Armed Attack Nuclear Energy Damage Response Emergency Technology Advisory Committee.

[7] A list of members of the Nuclear Reactor Safety Special Review Board is available in Japanese at: www.nsc.go.jp/shinsa/shidai/genshiro/genshiro206/siryo1-2.pdf (accessed February 28, 2012).

size of committee membership makes it difficult to fully utilize the expertise of each individual, and the culture was not conducive to voicing disagreement due to the pressure to comply with government policies. Though the reviews and discussions were thorough, each expert was tasked to focus on checking particular aspects or components of safety regulation, and the overall perspective on nuclear energy safety was thereby weakened.[8]

One of the bodies, the Emergency Technology Advisory Committee, was convened on the day of the Fukushima accident—but without Nuclear Safety Commission Chairman Madarame, who remained at the Prime Minister's Office. The committee met without liaising with any other committee, and without any type of crisis handbook or guide. Meanwhile, Madarame was isolated from other colleagues and administrative support, often needing to make judgments on nuclear reactor conditions or to make emergency responses without real-time information.[9] The only support he received was from Yutaka Kukita, deputy chairman of the Nuclear Safety Commission.[10] In the case of an emergency, the Nuclear Safety Commission should dispatch the Emergency Response Special Review Board members to the area of the accident; however, there is no evidence that this happened at Fukushima. In addition, Prime Minister Kan personally appointed a large number of Cabinet advisers and established an advisory team. Thus, the Nuclear Safety Commission—and particularly the Emergency Technology Advisory Committee—provided no advice to the Nuclear Emergency Response Headquarters, the central command for crisis response chaired by the prime minister and supported by NISA. The Emergency Technology Advisory Committee's function was evidently limited, rather, to providing advice to the local governments.

Safety regulation and accident preparedness

Historically, the Nuclear Safety Commission focused its energies and capabilities on improving the safety regulation of nuclear power facilities and assuring their operational safety. Conversely, it had paid little or no consideration to events or accidents that might show any need for strengthened safety regulations. Envisioning an emergency scenario, the commission had constructed the Emergency Technology Advisory Committee, but it did nothing to assume that it might actually be utilized, lest it be taken as an indication that the commission's safety regulations will and should prevent accidents to make sure that the myth of safety will be protected; this invited strident criticism by the antinuclear movement. The commission, therefore, was deeply devoted to

---

[8] It is assumed that the Emergency Technology Advisory Committee will be convened in the event of an emergency, with its members providing advice to the prime minister (as the chair of the Nuclear Emergency Response Headquarters) through the Nuclear Safety Commission chairman.

[9] Personal communication with NSC Chairman Madarame, December 17, 2011.

[10] Personal communication with NSC Chairman Madarame, December 17, 2011.

overregulating for preventing accidents without planning and preparing for possible accidents—not to mention avoiding any brand of related questions and criticisms.

Thus, the commission formulated extremely detailed nuclear power plant safety regulations—and a pattern emerged in which the Nuclear and Industrial Safety Agency and the Japan Nuclear Energy Safety Organization, as the organizations involved in the actual on-site procedures, mechanically applied paper-based documents as proof that safety had been secured.

The results of our investigation clearly showed that, under ordinary circumstances, the commission held a high level of expertise, but developed a bias for securing technological safety and became narrowly specialized in examining safety performance of finely compartmentalized technological fields and aspects of nuclear power. Initially, the government announced that the chairman of the Nuclear Safety Commission—in his role of consolidating many special review boards, expert committees, and subcommittees—would serve as a principal adviser to the prime minister in emergency situations, but no clear and specific authority or procedure was provided to achieve this advisory role.[11] Thus, during the accident at Fukushima, the Nuclear Safety Commission chairman was, in effect, left to advise the prime minister, departing substantially from the established system and leading to a situation in which it became necessary for the prime minister to seek a second opinion.[12]

Nuclear and Industrial Safety Agency

With an initial staff of 800—about 330 of whom were delegated jobs in the realm of nuclear power plant safety—the Nuclear and Industrial Safety Agency, or NISA, examined and inspected nuclear fuel refining and processing, interim storage, and waste materials facilities, as well as remained responsible for high-pressure gas security, mining safety, and other areas as it did before the reform. Not to mention, NISA opened twenty-one nuclear safety inspector offices at the country's nuclear plants and related facilities; between one and nine inspectors and senior specialists for nuclear emergency preparedness staffed each office.

NISA was responsible for nuclear emergency preparedness: Specifically, it was charged with disaster prevention and damage mitigation in the event of an accident, dispatching personnel to the accident site, gathering information to the Emergency Response Center, a part of NISA, and reporting to the government. In 2003, the Ministry of Economy, Trade, and Industry (METI) established the Japan Nuclear Energy Safety Organization (JNES), an incorporated administrative agency supervised by NISA, to implement nuclear

[11] This is particularly evident in the absence of any legal provision designating the Nuclear Safety Commission chairman as a member of the Nuclear Emergency Response Headquarters.
[12] Personal communication with former Prime Minister Naoto Kan, January 14, 2012.

power facility inspections, safety analysis and assessment, and emergency-response support. Employing about 420 workers, seventy-five of whom were full-time inspectors, the organization mainly provided technical support to NISA and, by substantially expanding its inspection and examination functions, compensated for ministry and agency deficiencies by accumulating industry experts through personnel transfers.

Since many of the inspectors had previously worked in companies manufacturing nuclear reactors and facilities, the organization was expected to be fully operational with experienced workers with expertise. In actuality, however, NISA and JNES could not coordinate effectively, and the organization failed to bring about substantially improved safety regulations. Yoshihiro Nishiwaki—who had worked for both the Agency of Natural Resources and Energy and JNES before becoming a guest professor at the University of Tokyo—found that, contrary to expectations, the communication and understanding of technical content at NISA only worsened when the highly specialized safety organization was established.[13] While the technical competence and knowledge of power plants resided in the Japan Nuclear Energy Safety Organization, NISA remained a strongly administrative, rather than a technical and regulatory body. Thus, despite the organization's vigorous research on safety, NISA did not have the know-how to properly incorporate the research results and proceeded without incorporating them into the regulatory process.[14]

NISA's code of conduct included scientific and rational determinations, transparency in operational performance, and fair and impartial performance. As one high-level METI official responsible for system design later noted: The Science and Technology Agency's approach, as was the case during the 1999 Tokai-mura accident, had been to withhold information that might arouse public concern, but NISA pushed to change this approach and share everything. That is, in the spirit of public disclosure, it set a goal to properly explain accidents, problems, and the methods of its assessments.[15] It set out to highlight even slight problems in nuclear power facility operations, and to be transparent with what organizations its personnel were affiliated.

Overall, however, in NISA's relations with the electric utilities, the interdependence remained mostly unchanged from the time of the Science and Technology Agency. In August 2002, NISA first publicly disclosed that the Tokyo Electric Power Company (TEPCO), one of the country's leading electric utilities, had for many years concealed problems by falsifying data and other means. Moreover, during the enquiry, NISA informed TEPCO of the whistle-blower's identity.[16] This public exposure provoked the opposition

---

[13] Personal communication with Professor Nishiwaki, December 16, 2011.
[14] Nishiwaki, Y. (2011) Changes in Japan's protections against severe accidents: Where did regulations go wrong? *Genshiryoku Eye*, September/October. (In Japanese.)
[15] Personal communication with a former METI member, October 28, 2011.
[16] Criticality.org (2013) Kei Sugaoka, the GE/Tepco Whistleblower. Available at: www.youtube.com/watch?v=fBjiLaVOsI4&feature=youtu.be (accessed August 31, 2013). (In Japanese.)

party to separate the regulatory organ from METI, which, itself, was engaged in the promotion of nuclear energy policy and planning. The government wanted to build NISA into an even stronger, singular regulatory organ—similar to the US Nuclear Regulatory Commission.[17] However, it took the disastrous Fukushima accident to realize the separation of nuclear regulator from METI.

NISA responded to the TEPCO case by forcing the company to shut down all seventeen nuclear reactors until it could comply with safety regulations. NISA vigorously inspected the reactors from 2003 and 2005, and then restarted each reactor one by one. NISA also increased the items for regular periodic inspection and strengthened penalties for infractions. To promote quality assurance, NISA introduced the Periodic Safety Management Examination system, which scrutinized corporate management policies and operational processes. Consequently, it became obligatory to report even spilled water and other minor errors as noncompliant. The objective of this was to attain a higher level of nuclear energy safety.[18] In actuality, however, the addition of new national periodic inspections, together with those previously conducted, greatly increased the paperwork for the electric utilities and plant-site personnel, causing nuclear plant staff to complain that the preparations for inspection were leaving less time to attend to the equipment. As described by Takuya Hattori, president of the Japan Atomic Industrial Forum, complaints arose in the nuclear energy industry that the content of these inspections were largely devoted to trivial matters that simply engendered a massive exercise in ticking boxes in documents—documents that lacked in any consideration of total safety, and could not be held to have increased the level of safety.[19]

Questions also arose within the ranks of the regulators, who were concerned with the reliance on plant operators to conduct exams. A former JNES inspector observed that the actual inspections were largely unsophisticated: Checking for compliance, he said, consisted of reviewing the procedures described in the inspection documents, which were simply copied from those published by electric utilities, and that the nuclear plant inspections were largely ceremonial in nature. The inspector said that these reviews, therefore, were merely a formality, and he expressed doubt as to whether these procedural examinations should be regarded as true inspections.[20] Another JNES official, who formerly worked in NISA, held a similar understanding and mentioned that, by the time

---

[17] On July 16, 2003, the Democratic Party proposed the establishment of an independent organ to strengthen safety regulation.

[18] Japan Electric Association (2009) *Quality assurance regulation for nuclear power plant safety*. Report, JEAC-4111. Tokyo: JEA. (In Japanese.)

[19] Personal communication with Japan Atomic Industrial Forum President Hattori, November 17, 2011.

[20] Personal communication with former Japan Nuclear Safety Organization personnel, November 2, 2011.

of the Fukushima accident, inspections involved such a mass of details that "you could not see the forest for the trees."[21]

In the wake of the 2002 TEPCO case, NISA became preoccupied with its response to a succession of smaller accidents[22] and made little progress in the realm of severe-accident risk and other risk-related questions. Professor Akira Omoto of the University of Tokyo has observed that, after the TEPCO scandal, the focus of safety regulation on issues of risk was not only blurred, but replaced by excessive spending on increasingly tough inspections on hardware, which gave the illusion of ensuring quality and compliance to prevent further accidents.[23]

Through mid-career hiring, NISA doubled the number of its non-career inspectors, from approximately fifty to 100. Over the course of the decade, however, the retention of career inspectors remained problematic. That is, they were vital to the agency's operation, but they are instructed and guided by career bureaucrats. Though these career bureaucrats have strong administrative discretion, they are not always educated or trained as nuclear technology experts. Furthermore, nearly all of them shuttled back and forth between the agency and METI every two or three years under the personnel rotation system. Consequently, NISA was unable to retain specialized administrators, and its regulatory policy lacked continuity. The agency's director general and vice director general posts were held by bureaucrats with purely administrative backgrounds. However, career bureaucrats with expertise in nuclear technology were very small in number. In fact, this knowledge deficit would continue all the way to 2011, when the disaster unfolded at the Fukushima Daiichi Nuclear Power Station. This crippling weakness in the agency's human resources was made apparent by its slow emergency response.

The Ministry of Education, Culture, Sports, Science, and Technology

The Science and Technology Agency merged together with the Ministry of Education to form the Ministry of Education, Culture, Sports, Science, and Technology, known as MEXT. Though the agency's Atomic Energy Bureau was divided and transferred to the Japan Atomic Energy Commission and NISA after the Tokai-mura accident in 1999, it retained its competence in nuclear development because MEXT controlled public-service corporations. As such, MEXT maintained the budget allocated for nuclear energy, as well as oversaw the positions destined for ministry-personnel transfers in the

---

[21] Personal communication with Japan Nuclear Safety Organization officer interview, November 7, 2011.
[22] Japan Nuclear Energy Safety Organization (2012) *Third-party investigative committee on inspections and other operations*. Report. Tokyo: JNES.
[23] Personal communication with Professor Omoto, January 13, 2012.

longstanding *amakudari*—the tradition of the country's senior bureaucrats to retire to high-profile positions in the private and public sectors.

Though, after the 2001 government reorganization, NISA and the Nuclear Safety Commission both became the central agencies for nuclear safety, and MEXT retained its function of handling nuclear safeguards, as well as serving as the authority for radiation protection, including monitoring, nuclear reactor regulation, and radiation survey. That is, safety regulation became even more complex. As before, both MEXT and METI held jurisdiction in the legal framework as provided under the Nuclear Regulation Act; however, NISA was reconfigured to perform the actual inspections, and the Nuclear Safety Commission was redefined to provide the safety guidelines and collect the inspection results.

The government also handed METI the Nuclear Power Engineering Corporation, which tested nuclear power generation equipment and instruments for safety, and the Japan Power Engineering and Inspection Corporation, which performed on-site inspections of nuclear plant equipment. METI's role with the Nuclear Power Engineering Corporation, however, overlapped with MEXT's inspection role.

MEXT also controlled the Japan Atomic Energy Agency, an outfit that was formed in 2005, when the Japan Atomic Energy Research Institute merged with the Japan Nuclear Cycle Development Institute. The Japan Atomic Energy Agency housed the Nuclear Emergency Assistance and Training Center, which, during a nuclear emergency, dispatches radiation-monitoring cars, equipment, and materials, and issues technical information and governmental and community advice on protective measures.

Two key organizations, MEXT and METI, existed under different hierarchies, separating their administrative structures and impeding communication between them: The Japan Atomic Energy Agency and the Nuclear Power Engineering Corporation were both under MEXT, while the Japan Nuclear Energy Safety Organization was under METI. As became evident during the accident at the Fukushima Daiichi Nuclear Power Station accident, these organizations could not perform effective emergency accident response by acting within the limits of their prescribed scopes of operation. It should be noted that, in part, this was due to the overarching absence of authority from the ministry, agency, and public-service corporation hierarchies; thus, no definitive emergency-response instructions were issued.

This complexity and overlap of government responsibilities over nuclear safety was definitely one of the most important causes of the Fukushima accident. Particularly, in terms of SPEEDI (the System for Prediction of Environment Emergency Dose Information), MEXT did not see itself, but saw NISA, as the primary responsible agency for severe-accident response; however, MEXT is the agency solely responsible for radiation safety.

TEPCO

During World War II, the national government controlled electric power in Japan. But in 1951, the Japan Electric Generation and Transmission Company was privatized, creating nine electric power companies, including TEPCO. In the process of privatization, a system took shape of regional monopolies, in which each electric power company controlled everything—from the generation and distribution to the sale of electric power within a particular region it supplied. This was a so-called vertical integration. The objective of the vertical integration was to secure the stability of supply under the severe lack of electricity after the end of World War II. Later, in 1972, when Japan regained control of Okinawa, the Okinawa Electric Power Company opened its doors, adding a tenth name to the roster of Japanese power companies.

Today, as the industry leader, TEPCO enjoys a rich network of personal connections throughout the government, the bureaucracy, and the financial world, and has consistently exerted a major influence on national government policy decisions. The firm boasts annual sales of $66.25 billion, total assets of $193.75 billion, and a total workforce of 38,671.[24] The company supplies power to Tokyo and eight prefectures; in fiscal year 2011, it sold a total of 268 billion kilowatt hours of electricity, approximately one-third of all power consumed in Japan. It ranks among the largest private-sector power companies in the world.[25]

In Japan, the top power company executives are central players in the regional economies they serve, and the companies retain strong influence through donations to political parties—not to mention, by employing former bureaucrats as company advisers and management. TEPCO's former CEOs, in particular, have served as chairmen or vice-chairmen of the Japan Business Federation (*Keidanren*),[26] ensuring the firm has solid portals into the heart of

---

[24] The sales, assets, and workforce figures refer to fiscal year 2011.

[25] In return for the obligation of supplying stable electric power throughout their assigned areas, TEPCO and the rest of Japan's ten electric power companies are allowed to pass electricity costs, fuel costs, personnel costs, costs of facility maintenance, and other expenses—plus profit—on to the consumer; this system is known as an "overall base-price scheme," which has ensured a certain level of profitability and steady growth. In the past, the massive scale of infrastructure investments allowed the government (particularly MITI) to stimulate the national economy by making investments in electric power infrastructure—for example, extending transmission lines or installing more electric towers.

[26] Known as *Keidanren*, the federation, which was established in 1964, is a representative economic organization that has 1,285 companies, 127 industrial associations, and 47 regional economic organizations (as of March 29, 2013). This is the largest business lobby among Japanese economic organizations in the country. *Keidanren* has been the most powerful lobbying body to promote business interests and has held strong links with politicians affiliated with the Liberal Democratic Party. However, since 2009, the Democratic Party of Japan-led government has tried to reduce the influence of *Keidanren*. Thus, there was a shaky relationship between TEPCO, one of the most powerful members of *Keidanren*, and the DPJ-led government under Prime Minister Kan; this may have contributed to the distrusting relationship between TEPCO and the government in response to the accident at Fukushima.

Japan's political and financial world. The federation—Japan's largest lobby that represents industry interests—has historically helped candidates get elected to Japan's House of Councilors, the Diet's upper house. For the 1998 elections, former TEPCO Executive Vice President Tokio Kano, who was also a powerful member of the federation, won a seat in the upper house with the strong backing of the Liberal Democratic Party (LDP). In 1974, Japan's electric power industry—shaken by criticism of "money-soaked elections" and by "Don't Pay One More Yen," an organized movement to oppose corporate contributions—forced TEPCO to stop contributing to political campaigns from its own corporate account; however, individual corporate executives of power companies continued to make personal donations, the amount of which was determined by that executive's rank in the corporate hierarchy. During a fourteen-year period between 1995 and 2009, contributions from TEPCO executives to LDP fund-raising organizations totaled at least $630,000 (in 2009 dollars), and at least 448 people contributed.[27]

By the end of August 2011, TEPCO employed more than fifty consultants and part-time employees who were former members of Japan's central bureaucracy, including former staffers of the Ministry of Economy, Trade, and Industry—the oversight agency that is only a short walk from the company's Tokyo headquarters—as well as the Ministry of Land, Infrastructure, Transport, and Tourism; the Foreign Ministry; the Ministry of Finance; the National Police Agency; and other government institutions. Former government officials who rose to the rank of company vice president included a former deputy finance minister, a former energy minister, and a former deputy energy minister; all remained on staff in an auditory or consulting capacity after stepping down from their posts.[28] Throughout the years, TEPCO has exerted a profound influence on the direction of national government policy with pride that the company, not the government, provides Japan's stable electricity.[29] In the past, company representatives have served as members of energy policy subcommittees, and in some cases TEPCO has prepared various industry status reports that are ultimately distributed by government offices.[30]

In addition, the company has technological and planning capabilities far superior to those of the Nuclear and Industrial Safety Agency or the Agency

---

[27] *Asahi Shimbun* (2011) Untitled. Morning Edition, October 8. (In Japanese.)
[28] *Asahi Shimbun* (2011) Untitled. Morning Edition, October 8. (In Japanese.)
[29] TEPCO's website boasts that the company's highly reliable equipment and advanced technology ensure that its average annual power outage is only three minutes per building, an achievement that ranks the company among the world's most stable power providers both in terms of the frequency and the duration of outages. Available at: www.tepco.co.jp/corporateinfo/company/annai/jigyou/index-j.html (accessed August 31, 2013). (In Japanese.)
[30] Japan's electric companies provide the government with information on the state of the nation's electricity supply and demand, the price of imported fuel, safety issues, information about foreign safety regulations, among many other things. Often times, this information is spun to favor Japanese electric companies; this can have the power of influencing and controlling the government agenda.

for Natural Resources and Energy; this gives the firm significant power to direct government policy in ways favorable to its own interests by making various proposals on deregulation issues and pointing out technical problems in other proposals authored by regulatory bodies. A high-ranking staffer in the Prime Minister's Office describes TEPCO's lobbying strength as "overwhelming," noting that the company possesses far more talent than government bureaucracies and that it routinely comes to the table with largely unassailable proposals prepared by extensive and diligent behind-the-scenes groundwork—a dynamic that makes it extremely difficult for the government to effectively regulate the company.[31]

TEPCO as an industry leader

TEPCO has been a champion of nuclear power and greatly involved in nuclear power policy—a field that, in Japan, has evolved under the "state-planned, privately operated" framework (*kokusaku minei*). As many countries throughout the world saw nuclear power as a panacea in the world of climate change and the way to reduce greenhouse gas emissions, Japan was no different. The December 1997 Kyoto Protocol, in fact, recommended that Japan construct or expand up to twenty nuclear plants by 2010 in an effort to achieve Japan's target for greenhouse gas reductions. This benchmark for expanded nuclear power was revised in 2001 to read "up to thirteen plants by fiscal year 2010." However, even this number was too high. Though the government had hopes to meet this—a scenario that would have had utility companies increasing nuclear plant construction at a speed that would have been all but realistic—only five new nuclear plants have been constructed since 1998. When local governments were in support of nuclear energy policies, there was a system that could provide subsidies to regions surrounding nuclear plants; however, even with these subsidies, there were limited places where nuclear power plants could be accepted and built.

In addition, various power companies, including TEPCO, used nuclear fuel taxes, nuclear facilities asset taxes, and various forms of kickbacks—for example, financial contributions and public facilities, such as sports stadiums—to local governments. Beginning in 1971, when unit 1 at TEPCO's Fukushima Daiichi Nuclear Power Station went on-line, the company continued, over sixteen years, to add a total of six units at that nuclear plant, followed by four units at the Fukushima Daini Nuclear Power Station, and seven units at the Kashiwazaki-Kariwa station in Niigata Prefecture. Nuclear power accounts for around 30 percent of the firm's total power production. For more than twenty years, TEPCO has contributed more than $500 million (in 2012 dollars) to local governments in exchange for hosting nuclear power plants—including $110 million (in 1997 dollars) in 1997 for a soccer facility constructed in Naraha in

---

[31] Personal communication with Prime Minister Kan's staff, January 24, 2012.

Fukushima Prefecture.³² Naraha, a city with an increasingly aging population, needed to attract young people and thought a soccer stadium was just the way to do so. Until the Fukushima disaster, the city hosted national sporting events and highly prestigious international teams; the Japanese people in Fukushima Prefecture considered TEPCO's huge contribution as one that distinctly improved the community. Today, however, the stadium is used as a base for emergency-response teams and as a storage facility for highly contaminated water.

Striking the balance: business operations and state-planned, privately operated

Beginning in 1986, real estate and stock prices in Japan were grossly inflated, creating a bubble economy until the early 1990s. When this bubble burst, discussions of deregulation began in response to public outcry that Japanese electricity fees were higher than in other countries. In 1995, for the first time in thirty-one years, the Electricity Business Act was revised to allow independent power producers to enter the market. Since March 2000, a partial deregulation of the retail electricity market has been underway, with a gradually increasing scope to include small- to mid-sized businesses.³³ TEPCO accepted this widening deregulation; however, in 2002, when it came to the separation of power production and transmission, TEPCO CEO Nobuya Minami insisted that "Japan's system of vertical integration of production and transmission is essential for ensuring a stable supply of electric power."³⁴ Electric industries, including TEPCO, thought deregulation would make it more difficult to expand the number of nuclear plants, not to mention to further develop the technologies and other business activities related to the nuclear fuel cycle. That is, nuclear plant construction requires massive up-front investment—namely, the cost of pluthermal programs and other business activities related to the nuclear fuel cycle; this, therefore, poses heavy economic risks for private-sector corporations. At that time, the industry as a whole fiercely opposed separating production and transmission and—with the support of the LDP, which then controlled the government—succeeded in ensuring that the initiative never came to fruition.³⁵ TEPCO's influence demonstrates how cozy the relationship was between government and industry under the government system.

---

³² *Asahi Shimbun* (2011) Untitled. Morning Edition, September 15. (In Japanese.)
³³ In April 2005, the scope of retail electricity deregulation was enlarged to include contracts for 50 kilowatts or more, encompassing supermarkets and small- to mid-sized buildings.
³⁴ *Mainichi Shimbun* (2002) Untitled. Morning Edition, December 27. (In Japanese.)
³⁵ In June 2002, Akira Amari, the chair of the LDP's General Energy Policy Subcommittee, together with other members of Japan's House of Representatives, proposed an Energy Policy Basic Law. This legislation recognized nuclear power as a core power source within the energy plan adopted by the Ministry of Economy, Trade, and Industry in October 2003, and the preservation of the existing system of vertical integration of production and transmission of electricity.

Within the electric power industry, TEPCO was unusually quick to foresee the coming of the age of electricity deregulation, and, in response, it pushed through certain managerial reforms. Former TEPCO CEO Hiroshi Araki, who was in office from 1993 to 1999, was fond of statements such as "Let's make this an ordinary company" and "Let's point our desks toward *Kabutocho* [Japan's Wall Street] when we work."[36] His goals of improving the bottom line, by cutting costs and streamlining operations, accelerated when talk of electricity deregulation began: Costs for repairs and maintenance were reduced by reevaluating supply vendors, among other ways; authority devolved from headquarters to branch offices; and new business investments—including communications, caregiving, and novel energy sources—were stepped up. The company's Nuclear Power Division introduced an independent profit system for individual power plants, spurring competition among plants. It was said that a one-day shutdown of a single nuclear reactor could increase operating expenses by $900,000 (in 1999 dollars), and the duration of routine inspections was shortened: Not only did the company make cost-saving changes like switching to a higher-efficiency fuel, it also requested that nuclear plant manufacturers, such as Toshiba and Hitachi, reduce costs.

But how did cost and reducing expenses factor into nuclear plant safety provisions? "A system that allows safety to be strictly assured in a sustainable way will also be a low-cost system from an economic perspective," said former TEPCO CEO Nobuya Minami. "Safety provisions coincide with measures to simplify and streamline operations, and I'm confident that we did all the things that I consider important from a safety standpoint."[37] Former TEPCO executives agree that safety provisions were never neglected: One told us, "When it came to important safety provisions, there was no resistance from the higher level," while another stated, "Even after 2000, safety provisions costing tens of billions of yen were implemented at the Fukushima Daiichi Nuclear Power Station to reduce earthquake risks."[38]

However, amid company-wide efforts to reduce costs, it wasn't possible to spend money on every last-minute safety measure, and the heads of the company's Nuclear Power Division were forced to make some difficult decisions to define priorities. Press reports have previously noted that, in 2006,

---

[36] Before Araki, the CEO role was held by a succession of people—Hisao Mizuno, Gaishi Hiraiwa, and Sho Nasu—who had risen through TEPCO's General Affairs Division and had strong ties to Japan's political and financial communities. Araki, who placed considerable emphasis on the importance of reform efforts, selected Nobuya Minami—who had worked for the company's planning division, which oversees efforts to diversify business—as his successor. Minami set out to cut costs by 20 percent within five years; he took steps to reduce the company's interest-bearing debt, which had climbed as high as $884 million (in 1999 dollars) in the late 1990s, and kept a promise to reduce electricity rates. Minami paved the way and was followed by two CEOs who also rose through that same division: Tsunehisa Katsumata (from 2002 to 2008, then from 2008 to 2012 as chairman) and Toshio Nishizawa (from 2011 to 2012).
[37] Personal communication with former TEPCO CEO Minami, November 15, 2011.
[38] Personal communication with former TEPCO employees, November 2011 and January 2012.

TEPCO considered a construction project at the Fukushima Daiichi Nuclear Power Station to improve power-supply interconnections to prevent power outages, but decided to postpone the work due to technical impediments, among other reasons.[39] One former TEPCO managing director disclosed to us that, under Araki's "ordinary company" push, "achievements in cost reduction became a benchmark for evaluating executive performance, which created some dangers on the safety side."[40] This source went on to note: "There was never a proper discussion within the company of precisely the extent to which a company that handles massive risks such as nuclear power, and a public utility that provides a critical life-and-death commodity such as electric power, can ever become an 'ordinary company.'"[41]

Ambiguities of responsibility created by privately administered policy

On December 2, 2011, TEPCO released an interim report[42] from its internal committee to investigate the Fukushima nuclear accident. The report included the company's estimates of tsunami severity and repeated TEPCO's past assertions that the massive tsunami that struck the Fukushima plant was "outside the realm of expectation" (*souteigai*). For example, TEPCO claimed that its tsunami simulations were "based on assumptions for which there was no concrete evidence" and that, because the models considered "the impact of a tsunami that significantly exceeds expectations, such an event would lie outside the range of assumptions used to prepare initiatives to respond to accidents." Regarding accident management, the report insisted that the responsibility was not TEPCO's alone, stating that "electric power companies and the national government have worked together to advance [accident-management] provisions; we report on the content of our provisions to the government, and we move forward only after receiving their approval and confirmation."

Hitotsubashi University Professor Takeo Kikkawa, who studies the history of management in the electric power industry, cited Japan's state-planned, privately operated system as a problematic aspect of the nation's nuclear power industry, noting that it has had deleterious effects on the robustness of power companies as private-sector corporations. Kikkawa acknowledged that several advantages exist to running a public utility company as a private-sector entity; such companies take advantage of the efficiency of a private-sector company, exhibit performance in excess of government-run entities, and bring their risk management capabilities to bear on safety problems, which may enhance safety

---

[39] *Asahi Shimbun* (2011) Untitled. Morning Edition, October 23. (In Japanese.)
[40] Personal communication with former TEPCO employee, November 1, 2011.
[41] Personal communication with former TEPCO employee, November 1, 2011.
[42] Tokyo Electric Power Company, Inc. (2011) *Fukushima nuclear accidents analysis report (interim report) supplement*. Report, December 2. Tokyo: TEPCO. Available at: www.tepco.co.jp/en/press/corp-com/release/betu11_e/images/111202e14.pdf (accessed August 31, 2013).

provisions. However, TEPCO's repeated invocation of the phrase "outside the realm of expectation" carries with it a certain nuance: *We were just upholding the safety standards the government told us to uphold. Therefore, there was nothing that could have been done about it. It's not our fault.* This, said Kikkawa, illustrates what is wrong with the private-sector operation of a public utility. "In theory, private-sector corporations should be able to apply their risk management skills to do a better job of ensuring high safety standards than the government could," he explained. "Unfortunately, in the case of the Fukushima Daiichi Nuclear Power Station, these private-sector advantages failed to materialize."[43]

TEPCO and tsunami precautions

TEPCO's simulations, which it released after the Fukushima accident, indicated that the company was aware of the possibility of a tsunami in excess of then-current expectations striking the Fukushima Daiichi Nuclear Power Station; it became clear that, although the company reported this finding to the Nuclear and Industrial Safety Agency, TEPCO failed to institute any precautionary measures.

Released in December 2011, the interim report by the Japanese government's inquiry commission on the Fukushima accident presented a detailed chronology of the tsunami simulations and internal corporate discussions surrounding them:

> In February 2002, TEPCO, using new tsunami assessment technology developed by the Japan Society of Civil Engineers (JSCE), raised the expected tsunami height at the Fukushima Daiichi Nuclear Power Station from the value of 3.1 meters, used during the plant's design phase, to a new value of 5.7 meters. TEPCO implemented several provisions to reflect this heightened expectation, including increasing the electric drive for the unit 6 emergency diesel generator seawater pump cooling system. However, experts subsequently went on to note the possibility of even higher tsunamis, and in July 2002 Japan's Headquarters for Earthquake Research Promotion announced its opinion that earthquakes and tsunamis were possible throughout a region running from the northern Sanriku coast to oceanic trenches off the coast of the Boso peninsula—a region that included coastal regions of Fukushima, for which there were no previous records. The results of new research on the massive Jogan tsunami, which struck the coast of Japan's Tohoku region in the year 869 AD, also became available during this time.
>
> TEPCO conducted simulations of tsunami heights based on a model in which waves originated off the Sanriku coast, and concluded that tsunamis

---

[43] Kikkawa, T. (2011) *TEPCO: The True Nature of Failure.* Tokyo: Toyo Keizai Shimbunsha. (In Japanese.)

of heights up to 52 feet were possible. In response to this finding, the company convened an in-house study group around June 2008. At this meeting, Sakae Muto, who was then assistant manager of the Nuclear Power and Plant Siting Division, and Masao Yoshida, who was then head of the Nuclear Power Plant Management Department, were informed by knowledgeable personnel that the construction of tsunami sea walls was estimated to take four years and cost several billion yen; Muto noted that the construction of such sea walls would not be regarded favorably by the regional population, which would conclude that TEPCO needed to sacrifice the local surroundings in order to protect the safety of the plant.[44] Muto and Yoshida, despite believing that "in reality, there isn't going to be any tsunami," requested the JSCE to reevaluate tsunami risks "just in case."[45] With the JSCE scheduled to release the results of this study in October 2012, TEPCO formed a "Working Group to Study Tsunami Precautions at the Fukushima Site" in August 2010, which was to study—confidentially, within the company—the construction that would be required to implement tsunami precautions. Among the proposed precautions to be considered were methods for waterproofing the electrical drive units for seawater pumps, relocations of the buildings in which pumps were housed, and the installation of guard walls within the power plant facility. However, barring some dramatic finding from the JSCE's report, these sorts of construction efforts were considered inessential; they were not reported to Akio Omori, then assistant manager of the Nuclear Power and Plant Siting Division, and the report of the Japanese government's Fukushima inquiry commission concluded that "there is no evidence that tsunami risks were treated as a serious problem for TEPCO."[46]

Toshiro Kitamura, a former top executive at the Japan Atomic Power Company (JAPC)—an electric power company specializing in nuclear power generation and wholesale to other utilities such as TEPCO—noted that TEPCO was a leading industry player whose actions exerted strong influence on the behavior of all Japanese power companies, and probably resisted incorporation of new knowledge. Meanwhile, Kitamura speculated, Japan's regulating agencies—

---

[44] At the time of the Fukushima accident, Muto was TEPCO's deputy CEO and Yoshida was the director of the plant and responsible for on-site accident management.

[45] Tokyo Electric Power Company, Inc. (2011) *Fukushima nuclear accidents analysis report (interim report) supplement*. Report, December 2. Tokyo: TEPCO. Available at: www.tepco.co.jp/en/press/corp-com/release/betu11_e/images/111202e14.pdf (accessed August 31, 2013).

[46] This content was excerpted and aggregated from the interim report of the Japanese government's Fukushima inquiry commission. Cabinet Office, Government of Japan (2011) *Interim report of the Government Investigation Committee on the Accident at TEPCO's Fukushima Nuclear Power Stations*. Report, Government Investigation Committee on the Accident at the Fukushima Nuclear Power Stations, December 26. Tokyo: Cabinet. Available at: http://jolisfukyu.tokai-sc.jaea.go.jp/ird/english/sanko/hokokusyo-jp-en.html.

understanding TEPCO's power and importance in the electricity supply—chose simply to accept TEPCO's decisions and act accordingly. JAPC, following the Ibaraki prefectural government's observations of tsunami risks, raised the height of tsunami sea walls at its Tokai Daini Nuclear Power Station in Tokai-mura; this action allowed the plant to avoid—albeit just barely—flood damage to equipment, thus drawing a stark contrast to the situation at TEPCO's Fukushima Daiichi Nuclear Power Station. Kitamura insists, "When it came to implementing tsunami defenses, TEPCO was in a far more advantageous position than JAPC both financially and in terms of human resources," but goes on to suggest that the large size of TEPCO's organization was perhaps an impediment: "Within the nuclear power industry, JAPC was a relatively small organization, and consequently the top executives had a direct connection to the perspective of workers at facility sites, which made the company more agile in both planning and execution."

TEPCO's cover-up scandal and the culture of concealment

In TEPCO's history, the major turning point for the company was undoubtedly in August 2002, when an internal whistle-blower came forward with information that, from the 1980s through the 1990s, the company had doctored self-inspections. It came to light that, during this time, twenty-nine inspection reports on thirteen nuclear reactor units concealed cracks and similar defects in components. An investigation revealed that, in 1991 and 1992, TEPCO had improperly injected air into unit 1's reactor containment vessel at the Fukushima Daiichi Nuclear Power Station during leak inspections in order to make leakage rates appear artificially low; as punishment, NISA ordered a year-long operational shutdown. The incident cost TEPCO dearly; in order to comply with regional governments' requests for safety checks, the company accelerated its schedule of routine nuclear plant inspections, shutting down all seventeen of its reactor units in the spring of 2003. The operating rate (facility utilization rate) of TEPCO's nuclear power plants, which had hovered near 80 percent in the late 1990s, fell to 26.3 percent in 2003. The incident prompted four employees—then-CEO Minami, chairman Hiroshi Araki, and advisers Gaishi Hiraiwa and Shou Nasu—to resign.

This scandal revealed a number of problems both in TEPCO's internal organizations and in Japan's regulatory system. Indeed, flaws in Japanese safety regulations created an environment that led TEPCO employees to doctor data. In the United States, so-called "sustainability standards" allow plants to continue operating in the presence of minor damage; in Japan, however, there were no such standards. Instead, plants required that components be "as good as new" at all times. For this reason, supervisors of maintenance and repair operations in TEPCO's Nuclear Power Division—in collaboration with parts vendors—effectively implemented the US standard on their own, falsely reporting "no irregularities" even for parts with cracks or other defects.

TEPCO's internal inquiry commission, which investigated this incident, concluded in its report that:

> Employees of the Nuclear Power Division failed to appreciate the societal weight of irregularities in nuclear power production, were overconfident in assuming that they knew more about nuclear power than anybody else, and came to believe—incorrectly—that, as long as there were no safety problems, there was no need to report irregularities.[47]

Still, this sort of culture of concealment was not exclusive to TEPCO. After the company's scandal came to light, data-doctoring incidents were discovered at several other power companies, as well: Tohoku Electric Power Company, Chubu Electric Power Company, Chugoku Electric Power Company, Shikoku Electric Power Company, Japan Atomic Power Company, to name a few.

Organizational structure and the culture of concealment

Former executives who supervised TEPCO's Nuclear Power Division cited the stratified structure of the organization as a factor behind the company's longstanding culture of concealment. To TEPCO, the administrative departments, such as the General Affairs and Planning Division, were more valued than the technical departments, like the Nuclear Power Division. Unlike other power companies, TEPCO has never had a CEO whose expertise was rooted in nuclear technology or engineering.[48] In short, the administrative divisions were the most highly regarded in the company and those who worked in these divisions were on the fast track to becoming the company's elites. In the past, executives with experience in managing nuclear plants had risen to the rank of vice president of the Nuclear Power Division, a position that oversaw the 3,000-plus employees (now known as the Nuclear Power and Plant Siting Division). Past CEOs who rose up through the General Affairs and Planning Divisions were largely incapable of understanding the details of business operations in the highly specialized Nuclear Power Division and were forced to entrust matters, such as safety provisions, to underlings familiar with nuclear power.[49] Former CEO Minami noted that the environment was such that non-experts dared not open their mouths, and explained, "Even among

---

[47] Tokyo Electric Power Company, Inc. (2002) *Investigation report on the items pointed out by GE on the periodic inspections and repair for our nuclear power plants.* Report. Tokyo: TEPCO.

[48] Makoto Yagi, who had headed Kansai Electric Power Company's Nuclear Power Division, became CEO of that company in June 2010. The Hokkaido, Hokuriku, and Kyushu electric power companies have all promoted executives from the technical side of the business to the rank of CEO.

[49] However, the CEOs did take on the responsibility of lobbying, and explaining decisions to, municipal authorities and prefectural governments located near nuclear plant sites.

the executives, managerial matters could really only be appreciated by a limited number of people."[50]

The Planning Division decided that business policy was simply to give orders to technical organizations, such as the Nuclear Power Division, and offer little or no opportunity for feedback. Even though the nuclear power business was a "cash cow" within TEPCO, any trouble at nuclear plants meant that responsibilities were shifted to various divisions—thus, there was constant pressure to maintain high uptime at nuclear plants and ensure that information was kept quiet on nuclear plant troubles that might provoke a reaction from local governments. Instead, the company attempted to make all decisions and implement all necessary responses on its own, a trend that pushed the company toward an increasingly closed and insular mind-set. Indeed, even within the Nuclear Power Division there was a failure to share information between the maintenance and inspection groups—which were heavily populated by electrical and mechanical experts—and the technology groups, staffed primarily by nuclear reactor experts; information was divided between the two branches, with data doctoring carried out by the maintenance departments. In the past, equipment manufacturers held the greatest stockpiles of information, but TEPCO had something of a master–servant relationship with Hitachi and Toshiba—two of the top vendors and producers of nuclear plants in the world—and it was exceedingly rare for vendors to raise issues.

One former TEPCO managing director told us that the company's problems stemmed from its siloed structure and stovepiped bureaucracy. Even at the company's frequent managing directors' meetings, executives rarely knew—and thus did not speak to—executives from outside their own divisions, save for the company's CEO or chairman; consequently, each division simply made its own proposals, with the Planning Division responsible for organizing them and assigning responsibility. One former executive put it this way: "Power companies are supposed to produce power and sell it—really a pretty simple proposition—and yet the planning department was always making grandiose statements and gestures. The company was not being managed by people with on-site knowledge and experience of power plants."[51]

Regulations, reforms, and the deterioration of a safety culture

Professor Akira Omoto of the University of Tokyo observed that, after the TEPCO scandal, the focus to regulate safety risks was lost and replaced by an excessive expenditure on quality assurance and compliance.[52] Both TEPCO and the Japanese government attempted to learn the lessons of the event by revisiting the regulatory and organizational structures that had been in place up to that

---

[50] Personal communication with former CEO Minami, November 15, 2011.
[51] Personal communication with former TEPCO employee, November 1, 2011.
[52] Personal communication with Professor Omoto, January 13, 2012.

point. Yet the resulting reforms had the paradoxical consequence of *reducing*, rather than *improving*, the safety of nuclear power plants, and may have been a distant cause of the disaster at the Fukushima Daiichi Nuclear Power Station.

The focus of government regulations on quality assurance and compliance not only had the consequence of delaying initiatives to assess safety risks such as the possibility of severe accidents, but also tied the hands of workers at nuclear plant sites. Omoto, who left TEPCO in 2003 and later served at the IAEA, recalls the frustration of a TEPCO employee who complained that the intensifying focus on document-based inspections meant that "people were so busy searching for old documents and making comparisons to other companies that they didn't have time to visit the actual plants. We were handcuffed by all the paperwork." TEPCO's official apology for its 2002 cover-up involved a promise to "overcome the insularity and secretiveness of the Nuclear Power Division," and the ensuing reforms of that division involved personnel changes including the selection of an executive from the Thermal Power Department to head the Nuclear Power and Plant Siting Division in October 2002.[53] Omoto said he was surprised by such a selection. "Before 2002, the tradition within TEPCO had been that the people with the deepest understanding of nuclear reactors would lead the technical divisions," Omoto said. "I think it was a mistake to put people with no technical skills in positions of responsibility over critical matters such as reactor cores, fuel, and radiation controls. I appreciate the importance of bringing a variety of different viewpoints to bear on nuclear power issues, but these personnel exchanges were going too far; it seems clear to me that they were causing a degradation in the safety culture."[54]

Generation gap in the awareness of safety risks

Until the 1980s, TEPCO was active and enthusiastic in its institution of safety provisions. However, in the 1990s, the company's expanded portfolio of nuclear reactors and the societal wave of opposition to nuclear power forced the company to respond to regional governments, and this—together with the pressure to cut costs—made the company more conservative. Repairs and inspections of nuclear plants were increasingly outsourced to contractors to restrain personnel costs, a fact that lessened awareness among executives of the on-site realities of nuclear plants. Because our requests for interviews with TEPCO executives in office at the time of the Fukushima disaster were denied, we cannot say to what extent TEPCO had been aware of the tsunami risks, but former TEPCO executives have suggested that peer pressure among colleagues

---

[53] This was the only time that an executive from the company's thermal power branch led the Nuclear Power and Plant Siting Division; the post was later held by Ichiro Takekuro, who was plant manager of the Kashiwazaki-Kariwa Nuclear Power Station and subsequently became a company fellow serving as an adviser to the CEO.

[54] Personal communication with Professor Omoto, January 13, 2012.

may have reduced awareness of safety concerns among managers and diminished sensitivity to risks.

"Until the 1980s, we had the sort of spirit that motivated us to do things of our own volition, even if the government didn't tell us to," said former TEPCO Vice President Toshiaki Enomoto, who served a very short term in 2002. "We studied safety design philosophy for nuclear plants and learned the lessons of past accidents, and our people were trained in an atmosphere of constant initiatives to improve and upgrade things. But in the 1990s, the core focus of nuclear power work shifted to maintenance and operational issues, and the practical aspects of safety design started to recede into the background." Enomoto went on to speculate that "these changes in day-to-day operations and in the nature of our human resources, I think, probably influenced safety assurance initiatives."

Enomoto, who resigned after the TEPCO scandal, says he was shocked to learn, following the Fukushima disaster, that TEPCO nuclear power executives had presented a paper at the 2006 International Conference on Nuclear Engineering in the United States on a probabilistic analysis of tsunami hazards at the Fukushima nuclear station—and that internal corporate simulations had demonstrated the possibility of a massive tsunami. "The instant that simulation was done, people should have been considering the possibility of a tsunami-triggered power outage at the Fukushima Daiichi Nuclear Power Station and brainstorming with people from the plant site to imagine what might happen," he said. "If that brainstorm had taken place, then at least some minimal protections might have been put in place to prevent a large-scale release of radioactive materials. You buy insurance precisely to account for possible gaps in your own knowledge and thinking. These are things that need to be done by plant operators, who have the primary responsibility for ensuring nuclear power safety; even more than an engineering decision, this is really a management decision."[55]

"Different generations have different thought processes and different senses of danger," explained Akira Omoto, University of Tokyo project professor and former TEPCO executive. "In the beginning, and in our generation, people were always asking themselves why things were designed in such-and-such a way, and there were still all sorts of case studies to which we could apply that type of thinking. However, as nuclear power became more and more a domestic Japanese product, instead of thinking about why a machine is designed in a particular way, TEPCO's operations and inspections staff just reached for the telephone and called to ask the manufacturer. There will always be flaws and gaps in the decision-making process. But I think that the root causes of the Fukushima disaster are things such as failures of imagination, overconfidence, and biases."[56]

---

[55] Personal communication with former TEPCO Vice President Enomoto, November and December 2011.
[56] Personal communication with Professor Omoto, January 13, 2012.

Chapter 5

International Safety

Amid rapid international growth in nuclear reactor construction in the 1970s, pressure mounted for the International Atomic Energy Agency (IAEA) to establish safety criteria for nuclear power plants. In 1974, the agency did just that and established its Nuclear Safety Standards.[1,2] But there was one twist: Though these international standards could be adopted by a country at its own discretion, they were not legally binding. At best, they amounted to a set of reference conditions for nations. This raised the question of how safety standards could be made effective without infringing upon the sovereignty of member nations. In response to this, the agency launched a peer-review system in 1983—the Operational Safety Review Team, or OSART—that each member nation, of its own volition, could choose to use. Upon a nation's request, an international inspection team would assess the operational safety of a nuclear power plant.[3] In 1989, the agency launched the International Regulatory Review Team, known as IRRT, which offered, to any requesting nation, peer reviews of that nation's nuclear safety regulations, together with expert advice, based on the Nuclear Safety Standards. In the following years, several similar programs were rolled out that evaluated the efficacy of a nation's regulatory platform for radiological safety and the security and safety of radioactive sources

---

[1] In 1979, after the investigation into the Three Mile Island accident, this program specified five major categories of safety standards and policies for nuclear power plants: government regulatory organizations, geographical safety of the nuclear power plant sites, design safety, operational safety, and quality assurance. The Nuclear Safety Standards program was thoroughly revisited in 1988 following the Chernobyl accident; in 1993, new safety principles, corresponding to upper-level regulations within safety standards, were ratified. This had the effect of completing standards into a three-layered structure: safety principles, safety requirements, and safety policy. In addition, standards documents, which previously had been prepared according to different processes for each field, were consolidated in 1996.

[2] Cabinet Office, Government of Japan (2001) *Nuclear power safety white paper*. Report. Tokyo: Cabinet. (In Japanese.)

[3] International Atomic Energy Agency (2013) OSART: Operational safety review teams. Brochure. Available at: www-ns.iaea.org/downloads/ni/s-reviews/osart/OSART_Brochure.pdf (accessed August 31, 2013).

(known as RaSSIA),[4] assessed the implementation status of IAEA radioactive-material transport regulations (known as TranSAS),[5] and appraised preparedness for nuclear accident or radiological emergencies (such as the Emergency Preparedness Review Service).[6] Furthermore, in 2006, the IAEA launched the International Regulatory Review Service, or IRRS, which both integrated and refined IRRT and RaSSIA.[7] The agency and the Convention on Nuclear Safety work together to improve the effectiveness of these peer-review mechanisms.[8] The convention calls for the establishment of regulatory bodies and procedures to ensure nuclear safety and requests public reports, including status checks on goal attainment, every three years.

Today's peer-review system is the product of a fairly active community of nuclear plant operators, who, having realized that a Chernobyl-scale accident had the potential to wreak apocalyptic havoc, convened a conference in Paris in October 1987 and solidified the conceptual framework for the World Association of Nuclear Operators, or WANO. Officially launched in 1989, the association peer reviews each participating plant every six years. Today, on an annual basis, WANO conducts thirty to forty reviews on organizational management, power plant operation and protection, and radiation defenses, among other things.

But, despite all of these international efforts, Japanese nuclear plant operators and regulatory bodies have not always worked earnestly to incorporate the many safety recommendations made by peer reviewers or the points raised by independent regulatory institutions.

Operational Safety Review Team

The IAEA conducted OSART reviews at multiple Japanese nuclear plants over the course of two decades, including units 3 and 4 at the Fukushima Daini Nuclear Power Station in 1992.[9] However, when it came to the topic of OSART, the relationship between Japan and the IAEA was not a good one. In 2002, at the time of TEPCO's cover-up scandal, IAEA Director General Mohamed ElBaradei offered Japan OSART assistance to investigate the matter,

---

[4] The Radiation Safety and Security Infrastructure Appraisal.
[5] The Transport Safety Appraisal Services.
[6] This is a service, offered by the IAEA, that evaluates a member state's preparedness for nuclear and radiological emergencies.
[7] IRRS offers comprehensive assessments of a nation's legal framework and organizational structure for nuclear power safety regulation; this new service is now operational.
[8] At the Third Review Meeting of the Convention on Nuclear Safety in March 2005, the importance of peer-review processes capable of clarifying the strengths and weaknesses in each nation's nuclear safety regulatory systems was emphasized.
[9] Units 3 and 4 of Kansai Electric Power Company's Takahama Nuclear Station in 1988; units 3 and 4 of TEPCO's Fukushima Daini Nuclear Power Station in 1992; units 3 and 4 of Chubu Electric Power Company's Hamaoka Nuclear Power Station in 1995; and TEPCO's Kashiwazaki-Kariwa Nuclear Power Station in 2004.

but Japan did not even respond to his offer.[10] Later that year, a trade magazine reported that, although OSART had made numerous recommendations following its 1992 inspection of the Fukushima Daini Nuclear Power Station, IAEA officials stated that TEPCO had refused to implement them.[11] Company officials deny this alleged negligence, but IAEA representatives also indicated that TEPCO's efforts to improve plant safety were insufficient. Furthermore, the agency said that Chubu Electric Power had failed to respond sufficiently to recommendations made after its 1995 inspection of units 3 and 4 of the power company's Hamaoka Nuclear Power Station.[12]

Plant operators felt as though the inspections that accompanied the peer-review process were excessive. In fact, multiple high-level executives operating the country's nuclear plants have suggested that the inspections that were introduced to comply with regulations in the late 1980s and early 1990s probably had the opposite effect: Ultimately, they spurred efforts to conceal operational details.[13]

Integrated Regulatory Review Service

The IAEA and Japanese officials failed to see eye to eye in the case of this program, as well. In 2007, the service conducted a five-day review of Japan's nuclear facilities and recommended five points for necessary improvement:[14]

1 The roles of both the Nuclear and Industrial Safety Agency (NISA) and the Nuclear Safety Commission—especially with regard to the nuclear safety guidelines—should be clearly identified;[15]

---

[10] *Nucleonics Week* (2002) IAEA aims for thaw with Japan on OSART, safety cooperation. *Nucleonics Week* 43(39), September 26.

[11] *Nucleonics Week* (2002) TEPCO says it has cooperated with IAEA on OSART findings. *Nucleonics Week* 43(41), October 10. Kansai Electric Power Company was cooperative during the investigation, while TEPCO's absolute refusals were the precise opposite. However, we were unable to verify any concrete examples to illustrate in what sense TEPCO refused to cooperate.

[12] *Nucleonics Week* (2002) TEPCO says it has cooperated with IAEA on OSART findings. *Nucleonics Week* 43(41), October 10. We were unable to confirm any concrete examples to illustrate in what sense Chubu Electric Power Company was uncooperative.

[13] US Nuclear Regulatory Commission (2002) Excessive Japanese requirements led to cover-ups, managers say. *Inside NRC* 24(22): 348–349.

[14] International Atomic Energy Agency (2007) *Integrated Regulatory Review Service to Japan: Report to the Japanese government*. Report. Vienna: IAEA.

[15] The first point, the need to clarify the roles of the Nuclear and Industrial Safety Agency and the Nuclear Safety Commission, was noted more pointedly in the main text of the "Report to the Japanese Government" that emerged from the IRRS: "The Nuclear Safety Commission is a council established in the Cabinet Office and supervises the Nuclear and Industrial Safety Agency, and the Nuclear and Industrial Safety Agency is the regulatory body. By the stipulation of law, JNES conducts some inspections. However, the organizational arrangements may cause complexity. The responsibilities for nuclear safety among these entities, although defined in the relevant laws, seem intertwined."

2   NISA should continue its efforts to address the human and organizational factors that could influence the operational safety of a nuclear power plant;
3   NISA should develop a strategic human resources management plan for the future;
4   NISA should foster relations with industry that are frank, open, yet formal and based on mutual respect—and fully recognize that the responsibilities and positions of the regulator and regulated industry fundamentally differ; and
5   NISA should continue developing its comprehensive management system.[16]

But Japanese officials did not take these sorts of observations very seriously. In fact, NISA followed up with a press release in which it cherry-picked and even stretched the reality of the improvements, effectively saying: Japan has a comprehensive national legal and governmental framework for nuclear safety in place; the current regulatory framework was recently amended and is continuing to evolve; NISA, as the regulatory body, plays a major role in directing and coordinating the evolution of the regulatory framework; and challenges have already been addressed to improve the relations among the Nuclear and Industrial Safety Agency, the nuclear industry, and stakeholders in order to establish better understanding and cooperation—and more work is underway.[17]

The agency went on to note that the "IRRS review team noted several positive examples, and gave recommendations and advice on areas in which improvements are necessary or desirable in order to further enhance the effectiveness of regulatory activity." This willingness to downplay the IRRS's call for urgent improvements—and, instead, to trumpet the service's praise for Japan's achievements—offers a telling glimpse into the attitude of Japan's regulatory authorities.

---

[16] Incidentally, to date, the points made by the IRRS reflect major regulatory trends among the world's primary nuclear power producing countries, and, as such, carry a certain authority as reasonable warnings. In practice, notwithstanding decisions as to whether to abandon or retain nuclear power production, disparate regulatory institutions in nuclear power producing countries have tended to merge into a single organization that includes oversight of nuclear safety, radiological protections, and other areas. For example, in Germany, the Federal Ministry for the Environment, Nature Conservation, and Nuclear Safety not only supervises radiological protections and nuclear safety, but also contains within it the Federal Office for Radiation Protection, established in 1989 to consolidate regulatory authority over all areas of radioactive waste management. In France, the Nuclear Safety Authority was established in 2006 as a unified organization, independent of government agencies, to house the dual functions of regulation and oversight of nuclear safety and radiological protections. Similarly, in Sweden, the country's Radiation Safety Authority was established in 2008 by merging two regulatory bodies charged with overseeing nuclear safety and radiological protection.

[17] Nuclear and Industrial Safety Agency (2008) Press release, March 14. Available at: http://warp.ndl.go.jp/info:ndljp/pid/286890/www.meti.go.jp/press/20080314007/irrs.pdf (accessed August 31, 2013). (In Japanese.)

The Nuclear Safety Commission's chairman even commented that "Japan's regulations are unusually outstanding even in comparison to international standards, and we are delighted to receive this highly positive assessment of the functional efficacy of Japan's system for ensuring nuclear safety."[18] This chairman's statement, while reaffirming the commission's faith in the regulations themselves, takes a less self-congratulatory posture in view of the IRRS's findings that Japan's nuclear regulatory framework was excessively complex:[19]

> The report notes that, from the outside, the interrelationship of the Nuclear and Industrial Safety Agency, the Nuclear Safety Commission, and other authorities appears complicated, with responsibilities seemingly intertwined. This point has been noted by a variety of others as well. We must take this opportunity to articulate more clearly the differences between the Nuclear and Industrial Safety Agency and the Nuclear Safety Commission.

The challenges of the peer-review model are, first, that it cannot override a nation's sovereignty in enforcing peer-reviewed recommendations, and, second, that it cannot control how the nation in question will respond to the results of the assessments. As such, when faced with questions and observations from other signatory countries on the need for independent regulatory institutions, Japan has displayed a posture of rigid adherence to its own views. In 2004—in response to Japan's third report for the Convention on Nuclear Safety—other signatory nations questioned whether the Nuclear and Industrial Safety Agency, as a subdivision of the Ministry of Economy, Trade, and Industry (METI), could adequately protect the independence of the agency as a regulatory body. Similarly, this inquiry report questioned if there might be a revolving door for personnel to oscillate between METI and NISA, and observed that such a cozy relationship would undermine the agency's independence as a regulatory body. However, Japan's response to these observations was merely to note that the agency and the ministry are separate entities under law—and, thus, preserving independence would not present any problem. Moreover, Japan, while acknowledging that there had been personnel movements between the two organizations, insisted that this had exerted no deleterious impact on NISA's ability to carry out its duties as an independent regulatory body.[20]

---

[18] Nuclear Safety Commission (2008) Chairman's comment regarding the results of the IAEA/IRRS assessment. Press release, March 17.
[19] Nuclear Safety Commission (2008) Chairman's comment regarding the results of the IAEA/IRRS assessment. Press release, March 17.
[20] Moreover, with regard to tsunami risks, the same national report asked what procedures and methods had been used to account for tsunami risks in the design of the plant. In response, Japan replied that it selected elevations for all nuclear power plants after it first measured the height of past tsunamis in the surrounding regions.

Four months after the Fukushima accident, in July 2011, the IAEA told Japan that, despite having been warned of the need to clarify the roles of the various regulatory institutions, the nation had failed to address this problem.[21] Michael Weightman, the head of an IAEA inspection team that visited Japan in June 2011 in the aftermath of the Fukushima accident, stated that "the complexity of Japan's systems and organizations might delay decision making in times of crisis."[22]

World Association of Nuclear Operators and the Nuclear Safety Network

In a committed effort to share wisdom and experience in the safe operation and maintenance of nuclear power plants, nuclear plant operators from around the world gathered in Moscow for WANO's first general meeting in May 1989. WANO instituted a peer-review meeting in 1991; after an experimental program was conducted in 1992 and 1993, the system was officially launched. Japan was inspired by WANO, and the nation drew on the organization's achievements in different ways. Instigated by the Tokai-mura nuclear accident in September 1999, nuclear plant operators, fuel processors, plant manufacturers, and research institutions came together three months later, in December, to form the Nuclear Safety Network (NS Net). Though this was a local and autonomous endeavor, led by the Federation of Electric Power Companies of Japan, it was largely inspired by WANO and drew from its established methodologies and accumulated data. Whereas WANO focused exclusively on nuclear power plants, NS Net focused on all Japan-based facilities involved in the nuclear fuel cycle.[23]

But WANO has not gone unused in Japan. In 2003, TEPCO, seeking to establish greater transparency in the wake of its cover-up scandal, turned to WANO to peer review its Fukushima Daiichi Nuclear Power Station in 2003, as well as the Fukushima Daini station in 2004. Today, WANO, as well as the NS Net, continue to be the peer-review systems that are most trusted in Japan.

---

[21] This is according to the former head of France's Nuclear Safety Authority, Andre-Claude Lacoste (who led the IRRS review of Japan). When he met with Goshi Hosono (then-special advisor to Prime Minister Kan and later named the nuclear disaster minister) in June 2011, Hosono indicated to Lacoste that the Japanese government had announced its intention to enhance the independence of its nuclear safety regulatory institutions.
[22] *Mainichi Shimbun* (2011) IAEA report on the Fukushima Daiichi Nuclear Power Station: Japan slow to make decisions during the crisis. June 18. (In Japanese.)
[23] The Japan Nuclear Technology Institute, which was established in 2005, assumed and advanced the work of NS Net and now conducts peer-review activities.

Early notification convention

In September 1986, four weeks after the Chernobyl accident, the IAEA adopted the Convention on Early Notification of a Nuclear Accident. In contrast to the Convention on Nuclear Safety, which focused on nuclear power plants, the treaty addressed any nuclear-related facility—whether for civilian or military use—where accidents had the potential to spread across national borders and required that the date, time, and location of such accidents be immediately communicated both to the IAEA and to any nation that might be at risk. In addition, the treaty mandated that all information necessary to implement safety precautions—including the cause of the accident, the volume of radioactive material emitted, and predictions for how that material might disperse—be communicated.[24]

Although Japan was a signatory state, it did not fully conduct the expected actions mandated in this treaty after the Fukushima disaster. In fact, on March 16, 2011, IAEA Director General Yukiya Amano complained that Japanese officials were not supplying enough information. Andreas Persbo, executive director of VERTIC, a nonprofit group that advises the IAEA, also referred to the treaty, stating that "Japanese officials themselves may not have [had] enough information to fulfill their obligation."[25]

Thus, the Japanese government turned to East Asian countries to learn from their existing early-notification protocols associated with nuclear accidents. Prime Minister Kan worked with Chinese Premier Wen Jiabao and with South Korean President Lee Myung-bak to communicate closely in order to construct a framework for early crisis warning notifications; ultimately, the three nations signed the Japan–China–Korea Nuclear Safety Initiative, a statement of intent to construct protocols to ensure rapid sharing of information in the event of a nuclear accident. The three nations promised to develop a mutually satisfactory approach to nuclear safety and regulation based on IAEA safety standards, to share experiences and best practices, to maintain the ability to prevent disasters and respond to emergencies, to build capacity, and to recognize the critical importance of transparency and open information in emergency situations.[26]

---

[24] International Atomic Energy Agency (1986) *Convention on early notification of a nuclear accident*. Report. Vienna: IAEA. Available at: www.iaea.org/Publications/Documents/Infcircs/Others/infcirc335.shtml (accessed August 31, 2013).

[25] Tirone, J. (2011) Nuclear watchdog says Japan falls short supplying information. *Bloomberg*, March 16. Available at: www.bloomberg.com/news/2011-03-16/nuclear-watchdog-says-japan-falls-short-supplying-information.html (accessed August 31, 2013).

[26] Nuclear and Industrial Safety Agency (2008) Press release, March 14. Available at: http://warp.ndl.go.jp/info:ndljp/pid/286890/www.meti.go.jp/press/20080314007/irrs.pdf (accessed August 31, 2013). (In Japanese.) The trilateral initiative among Japan, China, and South Korea suggests that perhaps it is possible for East Asian nations to develop a framework like the Western European Nuclear Regulators' Association (WENRA), which was created in 1999 as an informal framework through which European regulatory institutions could share best practices.

Problems related to sharing information on contaminated water

The Convention on Early Notification of a Nuclear Accident is an extremely terse document that leaves some room for interpretation and does not constrain the actions of signatory countries—a fact that became all too apparent on April 4, 2011, when TEPCO released water that contained a low concentration of radioactive contaminants into the ocean. Stored in the concentrated waste treatment facility at the Fukushima Daiichi Nuclear Power Station, the water had filled the facility almost to capacity. TEPCO said this release was an unavoidable and necessary measure in order to prevent collected water, which had a fairly high concentration of contaminants, from releasing from unit 2.[27]

Not only did the actual action—dumping the contaminated water into the ocean—not violate Japan's obligations under international law,[28] it also did not violate the convention, which only obliged Japan to notify countries of the nuclear accident that could lead to a release of radioactive material with the potential to pose radiation safety risks. On the one hand, this information was publically communicated at a press briefing. On the other hand, then-Foreign Minister Takeaki Matsumoto reported the event to the governments of various foreign countries and to the IAEA. "There is no significant health impact, and there is no immediate problem from the standpoint of our obligations under international law," he explained.[29]

Nonetheless, it is difficult to argue that the incident was not problematic on ethical and political grounds. From a diplomatic perspective, South Korea and Russia, the two neighboring countries with perhaps the greatest interest in the incident, are said to have been absent from the relevant briefings and received only delayed notifications. South Korea worked independently to gather its own information[30] and was not in attendance at the Japanese Foreign Ministry's briefings, where it was announced that the government approved the water release. The Korean Foreign Ministry expressed its strong grievance, thereby implying Korean doubt as to the Japanese fulfillment of the treaty. This incident raises questions warranting detailed future study, including not only the legality of the incident itself, but also the extent to which individual intent was communicated to counterparts. The notion that such important information would be communicated by statements at a briefing session, where attendance is voluntary, conveys a sense of inadequacy. The Foreign Ministry stated that it subsequently faxed notifications to the embassies of

---

[27] *Asahi Shimbun* (2011) Nuclear Safety Commission: Full transcript of Chief Cabinet Secretary Edano's press conference. April 4 and April 5. (In Japanese.)

[28] Such as the United Nations Convention on the Law of the Sea, which codifies general obligations to prevent ocean pollution, or the Convention on the Prevention of Marine Pollution by Dumping of Wastes and Other Matter, 1996 Protocol Thereto, i.e., the "London Protocol."

[29] *Sankei Shimbun* (2011) Japanese government pressured to explain ... dumping of contaminated water with no advance warning. April 6. (In Japanese.)

[30] Personal communication with an official of the South Korean Embassy in Tokyo, November 18, 2011.

various foreign countries, but these arrived after (if only a few minutes after) the release of water had begun.

In addition, prior to TEPCO dumping the water into the ocean, there is no indication of any discussions on whether ocean dumping defied international law. The Foreign Ministry, itself, did not know that this was even taking place until its liaison team, stationed at TEPCO's Joint Emergency Response Headquarters, reported that the decision had already been made to release the contaminated water into the ocean.

That is, given the limited amount of time available for decision making, it was understandably difficult for TEPCO to identify options other than ocean release of low-level contaminated water, and moreover the actions taken were not problematic from the standpoint of international law. Nonetheless, this is only because the contaminated water that was released had unexpectedly low concentrations of radioactive contaminants—and we cannot deny the possibility that, under slightly different conditions, with higher concentrations of radioactive contaminated water, the incident might have escalated into a full-blown violation of international law.

Chapter 6

Accident Preparedness and Operation

The implementation of safety measures and the evaluation of their effectiveness inevitably depend, to some extent, on the assumption of risk. Nuclear experts, including those responsible at the Tokyo Electric Power Company (TEPCO) and the regulatory authorities, often argue that the Fukushima accident was "beyond expectation." Nuclear experts, however, have never professed the ability to assess or consider *all* possibilities of an accident. Needless to say, there are innumerably many distinct events that could trigger an accident, as well as infinitely many different sequences and pathways through which such an event could escalate into a full-blown accident. Anticipating and instituting countermeasures for every possible contingency is probably impossible, and certainly impractical. In terms of the Fukushima accident, TEPCO indeed had assessed the plant's moderate safety measures. But the aim of a deterministic assessment is, as the International Atomic Energy Agency (IAEA) defines, "to specify and apply a set of conservative deterministic rules and requirements for the design and operation of facilities or for the planning and conduct of activities."[1] That is, it is necessary to collect information in order to institute moderate safety provisions—an objective that is not quite the same thing as proving definitively that all requisite safety measures are fully in place.

A conceptual paradigm that offers a realistic approach to this dilemma is the notion of the design-basis event, which represents a choice of a single characteristic phenomenon selected from the infinitude of possible accident sequences—these possibilities form the hypothetical accident sequence. (For example, a momentary and total pipe rupture is not a phenomenon that would be expected to occur in the real world, but such a hypothetical event offers a useful scenario within which to anticipate a large-scale leakage of cooling liquid from damaged pipes.) By considering a wide variety of design-basis events and ensuring that existing safety precautions are sufficiently conservative to address all hypothetical events, governments and the nuclear industry can reasonably expect to ensure adequate capacity for handling most accident sequences that

---

[1] International Atomic Energy Agency (2009) *Safety assessment for facilities and activities*. Report, IAEA Safety Standard Series, GSR-Part 4. Vienna: IAEA.

might occur in practice. This strategy of anticipating accidents at the design stage is the substantive theory underlying safety assessments—that is, the system must be capable of performing its task in the presence of any single failure.

Among the safety measures deemed most critical for nuclear power safety is the design, and functional verification, of engineered safety equipment to ensure that the containment walls shielding the reactor core block the spread of radioactive material in the event of a real-life design-basis event. Moreover, to ensure that damage to any one piece of equipment cannot jeopardize the longevity of such barrier, nuclear facility designs incorporate redundancy in safety installations.

To verify the effectiveness of safety measures, large-scale experiments using models of actual equipment and facilities are conducted systematically; these experiments assess parameters such as the performance of the emergency reactor core cooling system and the behavior of nuclear fuel during accidents. Moreover, advances in computational science have enabled numerical modeling and analysis of heat-flow dynamics, structural deformations, and other phenomena, and these calculations have begun to be used in assessing safety as well.

Following a wave of research that began in the second half of the 1960s, safety provisions, inspired by design-basis events, were applied to one piece of equipment after another. The fruits of this safety research were incorporated into the design and construction of new plants, and higher-reliability safety equipment replaced older, less reliable installations. In addition, design-basis events have been revisited in conjunction with advances in plant design.

Accidents that significantly exceed design-basis events: probabilistic safety assessments

In the 1970s, scientists began to note the dangers—specifically, the multiple failures that could happen in accidents—that significantly exceeded design-basis (so-called "severe accidents"). In 1975, physicist Norman Rasmussen and a group of experts in the United States published the first quantitative assessment of the risks of accidents that significantly exceed design expectations. Intended for the US Nuclear Regulatory Commission, the report—known as WASH-1400, or the Rasmussen report—identified the minimal set of simultaneous equipment failures that would cripple a plant's ability to protect against accidents and calculated the probability of this set of simultaneous equipment failures.[2] This technique, known as a probabilistic safety assessment, has been continually improved and remains in use around the world to this day.

---

[2] It must be noted that, after much criticism, the report overestimated the risk of a severe accident, and the commission withdrew its support of WASH-1400 in 1979. That said, though there were errors in Rasmussen's work, his method was influential in nuclear engineering, as well as in other fields. Rasmussen, N. (1979) *Reactor safety study (WASH-1400)*. Report, US Nuclear Regulatory Commission.

The most important lesson identified in the WASH-1400 report was that, out of all the risks facing nuclear power plants, the majority arose from accidents that exceeded design-basis. In addition, the report demonstrated that the probability of an event escalating to a severe accident was higher than had been previously estimated, and moreover that there existed events that could pose greater risks than an accident with a large loss of coolant, which had previously been considered the single most important contingency. Since WASH-1400, research on severe accidents has proceeded vigorously in the United States and was intensified by Three Mile Island.

The objectives of probabilistic safety assessments, according to the IAEA, are "to determine all significant contributing factors to the radiation risks arising from a facility or activity, and to evaluate the extent to which the overall design is well balanced and meets probabilistic safety criteria where these have been defined."[3] These assessments, the agency states, "may provide insights into system performance, reliability, interactions and weaknesses in the design, the application of defense in depth, and risks, that it may not be possible to derive from a deterministic analysis."[4] In the United States and other countries, probabilistic safety assessments have a history of serving precisely these roles. In Japan, however, probabilistic safety assessments were never promoted to the status of regulatory conditions.

Defense-in-depth

The assurance of safety at a nuclear reactor requires that extremely high-reliability mechanisms be in place to ensure that the following three capabilities are retained at all costs: the ability to control the nuclear reactions occurring inside the reactor (that is, to *stop* the reaction); the ability to remove heat from inside the reactor (to *cool* the reactor); and the ability to prevent the spread of radioactive material outside the reactor (to *contain* the radioactive material).

The engineering paradigm that attempts to ensure that all three of these functions are maintained in a highly reliable way is known as "defense-in-depth," which is the notion of multiple, simultaneous layers of safety precautions that ensure—even in the event that some subset of the preventative measures fail or are disabled—that the safety of the overall system is retained. Not only are nuclear power plants equipped with multiple layers of defenses, but they also incorporate countermeasures to ensure that, even if all the safety provisions in a plant's design fail simultaneously, human life and the environment are nonetheless protected from the effects of radiation.

---

[3] Rasmussen, N. (1979) *Reactor safety study (WASH -1400)*. Report, US Nuclear Regulatory Commission.

[4] Rasmussen, N. (1979) *Reactor safety study (WASH -1400)*. Report, US Nuclear Regulatory Commission.

There are five defense-in-depth categories, each existing independently of the others; in other words, the safety provisions of each layer are to be instituted as though they alone represent the final defense against catastrophe. Moreover, each category is to be implemented on the assumption that preceding provisions have entirely failed. It is only when all five categories are independently strengthened in accordance with this philosophy that a system can be said to exhibit defense-in-depth. The IAEA classifies the five levels of defense as follows:[5]

1 ***Prevention of abnormal operation and failures.*** There are two types of irregularities at nuclear power plants: internally induced irregularities (such as worker errors or equipment malfunctions) and externally induced irregularities (such as earthquakes and tsunamis, or events such as plane crashes). Important measures to prevent these issues include building plants in locations with minimal sensitivity to natural disasters and other external events, using only highly reliable equipment, and choosing a plant design and an operating environment that serve to inhibit worker errors.

2 ***Control of abnormal operation and detection of failures.*** Even with careful site selection, plant design, construction, and operation in place, it is impossible to prevent all irregularities. For this reason, nuclear power plants must be equipped with multiple methods of rapidly detecting irregularities and multiple protocols for taking appropriate actions to prevent them from turning into accidents.

   The Fukushima accident arose from an earthquake and tsunami—external irregularities that significantly exceeded design assumptions. Control rods were automatically inserted into the reactors immediately after the earthquake struck, so the *stop* function was upheld. Moreover, fail-safe mechanisms ensured that the main steam isolation valves were automatically closed; this ensured that, even in the event of any major pipe damage, cooling water containing radioactive material would not leak outside the containment vessel, thus upholding the *close* function. Furthermore, an operational mechanism for upholding the *cooling* function was activated—namely, the isolation condensers in unit 1 and the reactor core isolation cooling systems in units 2 and 3—and workers continuously and actively injected water into the reactor cores.

   However, the tsunami caused a loss of all AC and DC power, which also disabled seawater pumps; this ensured that the irregularity could not be contained, and so began one of the worst nuclear accidents in history.

3 ***Control of accidents within the design-basis.*** When the second level of countermeasures fails, an accident ensues. To mitigate the damage, another

---

[5] International Atomic Energy Agency (2000) *Safety of nuclear power plants: Design.* Report, IAEA Safety Standard Series, NS-R-1. Vienna: IAEA. Available at: www-pub.iaea.org/MTCD/publications/PDF/Pub1099_scr.pdf

level of specially designed equipment and support systems—such as electric power sources and heat sinks—are employed.

When disaster struck the Fukushima Daiichi Nuclear Power Station, however, nearly all support systems were either destroyed or rendered nonfunctional. Although attempts were made to restore electrical power at the power plant, it ultimately was not restored before the workers could contain the accident and radioactive material was emitted into the environment. Consequently, nuclear fuel tubes, reactor pressure vessels, and primary containment vessels all suffered major damage from decay heat emitted by nuclear fuel. Fuel pellets, thought to have partially melted, released large volumes of radioactive material, including iodine and cesium.

4 *Control of severe plant conditions, including prevention of accident progression and mitigation of the consequences of severe accidents.* Defense-in-depth considers techniques for responding to situations that cannot be addressed even with the full arsenal of methods prepared at the design stage. Such crisis-response techniques, known as accident management, must all be employed to control a situation.

Workers applied various techniques at Fukushima to calm the effects of the disaster—they injected boric acid into the reactors, used fire engines to inject water into the reactors, and vented the primary containment vessels to prevent a large break in these vessels. However, in addition to the damage caused by the earthquake and tsunami, a number of factors, including aftershocks, widened the scope of the accident, while hydrogen explosions further complicated matters; ultimately, the release of large quantities of radioactive material could not be prevented. Thus, iodine and cesium, which emitted from fuel pellets, escaped beyond the containment vessel and released into the atmosphere in the form of micron-sized particles.

5 *Mitigation of radiological consequences of significant releases of radioactive materials.* The fifth and final stage of defense-in-depth is the government's responsibility; the electric power companies are responsible for the first four stages. Even in cases in which a large-scale release of radioactive material cannot be avoided, mitigation of the radiological consequences can be incorporated into a design to protect public health. If radioactive material is released and spreads throughout the atmosphere, measures can be taken, such as citizen evacuation or indoor confinement. However, once radioactive material has fallen to the ground, measures like temporary displacements, decontamination, and food regulation become necessary. More than 100,000 citizens were forced to evacuate from the Fukushima Prefecture and the surrounding communities.

Misconceptions about defense-in-depth

Though the concept of defense-in-depth is based on the independency of each level, discussions of nuclear power's dangers are frequently based on the misconception that the emergency response set out in the fourth and fifth levels of defense-in-depth are necessary because the safety provisions of the first three levels are inadequate. Any such discussion is based on a fundamental misunderstanding of the engineering paradigm encapsulated in the defense-in-depth hierarchy.

Nonetheless, such misinterpretations are rampant even among nuclear power professionals. A typical example of this is the idea that the retention of all three safety functions—stopping, cooling, and containing—only exist in the second and the third levels of defense-in-depth. This misunderstanding was also present in the interim report of the government's Fukushima inquiry commission published in 2011.[6]

Power companies fear that, if they strengthen their accident-management provisions or their emergency-response protocols, local residents will become nervous and grow opposed to nuclear power. If a proper understanding of the defense-in-depth paradigm had been disseminated throughout Japanese society, these fears might never have developed. However, there is reason to doubt that the meaning and importance of defense-in-depth were fully appreciated even by nuclear power professionals; thus, the wider public never properly and fully understood the significance of the paradigm.

Japan's preparedness: roles and responsibilities of nuclear safety regulations

As we have laid out in previous chapters, TEPCO was primarily responsible for ensuring safety at the Fukushima Daiichi Nuclear Power Station; despite this, the company remained largely unprepared for accidents. And it was precisely this failure to institute preparedness against tsunami threats that stands as one of the primary causes of the disaster—but government regulations also did not serve their proper role. Media reports have suggested that the investigation project undertaken by Fukushima Prefecture showed twelve cases of thyroid cancer in people under 18 years old in July 2013;[7] however, some experts involved in this project denied that these cases were the consequence of the nuclear disaster. Nonetheless, the general public in the Fukushima Prefecture are concerned with the possible radiation effects. It is true that the safety precautions in place failed to prevent the radioactive contamination over a wide geographical area; for this reason there were, at the very least, clear regulatory deficiencies to protect the environment from radiation risks.

---

[6] Morokuzu, M. (2011) Now is the time for a deeper understanding of defense in depth. *Journal of the Atomic Energy Society of Japan (ATOMOΣ)* 53(12): 794–795. (In Japanese.)

[7] Fukushima Medical University (2013) *Survey results: Proceedings of the 11th prefectural oversight committee meeting for Fukushima Health Management Survey*. Report, June 5. Available at: www.fmu.ac.jp/radiationhealth/results/20130605.html (accessed August 30, 2013).

In the following sections, we investigate the role of safety regulations in the context of four of the most important factors underlying the complex Fukushima accident: the tsunami and earthquake defenses, the total loss of AC power, and the countermeasures against severe accidents. In each case, we review what preparedness was in place and where preparedness was inadequate.

Regulatory preparedness against tsunamis and the Fukushima accident

Between 1966 and 1972, TEPCO obtained its permit to establish and construct the Fukushima Daiichi Nuclear Power Station. It wasn't until 1970 that Japan's Atomic Energy Commission introduced guidelines for regulatory safety reviews, bringing natural phenomena, including tsunamis, into the conversation.[8] In fact, prior to this,[9] there were no clear standards regarding tsunami risks. Thus, TEPCO based its Fukushima plant design on records of earlier tsunamis: The company took the specified design height to be 3.122 meters (the highest tide level from the 1960 Valdivia earthquake and tsunami in Chile) above the base construction surface of the Onahama port.

In July 1993, an earthquake struck off the southwest coast of the Japanese island of Hokkaido; Okushiri, an island close to the epicenter of the quake, suffered heavy damage from the ensuing tsunami. In October of that year, the Ministry of International Trade and Industry (MITI) responded by ordering Japanese electric power companies to reassess the tsunami safety of existing power plants. For the Fukushima Daiichi and Daini stations, TEPCO scoured written records to obtain a data sample on historical tsunamis, then used simple predictive models to estimate the possible water heights of anticipated future tsunamis. The company submitted the results of these safety evaluations to the ministry in March 1994, and MITI officially accepted the report in June. Though the report mentions the Jogan earthquake and tsunami of 869 AD, it isolates no reason to believe that the height of the tsunami waves exceeded that

---

[8] These guidelines were largely inspired by construction-design criteria implemented by the US Nuclear Regulatory Commission three years prior, in 1967.

[9] In 1970, the Japan Atomic Energy Commission established its "Regulatory Guide for Reviewing Safety Design of Light Water Nuclear Power Reactor Facilities," which stated: "Systems and devices that, upon malfunction, could potentially become direct causes of accidents with serious safety ramifications must be designed to withstand the most severe forces of nature that may be anticipated based on the prevailing natural conditions at the plant site and in the surrounding regions, as determined by considering past records. ... In the event of an accident with serious safety ramifications, or in any other situation in which the nuclear reactor must be shut down with absolute certainty, systems and devices that (for safety reasons) are important or essential for reducing or containing the consequences of an accident must be designed to withstand, and retain functionality in the face of, not only the most severe forces of nature that may be anticipated based on the prevailing natural conditions at the plant site and in the surrounding regions (as determined by considering past records) but also the simultaneous impact of any accident conditions that may be present."

of the 8.1-magnitude Keicho Sanriku earthquake in the year 1611. Moreover, the report estimates that the wave height in the earthquake and tsunami in Chile exceeded that of Keicho Sanriku.

The Nuclear Safety Commission began revising the Seismic Design Regulatory Guide in June 2001[10] and completed it in September 2006 to prescribe that, "facilities are to be designed with sufficient consideration so that their safety functions shall not be significantly affected by tsunamis anticipated to occur but very rarely during the operation of the facilities."[11] However, whereas the guidelines detailed methods for addressing seismic motion, they provided no specific procedures for estimating tsunami heights. During this five-year revision stage, numerous debates emerged. In fact, recent analysis of these deliberations shows that the commission spent 3.4 percent of its time discussing events accompanying earthquakes, while it devoted only 0.04 percent of its time evaluating tsunami risks.[12] Indeed, the review subcommittee for the guide included numerous experts in earthquakes and seismic design, but not one of the committee members could be considered a tsunami expert.

The revised Seismic Design Regulatory Guide and the inclusion of tsunami risks

Though, as we noted, tsunamis were listed among the "events accompanying earthquakes" in the revised regulatory guide, we challenge the adequacy of that classification.

During the five years that it took to review and revise the regulatory guide, it is important to note that several incidents unfolded—namely, the August 2005 earthquake in the Miyagi Prefecture that exceeded design-basis, as well as a 2006 ruling by the Kanazawa District Court that the Hokuriku Electric Power Company did not sufficiently consider the active fault lines when it constructed unit 2 of its Shika Nuclear Power Station—that heightened the focus on seismic activity. The impact of these events, as well as the discussion they engendered, is clearly visible in several passages within the revised "Seismic Design Regulatory Guide." The revision not only emphasizes the importance of properly acknowledging the inherent uncertainty of standard seismic ground

---

[10] Established in 1981, the guide determined design-basis seismic motion. Nuclear Safety Commission (2006) *Regulatory guide for reviewing seismic design of nuclear power reactor facilities*. Report, September 19. Tokyo: NSC. Available at: pbadupws.nrc.gov/docs/ML0803/ML08031 0851.pdf

[11] This is found in the section: "Consideration of Events Accompanying Earthquakes." Nuclear Safety Commission (2006) *Regulatory guide for reviewing seismic design of nuclear power reactor facilities*. Report, September 19. Tokyo: NSC. Available at: pbadupws.nrc.gov/docs/ML0803/ML0803 10851.pdf

[12] Tsuchiya T (2011) Issues on earthquake and tsunami risk assessments and seismic design—interim report of the expert hearing. In: *International symposium on joint fact finding: The possibility of earthquake risk analysis for nuclear power plants*. Tokyo: Tokyo University Policy Alternatives Research Institute. (In Japanese.)

motion, but also notes the possibility that seismic motion could exceed design-basis levels; the guide notes that this "residual risk"[13] must be acknowledged by nuclear utilities, which are required to minimize the risk as much as reasonably practicable.

During safety reviews, the regulatory authority was to consider the probability of exceeding standard seismic ground motion. In addition, when the Nuclear Safety Commission adopted the revised regulatory guide, it recommended that nuclear utilities actively use probabilistic safety analysis.[14] Thereafter, nuclear plant operators understood they were to voluntarily implement this as part of their seismic back-checks in order to verify the safety margin.

Notably, some experts we interviewed pointed out that the severity of earthquake and tsunami are not proportional.[15] The significance of this fact is that the allowed design safety margin to ensure that tsunamis do not escalate into damaging events is narrower than for earthquakes. Consequently, although it is important to improve the accuracy of predictions for tsunami wave heights, it is even more important to account for the residual risk. For example, plant operators must consider not only preparedness against tsunami height, such as the choice of a high-elevation site or the installation of embankment, but also design precautions against flooding, such as both water-resistant buildings and equipment, as well as redundancy in electrical switchboards. Moreover, the accumulation of academic knowledge of tsunamis is much less extensive than the present scientific understanding of earthquakes, and it is difficult even to estimate tsunami heights with any degree of certitude.

It is clear that, despite the fact that earthquakes and tsunamis pose very different risks to the safety of nuclear power plants, tsunamis were downplayed as merely one of many types of events that may accompany an earthquake—and no regulations were established to urge nuclear plant operators to institute safety provisions taking proper account of such tsunami risks.

---

[13] "Residual risk" is defined as "various risks that a facility will be impacted by seismic motion in excess of design-basis level in such a way as to provoke (1) events resulting in serious damage to a facility, (2) events resulting in the release of large quantities of radioactive materials from a facility, and/or (3) the possibility of harmful radiation exposure to the surrounding population as a result of such an event." Nuclear Safety Commission (2006) Section 1: Basic policies (commentary). In: *Regulatory guide for reviewing seismic design of nuclear power reactor facilities*. Report, September 19. Tokyo: NSC. Available at: pbadupws.nrc.gov/docs/ML0803/ML080310851.pdf.
[14] Nuclear Safety Commission (2006) Decision No. 60, item 18. In: *Regulatory guide for reviewing seismic design of nuclear power reactor facilities*. Report, September 19. Tokyo: NSC. Available at: pbadupws.nrc.gov/docs/ML0803/ML080310851.pdf
[15] Omoto, A., Juraku, K. and Tanaka, S. (2011) Why was the accident not prevented? In: *University of Tokyo global COE program GoNERI symposium 2011: Rethinking nuclear power education and research after the accident at TEPCO's Fukushima Daiichi Nuclear Power Station*. Tokyo, Japan. (In Japanese.)

Earthquakes, tsunamis, and the regulatory community

Since 2000, tsunami simulation technology has finally begun to progress. Until then, however, tsunami research consisted largely of efforts by a select, yet small, number of general researchers—rather than that of actual earthquake researchers—operating with limited resources, and using sedimentological studies and similar techniques, to estimate the maximum height of historical tsunamis in specific areas. Such research could barely deliver clear answers to the sort of questions required to assess the safety of nuclear power plants—questions such as: How tall will the one-in-every-10,000-years tsunami be? Similarly, Shunsuke Kondo, chairman of the Japan Atomic Energy Commission, noted that nuclear safety experts, who were reluctant to conduct probabilistic safety analysis studies, did not precisely communicate to tsunami researchers the necessary information.[16]

Even within the Atomic Energy Society of Japan, methods of probabilistic safety analysis for earthquakes exhibited steady progress, but tsunami investigations had only just begun to scratch the surface. In April 2010, the society laid out its basic approach to earthquake safety, while merely acknowledging other consequent disasters:

> As for fires, floods, tsunamis, and other events that may accompany an earthquake, each such event must be separately analyzed to develop an understanding of, and a logical approach for addressing, its potential impact on nuclear plant safety; the results of these analyses must serve as a basis for further developments in the paradigm of seismic safety.[17]

Still, the paucity of academic understanding of tsunamis did not mean that there were no opportunities for the regulatory community to appreciate the importance of tsunami risks or to revisit anticipated tsunami heights. For example, a nuclear reactor in the Indian city of Chennai (Madras) was damaged during the Sumatra-Andaman earthquake in December 2004; seawater pump motors were submerged, and the plant was shut down.[18] This incident could have been an opportunity to codify substantive regulatory guidelines for

---

[16] Kondo, S. (2011) *Where Japan is and where Japan will go: Update of the Fukushima accident and the deliberation of post-Fukushima nuclear energy policy in Japan.* Presentation, Japan Atomic Energy Commission, December 2. Available at: www.aec.go.jp/jicst/NC/about/kettei/111202b.pdf (accessed August 31, 2013).

[17] Atomic Energy Society of Japan (2010) Meeting notes from Special Investigatory Committee on Earthquake Safety. In: *The logic of seismic safety in the design and evaluation of nuclear power plants*, Section 7: Summary and future work. Tokyo: AESJ.

[18] International Atomic Energy Agency (2013) The international nuclear event scale. Available at: www-ns.iaea.org/tech-areas/emergency/ines.asp. See also: Japan Nuclear Energy Safety Organization (2013) Seawater flooding of pump building triggers shutdown of Kalpakkam unit 2 for safety purposes. Available at: www.atomdb.jnes.go.jp/content/000023842.pdf (accessed August 31, 2013). (In Japanese.)

tsunami preparedness, but there is no evidence that Japanese regulators paid careful attention to the event.

In addition, since 2008, the Japan Nuclear Energy Safety Organization has spearheaded an initiative to improve probabilistic safety research and analysis on earthquakes. This effort, though still under development, has emphasized that tsunami risks were high, but there is no evidence that the Nuclear and Industrial Safety Agency will incorporate the results of this research into its regulatory activities.

In 2009, the Nuclear and Industrial Safety Agency; the joint working group of earthquake, tsunami, and geology (under the Advisory Committee for Natural Resources and Energy); and the Ministry of Economy, Trade, and Industry accepted TEPCO's interim report that omitted any reference to the Jogan tsunami, and the joint working group said that it would consider including the natural disaster in the final report instead.[19] Similarly, in August 2010 and March 2011, the Nuclear and Industrial Safety Agency conducted hearings on TEPCO's tsunami provisions. At the hearing in March 2011, the agency's inspectors were briefed by a TEPCO representative who acknowledged the possibility that a tsunami—eight meters or greater—could submerge electric-powered pumps, causing a loss of reactor cooling capacity; however, the severity of this risk appears not to have been appreciated by the inspectors, who did not order the power company to implement any additional precautions.[20]

Although local governments did not incorporate tsunami predictions in nuclear safety regulations, there is one example in which a power company used predictions as an opportunity to implement tsunami countermeasures. In October 2007, Ibaraki Prefecture announced forecasts and predictions for tsunamis and the ensuing floods—this accelerated the implementation of tsunami protections at the Tokai Daini Nuclear Power Station operated by the Japan Atomic Power Company, which ultimately mitigated the damage suffered by the plant in the 2011 Tohoku earthquake and tsunami. At the Tokai Daini station, seawater pumps within the residual heat removal system had originally been protected by 4.9-meter-high walls to protect against tsunamis. The height of these walls was based on the Japan Atomic Power Company's reading of the Japan Society of Civil Engineers' tsunami assessments, which used the 1677 Empo Boso-oki earthquake as a basis for estimating a tsunami height of 4.86 meters. However, in its own assessment,[21] the Ibaraki

---

[19] Further, despite the recognition of the 869 AD tsunami, no substantive countermeasures were put in place.
[20] Cabinet Office, Government of Japan (2011) *Interim report of the Government Investigation Committee on the Accident at TEPCO's Fukushima Nuclear Power Stations*. Report, Government Investigation Committee on the Accident at the Fukushima Nuclear Power Stations, December 26. Tokyo: Cabinet. Available at: http://jolisfukyu.tokai-sc.jaea.go.jp/ird/english/sanko/hokokusyo-jp-en.html
[21] Titled the "Map of Possible Tsunami Inundation of Ibaraki Prefecture."

prefectural government used the same earthquake to estimate wave heights—between 2 and 7 meters—along the Ibaraki coast. Based on this input, the power company independently revisited its tsunami safety analyses and revised the tsunami height estimate to be 5.72 meters. Based on this new finding, the company then set out to build new walls 6.1 meters tall: That is, the tsunami that struck the plant following the Tohoku earthquake was 5.4 meters high. Because this exceeded the height of the original walls, there was some partial flooding of the seawater pump area, and some of the flooded pumps were incapacitated; however, the areas that were protected by the newly constructed walls remained safe from flooding, thus avoiding a loss of seawater cooling capacity.[22] Of course, regional governments do not have regulatory authority over nuclear safety, but in this case, the specialized knowledge possessed by Ibaraki Prefecture helped to accelerate the power company's tsunami preparation, thereby preventing a loss of reactor cooling capacity and any serious potential ramifications.

Although power companies did take some voluntary steps based on the tsunami simulations methods that were emerging in academic research, the fact is that today academic understanding of tsunamis remains incomplete, and the failures (or nonexistence) of communication among nuclear safety experts and tsunami researchers all contributed to the inadequacy of preparedness against the 2011 tsunami.

Though the question of whether a tsunami of such magnitude could possibly have been anticipated remains an extremely difficult one to answer, this in no way implies that preparedness against tsunamis could not have been implemented. Nuclear plant operators and safety regulators alike recognized that tsunami risks were not negligible, and the failure of regulatory agencies to incorporate adequate recognition of these risks into regulatory policy must be noted. Moreover, even if tsunami height predictions had high uncertainty (indeed, precisely because such predictions were so inherently imprecise), plant operators must account for and protect against the residual risk of tsunamis that exceed expectations—and the regulatory framework must actively promote and enforce these protections. This was an imperative that Japan's regulatory system failed to uphold.

Regulatory preparedness for floods

In many European countries, regulations exist to protect nuclear power plants from floods. Mainly, this is inspired by the fact that a great number of European nuclear facilities are located near rivers or lakes, and flooding is a major external

---

[22] Japan Atomic Power Company (2011) *On the status of the Tokai Daini Power Plant and its safety precautions after the Tohoku Pacific earthquake.* Report (current as of the end of June 2011).

risk factor; thus, throughout the continent, nations institute protective measures such as waterproofing reactor buildings and critical equipment.[23]

There is some evidence that Japan turned to Europe to learn from its experience on flood protections. For example, in 2007, the Japan Nuclear Energy Safety Organization (JNES) conducted an accident sequence precursor analysis, which drew from an earlier event at a French nuclear power plant in which floods caused a power outage. The results of this analysis suggested that, as was the case in France, the conditional probability of reactor core damage from flooding was high in Japan as well, particularly for boiling-water reactors. However, the lessons of these experiences in foreign countries were never reflected in the actual contents of nuclear safety regulations. Of course, river or lake flooding is much different than a tsunami in terms of energy, height, frequency, and time of occurrence—all of which are notoriously difficult to predict. Nonetheless, the possibility of damage due to the submersion of critical equipment is a feature common to both types of flooding, and the fact that regulatory agencies did not note such similarities highlights the inability of Japan's regulatory community to acknowledge real risks. Clearly, the results from JNES's investigations went ignored and were not applied to regulatory policy, suggesting an organizational and systemic failure—namely, a breakdown of technical communication[24] between the Nuclear and Industrial Safety Agency, which oversaw the enforcement of regulations, and JNES, in which expertise was concentrated.[25]

Loss of power: station blackouts

During the Fukushima accident, units 1, 2, and 3 suffered a total loss of AC power—that is, station blackout[26]—triggered by the earthquake and tsunami; the power outage lasted for several days and ultimately resulted in damage to the reactor cores of all three units. Though units 4 and 5 also experienced

---

[23] In December 1999, a surge in the water level at Gironde Estuary flooded the Blayais Nuclear Power Plant in France, whereby the external power supply was lost and pumps, electrical wiring, and other equipment were submerged; the accident incapacitated safety systems. See: International Atomic Energy Agency (2013) The international nuclear event scale. Available at: www-ns.iaea.org/tech-areas/emergency/ines.asp (accessed August 31, 2013).

[24] Nishiwaki, Y. (2011) Changes in Japan's protections against severe accidents: Where did regulations go wrong? *Genshiryoku Eye*, September/October. (In Japanese.)

[25] Indeed, the establishment of JNES as a technical support organization could accelerate the loss of in-house technical expertise from the Nuclear and Industrial Safety Agency.

[26] A station blackout, at a light water nuclear power plant, is defined as a "state in which the supply of power from all external AC power supplies and from all on-site emergency AC power supplies is lost." After the Fukushima accident and the subsequent investigation, this definition was clarified to read as follows: "The term 'station blackout' encompasses multiple scenarios, including loss of externally supplied electrical power due to a failure of transmission lines or a similar event, the failure of emergency diesel generators to start up, and a malfunction in the facility's central AC bus or power supply panels." In: *On studies of regulatory guides for reviewing safety design of light water nuclear power reactor facilities: A list of technical conditions for provisions to address total loss of AC power* (draft).

station blackout conditions, they were spared any core damage, as the former was down for periodic inspections and the latter was able to use electricity from unit 6's emergency diesel generator.

In 1990, the Nuclear Safety Commission revised its regulatory provisions on station blackouts.[27] In fact, one of its guidelines clearly stated: "The nuclear reactor facilities shall be designed so that safe shutdown and proper cooling of the reactor after shutting down can be ensured in case of a short-term total AC power loss." The guide goes on to comment that "[n]o particular considerations are necessary against a long-term total AC power loss because the repair of troubled power transmission lines or emergency AC power systems can be expected in such case. The assumption of a total AC power loss is not necessary if the emergency AC power system is reliable enough by means of system arrangement or management (such as maintaining the system in operation at all times)." The precise meaning of "short-term" is not clearly defined in the guidelines, but since the adoption of these station blackout policies in 1977,[28] it has been traditionally interpreted during safety reviews as meaning "thirty minutes or less." Although the question of why these guidelines were restricted to short-term station blackouts is one for which no rigorous answer is available, the subcommittee under the Nuclear Safety Commission speculated that, "based on the recognition that external power sources and emergency diesel generators are found to be highly reliable in Japan, the decision was made to regard the likelihood of a long-term station blackout as sufficiently low in Japan."[29]

Upon the revision of the commission's guidelines, the appropriateness of restricting consideration to the case of short-term station blackouts did not go entirely unquestioned. For example, the working group under the Nuclear Safety Commission[30] compared Japanese regulations against the US Nuclear

---

[27] The guidelines are known as the "Regulatory Guide for Reviewing Safety Design of Light Water Nuclear Power Reactor Facilities," which were first published in 1977 by the Atomic Energy Commission; the Nuclear Safety Commission revised them in 1990. Nuclear Safety Commission (1990) *Inspection guidelines for safety design of light-water type nuclear reactor for power generation*. Report, Ministry of Education, Culture, Sports, Science, and Technology. Tokyo: NSC.

[28] Japan Atomic Energy Commission (1977) Item 9: Design requirements to protect against power outages. In: *Safe design inspection guidelines*. The content of this item is continued in item 27 of the guidelines. Note that this was before the creation of the Nuclear Safety Commission, so the design inspection guidelines were formulated by JAEC.

[29] Nuclear Safety Commission (2011) Item 27: Design considerations for protection against power outages. In: *Report on Investigations of Total AC Power Loss* (draft). Report, subcommittee item 4-1-1. Available at: www.nsc.go.jp/senmon/shidai/anzen_sekkei/anzen_sekkei4/siryo4-1-1.pdf (accessed February 28, 2012). (In Japanese.)

[30] Nuclear Safety Commission (1993) *The event of total loss of AC power at a nuclear power plant*. Report, Total AC Power Loss Event Working Group of the Deliberation Committee on Analysis and Evaluation of Accidents and Failures in Nuclear Installations.

Regulatory Commission's station blackout regulations;[31] the group investigated questions such as the probability of station blackouts at representative Japanese nuclear plants and the length of time that batteries and cooling water supplies would last in such an event. The group found that, despite the fact that Japanese regulatory guidelines only required preparation for station blackouts lasting thirty minutes or less, the actual length of time that Japanese reactors could survive a power outage was five hours or more for the case of pressurized-water reactors and eight hours or more for the case of boiling-water reactors; these figures were sufficient to satisfy even US station blackout regulations. However, whereas US station blackout regulations require that geographic characteristics be taken into account when assessing expectations for external events (such as hurricanes or tornados), in Japan, even after the studies noted above, adequate consideration was never given to the possibility of station blackouts caused by external events.[32]

The station blackout during the Fukushima accident was caused by external events, namely, an earthquake and tsunami. First, the earthquake, and the landslides it provoked, damaged electrical transmission lines, switching stations, transformers, and other infrastructure; all external power supplied by six units at TEPCO's Shin-Fukushima Transformer Substation, and by one unit at TEPCO's Tomioka Transformer Substation, was lost. Immediately thereafter, the emergency diesel generators installed at the Fukushima Daiichi Nuclear Power Station sprang to life, ensuring a temporary retention of AC power; however, the tsunami that struck the plant some forty minutes later submerged some of the diesel generators, the cooling system for the generators (which were water-cooled models), and the power switchboard, plunging units 1 through 5 into station blackout conditions. The countermeasures instituted by the plant operator had only envisioned station blackouts caused by internal events. The primary countermeasures included steps such as restoring functionality to malfunctioning diesel generators and relaying electric power from neighboring reactors, but the former of these was inapplicable because the generators had been incapacitated by the tsunami, while the latter was inapplicable because multiple reactors suffered station blackouts almost simultaneously. (However, because unit 6's emergency diesel generators, which were air-cooled models, escaped flooding, a power connection established between units 5 and 6 allowed a successful restoration of functional capacity to critical safety equipment in unit 5.)

---

[31] Code of Federal Regulations (US): Title 10, Part 50.63.

[32] Nuclear Safety Commission (2011) *The 7th Nuclear Safety Commission Subcommittee for Investigating Safety Design regulatory guide*. Report, November 16. Available at: www.nsc.go.jp/senmon/shidai/anzen_sekkei/anzen_sekkei8/siryo1-2.pdf (accessed February 28, 2012). (In Japanese.)

Lessons learned from previous station blackouts

In March 2001, dense, salt-rich fog rolled in from the Pacific Ocean into Taiwan's Maanshan Nuclear Power Station and corroded the electrical insulation, stopping all supply of power to two external lines. Designed so that, in the event of a loss of external power, the plant's two emergency diesel generators would start up, this was not the case in practice, when one generator failed due to an Earth fault in a switchboard—the connection loss between electrical devices and the ground, caused by a failure of electrical insulation—while the other generator simply failed, pitching the plant into a station blackout. However, because the plant was able to make use of DC power, it was immediately possible to cool the reactor cores using the backup cooling system and other methods. Moreover, an additional emergency diesel generator was on hand for the two units; by connecting this generator to one of the systems, plant workers succeeded in ending the station blackout after approximately two hours. The backdrop for this incident is the fact that external power supplies in Taiwan were traditionally recognized as unreliable, and for this reason considerable emphasis had been placed on the redundancy in the emergency generators that were available.[33]

Through this event, Japan should have learned that damage to electrical switchboards, power supply motherboards, or similar equipment can cause a station blackout even if generators or external power supplies themselves are functioning properly.[34] Japan's Nuclear Safety Commission investigated the Taiwanese incident, but a comment made by a Nuclear and Industrial Safety Agency representative highlights the prevailing attitude at that time:

> Generally speaking, boiling-water reactors can last at least eight hours or so. For pressurized-water reactors, assuming certain measures have been taken, the state can be maintained for around five hours. During that time, external power supplies will be restored. In Japan, shutdowns in the electrical transmission system have generally only lasted for thirty minutes or so. On top of this, as I mentioned before, if we take into account repairs of diesel generators and those sorts of considerations, then it seems clear that there is plenty of leeway.[35]

This statement and the absence of any further discussion of this point demonstrate that no substantive lessons were learned from the Taiwanese case.

---

[33] Atomic Energy Council (2001) *The station blackout incident of the Maanshan NPP unit 1*. Report, April 18. Available at: www.aec.gov.tw/webpage/UploadFiles/report_file/1032313985318Eng.pdf (accessed February 28, 2012).

[34] Okamoto, K. (2012) On the possibility of total power loss. *Journal of the Atomic Energy Society of Japan (ΑΤΟΜΟΣ)* 54(1): 27–31. (In Japanese.)

[35] Statement from a Nuclear and Industrial Safety Agency speaker recorded in the minutes of the 47th meeting of the Nuclear Safety Commission (July 2, 2001). The comment was made to a colleague who worked for the commission.

Six years after the event in Taiwan, Japan had the chance to learn various lessons about station blackouts in its very own backyard when, in 2007, an earthquake rattled the Kashiwazaki-Kariwa Nuclear Power Station in the Niigata Prefecture—including the importance of having chemical fire-fighting trucks on hand and earthquake-resistant building for critical operations. These lessons, particularly the latter, did in fact serve valuable roles in the response to the accident at the Fukushima Daiichi Nuclear Power Station. However, the fact that a fire in an electrical transformer broke out in unit 3 at Kashiwazaki-Kariwa—right after the Chuetsu earthquake—should have served to highlight the fact that differences in the degree of earthquake-resistance between various pieces of equipment at a nuclear plant can pose an important vulnerability. Even if the nuclear reactor itself is highly quake-resistant, and even if all critical safety equipment within it is protected from earthquake damage, damage to other, less quake-resistant equipment, such as electrical connectivity equipment, can threaten the safety of the entire reactor. In reality, this experience did not motivate the nuclear utilities and the regulatory bodies to revisit the concepts of seismic design or earthquake-resistant infrastructure; the only outcome was that of more thorough seismic back-checks on all the nuclear power plants.

Through these two examples, it is clear that Japanese nuclear safety regulations adequately addressed precautions against *internally triggered* station blackouts, but insufficiently considered preparedness against *externally triggered* station blackouts. That is, when retracing the steps that led to this situation, it is clear that the "Regulatory Guide for Reviewing Safety Design" only considered short-term station blackouts, which may well have constrained the range of hypothetical situations envisioned by plant designers and, thus, choked efforts to investigate blackouts triggered by external events.[36]

Regulatory preparedness against severe accidents and the Fukushima accident

The Nuclear Safety Commission defines the term "severe accident" as an "event that significantly exceeds design-basis events, during which the response methods anticipated by design safety evaluations are unable to provide adequate reactor cooling capacity or to control the rate of reactions, resulting in severe

---

[36] At the time the policy guidelines were ratified, "the power companies had strong powers of persuasion over the adoption of draft policies," stated Yoshihiko Sasaki, the first chairman of the Nuclear and Industrial Safety Agency. (Sasaki, Y. [2012] Sociotechnology and the Fukushima Daiichi Nuclear Power Station accident. In: *8th annual sociotechnology research symposium*, January 28.) Indeed, one can only speculate as to the connection between the cozy relationship between regulators and regulated parties and the restriction of the policy guidelines to the consideration to "brief" station blackouts.

damage to a reactor core."³⁷ That is, equipment is designed with some additional tolerance beyond the range strictly required by the consideration of design-basis events. An event that significantly exceeds design-basis events is one that surpasses even this additional range of tolerance³⁸—and Fukushima was the very epitome of such an event. Precisely as stated in the definition, the response methods anticipated by design safety evaluations were unable to provide adequate cooling capacity for reactor cores, and the event resulted in severe damage to reactor cores.

But the Fukushima accident was not the first time that an event had escalated to the level of a severe accident. Even when limited to commercial nuclear power stations, the 1979 accident at Three Mile Island in the United States and the 1986 accident at Chernobyl in the former Soviet Union were both severe accidents. These experiences provided opportunities for the world's community of nuclear powered nations, including Japan, to advance their understanding of severe accidents and the appropriate countermeasures and to prepare regulatory preparedness against them.

Researched and implemented countermeasures against severe accidents qualify as accident management—that is, procedures such as preparing detailed manuals and equipment configurations to address specific severe scenarios.³⁹ In Japan, such measures were not required by regulations, but instead were considered to lie within the domain of voluntary activities entrusted to the discretion of plant operators. In the remainder of this chapter, we explore the ways in which severe-accident precautions were implemented in Japan by investigating their historical context.

Accident at Three Mile Island and the subsequent regulatory response

The 1979 accident at the Three Mile Island Nuclear Generating Station in the United States, despite the fact that it ultimately exposed the surrounding environment to only a tiny quantity of radiation, was nonetheless a major accident that resulted in reactor core damage. Since the accident, regulatory bodies in countries around the world have identified the prevention of reactor core damage as the primary goal of safety policy and have launched numerous initiatives to this

---

[37] Nuclear Safety Commission (1990) *Interim report of the common issues discussion group of Nuclear Safety Commission's special committee on safety standards and guides*. Report, February 19. Tokyo: NSC.
[38] Sato, K. (2011) *The Logic of Nuclear Safety* (revised edition). Tokyo: Nikkan Kogyo Shimbunsha. (In Japanese.)
[39] The "Accident Management as a Countermeasure against Severe Accidents at Light-Water Nuclear Reactor Facilities," adopted by the Nuclear Safety Commission on May 28, 1992, defines accident management as a "set of measures to prevent incidents beyond design-basis events, and/or incidents posing a risk of severe damage to reactor cores, from escalating into severe accidents or to mitigate the impact of severe accidents."

end. For example, the incident forced the Nuclear Regulatory Commission, the US regulatory authority, to acknowledge the reality that severe accidents beyond design-basis events can happen in practice—this recognition led the agency to invest efforts in probabilistic safety analysis and other methods of safety analysis to understand how to prevent or mitigate the consequences of such accidents.

In Japan, however, the Nuclear Safety Commission, which had only been established in 1978, convened a subcommittee to look into Three Mile Island. The subcommittee investigated and analyzed a wide range of areas, including nine issues related to standards, four on safety reviews, seven on design, ten on operation and maintenance, ten on disaster prevention, and twelve on safety research.[40] Beginning in May of the following year, these observations were incorporated into the commission's basic policy and its annual programs.[41]

But like Fukushima, the Pennsylvania accident was also tightly intertwined with the myth of safety. Until then—despite the public release of the 1975 WASH-1400 report and similar studies—the global nuclear power community had been sustained by an almost religious faith, which maintained that an accident beyond design-basis was theoretically possible, but in practice would never actually happen.[42] Indeed, the US report of the President's Commission on the Three Mile Island accident (the Kemeny report) states that: "After many years of operation of nuclear power plants, with no evidence that any member of the general public has been hurt, the belief that nuclear power plants are sufficiently safe grew into a conviction."[43] If we assume that this description applied to the worldwide nuclear power community as well, then the Three Mile Island accident clearly played a role in puncturing the myth of safety that had prevailed throughout the nuclear power world.

Chernobyl and regulatory response

The accident at the Chernobyl Nuclear Power Station, despite the fact that it involved an RBMK reactor—a light-water cooled graphite moderated reactor unique to the Soviet Union—was triggered by repeated regulatory violations by operating personnel; nonetheless, it was demonstrated that a commercial reactor was capable of releasing radioactive contamination across an extremely wide geographical area. The IAEA's International Nuclear Safety Group

---

[40] *ATOMICA* (1998) Japan's response to the Three Mile Island accident. Available at: www.rist.or.jp/atomica/data/dat_detail.php?Title_No=02-07-04-06 (accessed August 31, 2013). (In Japanese.)

[41] The subcommittee's report, dated September 13, 1979, is titled "Issues to be Reflected in Japan's Safety Assurance Measures."

[42] Sato (2011), *op. cit.*

[43] President's Commission on the Accident at Three Mile Island (1979) *Report of the President's Commission on the Accident at Three Mile Island: The need for change, the legacy of Three Mile Island.* Report, October. US Government Printing Office: Washington, DC. Available at: www.threemileisland.org/downloads/188.pdf

promotes the phrase "safety culture," which it defines as the "assembly of characteristics and attitudes in organizations and individuals which establishes that, as an overriding priority, nuclear plant safety issues receive the attention warranted by their significance." This notion has come to be emphasized as an essential ingredient for the assurance of nuclear safety.

The world had already begun to study the causes of severe-accident phenomena and to conduct probabilistic safety analysis research after the Three Mile Island accident. But that accident accelerated those trends, and accident measures gradually came to be established in various developed countries around the world beginning in the second half of the 1980s.

For example, in 1988 in the United States, nuclear plant operators were required to conduct individual plant examinations in search of vulnerabilities that might lead to severe accidents;[44] after 1991, they were required to conduct probabilistic safety analysis of external events. In 1994, in light of these trends, the Nuclear Energy Institute, a US nuclear power lobbying group, created severe-accident management guidelines, which consisted of individual plant examinations of both internal events and external events and containment performance improvements. Thus, power companies were forcibly required to comply with these guidelines, and by 1999, all nuclear plant operators had established severe-accident measures.

Throughout Europe, the Chernobyl disaster—and the radioactive contamination it caused—encouraged nations to recognize that protecting the environment from radiation risks was among the most important priorities in regulating nuclear power plants. Consequently, various measures were implemented, such as the installation of filters on containment ventilation systems. For example, in Germany, the government-formed Nuclear Safety Standards Commission recommended in 1986 that filtered containment venting systems be installed, so existing reactors were retrofitted to comply. By 1989 in France, containment ventilation systems equipped with sand filters had been installed at all the commercial nuclear power stations.

Likewise, Japan's regulatory agencies began to investigate severe-accident measures. In 1986, the same year Germany highlighted safety concerns, Japan's Ministry of International Trade and Industry (MITI), which held regulatory safety authority over commercial nuclear reactors, announced what was known as "Safety 21"—specific policies that upgraded safety regulations, based on technical standards grounded in the latest scientific knowledge; improved safety activities by the nuclear utilities, including training of operational and maintenance personnel; promoted research and technology development for enhancing safety; and promoted international collaborations.[45] Also included in

---

[44] This was based on the Nuclear Regulatory Commission's 1985 Severe Accident Policy Statement.
[45] Published in August of that year, the statement was titled "About the Enhancement of Safety Measures at Nuclear Power Plants (Safety 21)."

these policies were efforts to simulate research on severe accidents,[46] to work on developing emergency operation manuals that evolved into today's accident-measure procedures, and to create a predecessor of today's ERSS for predicting, detecting, and warning of incidents before they escalate into severe accidents.[47]

The Nuclear Safety Commission established a group to study severe-accident measures, probabilistic safety analysis methods, and other research methods that underlie them.[48] When, in May 1992, the group presented its findings, the commission recommended that "nuclear operators ... voluntarily prepare effective accident-management measures and ensure that they will be able to implement them properly in an emergency."[49]

In July 1992, MITI requested plant operators to implement accident measures not as a regulatory mandate, but rather as voluntary measures to be carried out by power companies at their discretion. This decision, the ministry reasoned, was because the safety of Japan's reactor facilities is sufficiently ensured by the philosophy of defense-in-depth; thus, the probability of severe accidents is so small that, from an engineering perspective, severe accidents can be said never to happen in reality. The ministry went on to say that it deferred this responsibility to the power companies because accident measures are based on the technical expertise of electric power companies to make this small risk even smaller. That is, it is desirable for electric power companies to put their own technical expertise to use in flexible and adaptive ways.[50]

MITI expected nuclear plant operators to implement levels of probabilistic safety analysis and establish accident-management measures.[51] In addition to these measures, MITI specifically required that plant operators assess their accident-management measures during periodic safety reviews, continually improve the accuracy of their probabilistic safety analysis methods and study for expanding their scope, and build databases of failure rates and other statistics on various types of equipment. At MITI's request, nuclear operators were required to report the

---

[46] Written Japanese documents from the 1980s use a combination of Chinese characters to express the phrase "severe accidents" that differs slightly from the combination used today.
[47] Nishiwaki, Y. (2011) Changes in Japan's protections against severe accidents: Where did regulations go wrong? *Genshiryoku Eye*, September/October. (In Japanese.)
[48] It was named the Common Issues Discussion Group.
[49] Nuclear Safety Commission (1992) *Accident management for the severe accidents at light water nuclear power reactor facilities*. Report, May 28.
[50] Ministry of Economy, Trade, and Industry (1992) The status of accident management measures within the safety regulatory system. Press release, July.
[51] Probabilistic safety analysis is typically categorized into three levels: Level 1 assesses the probability of serious damage to the reactor core and is the most important of the three levels, as it primarily assesses the possibility that an accident will exceed the range of design-basis and escalate into a severe accident, establishes accident-management measures to prevent severe accidents, and assesses the efficacy of those measures; Level 2 considers the findings in Level 1 and, based on that, assesses the probability that the reactor containment vessel will be incapacitated, which could lead to a release of radioactive materials into the environment; Level 3, then, takes the findings in Level 2 to assess the impact if radioactive materials were to be released in the surrounding area and environment. MITI encouraged the implementation of levels 1 and 2.

results of their probabilistic safety analyses, as well as the consequent accident measures to be implemented. MITI, then, would evaluate the technical feasibility of the measures. Thus, we see that, in post-Chernobyl Japan, the accident-management measures to address the risk of severe accidents was entrusted to power companies to implement at their own discretion, with regulatory agencies reviewing the adequacy of the efforts. Moreover, as of 1992, probabilistic safety analyses were only required to address internal events. As power companies and regulatory bodies continued to conduct research on probabilistic safety analysis techniques, however, their goal included external events, as well.

Plant operators and accident-management measures

In March 1994, power companies heeded MITI's request and presented their probabilistic safety analyses and their accident-management proposals. Seven months later, MITI released a report encouraging plant operators to establish, by the year 2000, accident-management measures at all nuclear power plants in operation or under construction.[52] Following orders, plant operators had such measures in place at nuclear power facilities by the end of March 2002. The next month, in April, the Nuclear and Industrial Safety Agency, which had only just been established in 2001, published a document that detailed the role of—and the specific measures that must be taken by—plant operators when implementing the containment ventilation system at boiling-water reactor plants, an item which was considered to lie within the realm of accident management.[53]

Thereafter, plant operators used probabilistic safety analysis to conduct individual plant examinations and to assess the efficacy of accident management; by March 2004, they had submitted for review their evaluations to the agency and the Nuclear Safety Commission—both of which were satisfied with the analyses.[54]

Regulations and severe-accident measures

Though it seems that plant operators were moving swiftly forward with strong accident-management measures—while regulatory agencies were in lockstep with appropriate monitoring and evaluation responses—a closer look at the situation reveals several problematic developments.

---

[52] Ministry of Economy, Trade, and Industry (1994) Report on the review of accident management measures to be developed at light water nuclear power reactor facilities. Press release, October.
[53] The document was titled "Basic Requirements to Be Satisfied in the Development of Accident Management Measures."
[54] The agency and commission reviewed these reports on the basis of the Nuclear and Industrial Safety Agency's basic requirements.

First, the measures studied and implemented by plant operators had traditionally been limited to consideration of internally, not externally, triggered events. Though probabilistic safety analysis is an essential component to address severe accidents triggered by external events, no reliable assessment methods had been developed; only policy promises had been made to continue research on probabilistic safety analysis with the eventual goal of expanding its scope to cover external events. However, severe-accident measures in the 2000s were essentially identical to the content of the accident-management reports submitted by plant operators in 1994, with no obvious indications that progress had been made. Yoshihiro Nishiwaki, visiting professor at the University of Tokyo, suggested that, once the government put in place the 1992 resolution on accident-management policy, the psychology around safety shifted for the worst—that is, from that moment onward, utilities purely emphasized regulation compliance and were barely motivated to cultivate their own voluntary actions beyond what the resolution stated. Furthermore, the desire for new probabilistic safety analysis methods and better severe-accident precautions seemed to ebb, and progress ground to a halt.[55] He also noted that the regulatory community did not rigorously review the results of the plant operators' individual plant examinations.

In 1992, MITI ordered plant operators to conduct periodic safety reviews of nuclear power plants approximately every ten years to assess their safety activities during the operation period. However, when—more than a decade later, in 2003—the Ministry of Economy, Trade, and Industry (METI) made it mandatory by ministerial ordinance for commercial operators to conduct periodic safety reviews, they excluded countermeasures against severe accidents from the Nuclear and Industrial Safety Agency's regulatory reviews.[56] The general understanding was that not enough technical expertise could be mustered to justify the promotion of probabilistic safety analysis (the backbone of the countermeasures against severe accidents) to the status of a legal requirement.[57] Consequently, there was no periodic assessment by regulatory authorities of the implementation status of countermeasures against severe accidents; thus, the expansion of probabilistic safety analysis—and the implementation of accident-management measures based on them—stalled.

Aside from this, the Safety 21 program called for further research into severe accidents and related topics, which the Agency for Natural Resources and Energy (ANRE) carried out through an institution known as the Nuclear

---

[55] Nishiwaki, Y. (2011) *Rethinking the nature of the nuclear safety agencies and the regulatory system for nuclear reactors*. Report, JNES Technical Information Seminar Materials. Tokyo: JNES.
[56] "Rules for the Installation, Operation of Commercial Power Reactors" were revised in October 2003 and were law under METI.
[57] Hirano, M. (2011) The history of countermeasures against severe accident and "residual risk." *Journal of the Atomic Energy Society of Japan (ΑΤΟΜΟΣ)* 53(11): 22–28. (In Japanese.)

Power Engineering Corporation.[58] However, safety research itself in Japan entered a period of decline in the second half of the 1990s, and its scope gradually diminished. One factor behind this trend was that the Japan Atomic Energy Research Institute, the primary institution for severe-accident research, existed as a special entity under the auspices of the Japan Science and Technology Agency; in practice, this meant that institutional barriers existed between the energy institute and ANRE, a factor which has been noted as one reason for the inability of the two bodies to collaborate on safety research.[59]

Though plant operators, at their own discretion, voluntarily initiated their own accident-management measures, the scope of these measures was never expanded to encompass precautions against external events. Of course, even if research had continued throughout the 1990s and beyond, and even if proper countermeasures had been established, the fact remains that academic understanding of probabilistic safety analysis for tsunami risk was insufficient; thus, it seems highly unlikely that the events triggered by the Tohoku earthquake and tsunami could have been entirely prevented from escalating into a severe accident. Nonetheless, by instituting advance countermeasures against severe accidents that adequately accounted for the impact of external events, it most likely would have been possible to reduce the amount of time required to establish a replacement water injection system and to vent the containment vessel, thereby mitigating the impact of the severe accident. According to this line of thinking, the primary problem with Japan's safety regulations was not so much that they failed to establish regulatory mandates for accident-management measures—but that they failed to create an environment in which the scope of these existing measures could be expanded.

Structural strength and the mishandling of risk

While Japan's nuclear safety regulations placed considerable emphasis on structural enhancements to facilities and equipment, they undervalued the importance of system functionality and analysis. Even after the Three Mile Island accident, Japan's nuclear safety regulations did not emphasize the importance of safety analysis, as other countries did, but instead chose to double down on its traditional focus by tightening safety regulations governing structural strength of the facilities—to this day, Japanese safety regulations attach little importance to system functionality and analysis. Whereas construction approvals require extremely detailed analysis of the structural

---

[58] This was founded in 1976 as the Nuclear Power Engineering Center under the initiative of academia and private companies; its purpose was to advance the performance and public acceptance of commercial nuclear power plants through engineering tests, safety analysis, information acquisition and analyses, and public relations activities.

[59] Nishiwaki, Y. (2011) *Rethinking the nature of the nuclear safety agencies and the regulatory system for nuclear reactors*. Report, JNES Technical Information Seminar Materials. Tokyo: JNES.

soundness of buildings and equipment, Japan's regulations have lagged behind the world and global trends.[60]

Though Finland, Sweden, and other European nations started using risk information in the 1980s and the US Nuclear Regulatory Commission included probabilistic risk assessments in 1995,[61] Japan's Nuclear Safety Commission did not follow suit until November 2003.[62] Similarly, Japan was slower than other nations in its effort to enact safety goals, the purpose of which is to acknowledge the impossibility of achieving absolute safety at a nuclear reactor facility and, instead, to seek a quantitative response to the critical nuclear safety question of what is "safe enough." In December 2003, Japan's Nuclear Safety Commission proposed qualitative goals to the public to control the risk of accidents and quantitative goals to establish specific criteria.[63] The clarification of safety goals can be important indicators of the efficacy of accident-management countermeasures against both severe accidents and the residual risk of earthquakes and other events. For example, in 1986, the United States released a policy statement on safety goals,[64] which was subsequently used as a benchmark for "backfit" procedures based on cost–benefit analyses.

In June 2010, the Nuclear and Industrial Safety Agency announced that, as part of its regulatory requirements, it would enforce severe-accident countermeasures.[65] Similarly, in December 2010, the Nuclear Safety Commission decided that its basic policy would clarify safety goals and utilize risk information in nuclear safety regulation. Although the importance of probabilistic safety goals in improving countermeasures against severe accidents

---

[60] Nishiwaki, Y. (2007) Issues and observations regarding regulations of nuclear power plants. *Transactions of the Atomic Energy Society of Japan* 6(3): 239–252. (In Japanese.)
[61] US Nuclear Regulatory Commission (1995) *Final policy statement on the use of probabilistic risk assessment methods in nuclear regulatory activities*. Report, SECY-95-126, August 16. Available at: www.nrc.gov/reading-rm/doc-collections/commission/policy/60fr42622.pdf
[62] It was in this month that the commission published: "Basic Policies on Introducing Nuclear Safety Regulation Using Risk Information."
[63] The commission proposed the following two safety goals. (1) Qualitative goal: The use of nuclear energy should not meaningfully increase the risk of damage to the public's health. (2) Quantitative goal: The mean value of acute fatality risk by radiation exposure, resulting from a nuclear power accident, to individuals who live in the vicinity of the nuclear installation should not exceed the probability of about 1×10^{-6} per year. And, for those who live near a nuclear installation, the mean value of fatality risk by latent cancer caused by radiation exposure, resulting from a nuclear power accident, should not exceed the probability of approximately 1×10^{-6} per year. In addition, the commission's proposal stated that, as specific performance goals for nuclear power facilities, the core damage frequency must not occur more than once every 10,000 years, and the containment failure frequency should not fail more than one time every 100,000 years; these two numerical performance goals, the commission stated, must be achieved simultaneously.
[64] US Nuclear Regulatory Commission (1986) *Safety goals for the operation of nuclear power plants*. Report, 51-FR-28044, August 4. Available at: www.nrc.gov/reading-rm/doc-collections/commission/policy/51fr30028.pdf
[65] Nuclear and Industrial Safety Agency (2010) Nuclear and Industrial Safety Agency's mission and action plans. Available at: www.nisa.meti.go.jp/oshirase/2010/files/220617-6-1.pdf (accessed February 28, 2012). (In Japanese.)

has long been noted,[66] as of this book's publication date, the proposed safety goals in Japan mentioned above remain at the draft stage—and the Nuclear Safety Commission has not yet written a resolution that incorporates them into the regulatory framework.

If, as the IAEA stipulates, safety regulations are to establish both standards and a regulatory framework to inspire safety and protection of society, then it is difficult to argue that Japan's safety regulations were successful. The crucial turning point after which Japanese risk assessment and management began to lag behind those of other nations was the accident at Three Mile Island, which offered lessons that Japan failed to put to practical use. If, starting in the early 1980s, Japan had broadened its focus away from the exclusive consideration of structural strength and had actively embraced probabilistic methods of safety analysis, then—although of course we cannot say with certainty that the Fukushima Daiichi Nuclear Power Station accident could have been prevented—it seems likely that more attention would have been paid to the possibility of incidents that exceed design-basis events.

SPEEDI

After the Three Mile Island accident in 1979 the Japanese government began to develop a system that would assess and predict the diffusion of radioactive materials in the event of a nuclear accident. The result—System for Prediction of Environmental Emergency Dose Information (SPEEDI)—went into operation in 1986, and has been consistently upgraded since then.

During a nuclear emergency, experts at SPEEDI's central information processing facility perform rapid computations of airborne radioactive materials concentrations and radiation doses in the surrounding environment. Their computations—based on meteorological, topographical, and release-source information—predict the movements of the radioactive plume. The resulting maps are meant to be used to determine evacuation zones and other protective measures.

In the Fukushima Daiichi emergency, however, the SPEEDI predictions failed to reach the Cabinet leaders in a timely manner, and were thus not utilized in decisions on evacuation instructions led by the Cabinet. This has become the target of strong criticism, centering on the possibility that some cases of exposure could have been avoided if the SPEEDI predictions had been utilized in evacuation instructions or made public at an earlier stage.

The SPEEDI system was switched to emergency mode on the afternoon of March 11, and the Nuclear Safety Technology Center began predictive

---

[66] See, for example: Abe, S. (2011) Some points regarding severe accidents and safety goals. In: *2010 Fall Planning Session of the Atomic Energy Safety Subcommittee of the Atomic Energy Society of Japan*, September 17. Available at: www.soc.nii.ac.jp/aesj/division/safety/H221021siryou2.pdf (accessed February 28, 2012). (In Japanese.)

calculations. Unfortunately, SPEEDI didn't have all the information inputs that it requires for a full calculation. Typically another system called the Emergency Response Support System (ERSS) would provide detailed information about the radioactive release (including such factors as the time, duration, and rate of the release; which radioactive isotopes were included; and so on); the ERSS gathers that information from the operator of the nuclear power station. However, the earthquake on March 11 disrupted data transmission networks, and SPEEDI could not get the necessary information from ERSS.

In the absence of hard figures relating to the release of radiation from Fukushima Daiichi, SPEEDI computations were done using hypothetical values. For example, some computations were based on the assumption of a release of radioactive noble gas or iodine at 1 becquerel per hour for one hour. Between March 11 and 16, nearly 100 other calculations were performed at the request of various government agencies; the resulting predictions were used in discussions on venting, damage to the reactors' containment vessels, and the selection process of the radiation monitoring sites.

During the initial period of their leadership of the Fukushima nuclear emergency response, Prime Minister Kan and his top Cabinet leaders did not know that SPEEDI was in full operation—in fact, for a considerable period, they did not know that SPEEDI even existed. METI Minister Kaieda later said, "In what may be termed the 'high command' of the nuclear emergency response, SPEEDI was unfortunately not brought up. … If we had known of it, someone would certainly have asked where the SPEEDI data was and called for it to be brought in immediately, but we didn't know about it and therefore no one asked."[67] Chief Cabinet Secretary Edano has said he first learned of the existence of SPEEDI from the mass media on March 15, but was told that it could not be used because release source information could not be obtained.

Once top officials were aware of SPEEDI, it still took some time for its calculations to be useful. On March 16 the Nuclear Safety Commission took charge of SPEEDI's management and operation, and prepared to assess the scope of the radioactive release based on readings from environmental monitors. But although some environmental monitoring had been conducted around the Fukushima Daiichi Nuclear Power Station on March 12 and 13, that data was left behind at the local off-site center when personnel were forced to relocate to the Fukushima prefectural office. No relevant data could be obtained from March 17 to 19 because the wind blew the plant's emissions out over the sea. When the wind shifted toward land on March 20, data could be collected, and SPEEDI's operators could finally calculate predicted exposure rates. The Nuclear Safety Commission presented the results to the top decision makers, including Chief Cabinet Secretary Edano and Prime Minister Kan, on March 23. However, in order to avoid inciting panic among evacuees, the forecast to

---

[67] Personal communication with former METI Minister Banri Kaieda, October 1, 2011.

include Iitate-mura was not used in the planning of extended evacuation zones. Recalling this decision, Deputy Cabinet Secretary Tetsuro Fukuyama said that it is a politician's job—not that of a computer, regardless of how sophisticated it is—to make a critical decision that could put thousands of lives at stake.

Opinions differ on whether SPEEDI calculations could have provided more precise and accurate evacuation instructions if the government had access to them earlier. Some officials believe that, even though the earliest SPEEDI calculations lacked real data about the amount of radioactive release, their predictions about wind direction and plume dispersal would still have been useful in planning the evacuations. For example, METI Minister Kaieda has said that he later felt abashed when he reviewed the SPEEDI results and found that they were in fairly close agreement with the "hot spots" that by then had become a subject of controversy—places with high levels of contamination that were outside the initial evacuation zones.[68] Other officials disagree, and say that the concentric circles of evacuation zones would have been adopted even if SPEEDI's calculations had been available, because those circles offered the broadest protection during a rapidly changing crisis.

Off-site centers

The national government originally conceived off-site centers to be local bases with the necessary equipment for effective response in nuclear emergencies; such configurations were based on the lessons drawn from the 1999 Tokai-mura criticality accident. That incident had shown the central importance of information sharing and coordination among the representatives from national, prefectural, and local community levels.

Each off-site center has the facilities and equipment to serve as an emergency response headquarters during a nuclear accident, and to coordinate with the nuclear operator, local government, and national government. On receiving notification under the Nuclear Emergency Act, local officials initiate activity at the off-site center. They gather information from the nuclear operator and the local government and facilitate deliberations over response measures.

The Fukushima Prefecture Nuclear Emergency Response Center was situated only 5 kilometers from the Fukushima Daiichi station in the town of Okuma. Like other off-site centers, the Fukushima facility included the latest communication systems, nuclear emergency response support systems, a decontamination room in case of exposure, radiation measurement instruments, and other equipment. However, the Fukushima off-site center, along with most others across the nation, had been cited in 2009 by the government for failing to install a ventilation system to reduce the radiation dose for personnel

---

[68] Personal communication with former METI Minister Banri Kaieda, October 1, 2011.

inside the building in case of a high-radiation environment outside. This deficiency was not corrected before the Fukushima Daiichi accident.

During the crisis, the Fukushima off-site center did not serve its function as a coordinating center for the emergency response. The Okuma center lost electricity following the earthquake, and its emergency power system also failed. The telephone company's nearby base station was also damaged, and on March 12 the off-site center lost access to public phone lines—and with them, to email, Internet, the national government's teleconferencing system, and the SPEEDI system. Only satellite phones remained to link the off-site center to the outside. An alternative facility at the Fukushima Prefecture Minamisoma Joint Government Building was damaged as well. On March 14, following the explosion of Fukushima Daiichi's unit 3 building, radiation levels began to rise inside the Fukushima off-site center. That night officials made the decision to relocate the response center to the Fukushima prefectural office, which had no communications or radiation monitoring systems.

In addition to the lack of infrastructure at this ad-hoc response center, the management structure was inadequate as well. This was in part because local officials were spread out through the prefecture, coping with the extensive and widespread damage wrought by the earthquake, tsunami, and nuclear disaster. It also took some time for specialists with expertise in responding to nuclear accidents to arrive on site. What's more, some personnel who were available to manage the accident response weren't adequately trained in response methods or weren't able to use the latest equipment. All these factors seriously impeded local information gathering and coordination, and made rapid decision-making regarding response measures impossible.

As the Fukushima off-site center was not able to fulfill its function of coordinating the accident response, the central government established the joint government-TEPCO response center to take over that duty. As a result of this reorganization, the local response agencies were often unable to coordinate their activities. In particular, the local government's evacuation efforts were carried out with an extreme lack of information.

Chapter 7

Impact of Radioactive Material Released into the Environment

The amount of radioactive material released into the environment from the Fukushima Daiichi Nuclear Power Station accident is roughly one-tenth of that which was released during Chernobyl. The radioactive emissions contaminated soil, seawater, and various natural resources, as well as food, drinking water, and other consumables—and necessitated environmental restoration and proper treatment of waste products before citizens could return to their homes.

To exacerbate these problems, present-day scientific understanding of the impact of low-level radiation exposure is incomplete—a fact that must certainly be cited among the causes of the societal anxiety created by the Fukushima disaster. Although low-level radiation exposure has been studied for many years, as yet no comprehensive understanding of its effects on the human body has been achieved.

Implementation of land monitoring

The Emergency Operation Center in Japan's Ministry of Education, Culture, Sports, Science, and Technology (MEXT) created maps showing the geographical distribution of radiation levels in the aftermath of the Fukushima accident; the center designed these maps in order to assess and lift evacuation orders in particular regions, in accordance with victim-assistance and monitoring plans that were established within days and weeks, respectively, of the disaster.[1,2]

---

[1] Ministry of Education, Culture, Sports, Science, and Technology (2011) A forum to discuss the creation of maps illustrating the geographical distribution of radiation levels and other issues. Press release, Nuclear Accident Response Center, May 16. Available at: www.mext.go.jp/b_menu/shingi/chousa/gijyutu/017/gaiyo/1307559.htm (accessed June 21, 2013).

[2] Ministry of Education, Culture, Sports, Science, and Technology (2011) Plan to strengthen environmental monitoring. Press release, Nuclear Accident Response Center, April 22. Ministry of Education, Culture, Sports, Science, and Technology (2011) Short-term policy initiatives to assist victims of the nuclear accident. Press release, Nuclear Accident Response Center, May 17.

Aircraft monitored a 100-kilometer radius around the Fukushima station (and a 120-kilometer region south of the plant), as well as over neighboring prefectures, to both assess the impact of radioactive materials and radiation levels and to ascertain what the short- and long-term consequences of radioactive materials would be in the evacuated areas. Figure 10 shows the radiation levels in the air (1 meter above ground), as well as the total quantity of cesium 134 and cesium 137 embedded in the ground.[3]

According to MEXT, various plutonium and strontium isotopes were in measurable quantities, which the ministry attributes to the Fukushima accident. However, in each of the locations where the maximum ground-embedded quantities of plutonium 238, 239, and 240 and strontium 89 and 90 were measured, the fifty-year effective accumulated radiation dose was less than that for cesium 134 (71 millisieverts) and cesium 137 (2 sieverts); consequently, the quantities of cesium found in the soil have been used as the most relevant measurement for assessing radiation exposure and formulating decontamination procedures.

Implementation of seawater monitoring

To assess the state of radioactive material emissions, MEXT started to conduct ocean monitoring within days after the Fukushima disaster. At points approximately 30 kilometers offshore—a distance that allows atmospheric radiation levels to be measured while ensuring the health of the ship's crew—seawater samples were collected from eight sites approximately 10 kilometers apart. For example, on March 23 at site 1, off the coast of the Fukushima Daiichi Nuclear Power Station, the maximum detected levels were 76.8 becquerels per liter of water for iodine 131 and 24.1 becquerels per liter for cesium; these values are higher than normal conditions, but lower than the provisional restrictions of tap water.[4] MEXT monitored air radiation levels at sea, as well as the radioactivity concentration of ocean dust. Seafloors were monitored in Miyagi, Fukushima, and Ibaraki prefectures, where the ministry detected low levels of iodine 131, cesium 134, and cesium 137—all thought to be a result of Fukushima. Since April 2011, the Japan Agency for Marine-Earth Science and Technology has used numerical oceanic prediction systems to conduct simulations of the dispersion of radioactivity concentrations, as well as other quantities, on the ocean surface.[5]

---

[3] Ministry of Education, Culture, Sports, Science, and Technology (2011) Monitoring information of environmental radioactivity level. Available at: http://radioactivity.mext.go.jp/ja/1910/2011/10/17485.pdf (accessed February 28, 2012). (In Japanese.)

[4] Before the accident, provisional restrictions were in place for tap water with 300 becquerels per kilogram of iodine and 200 becquerels per kilogram for cesium. For breast milk, the limit was 100 becquerels per kilogram.

[5] Japan Coastal Ocean Predictability Experiment (2012) Fukushima radionuclide dispersion simulation in the ocean using JCOPE. Available at: www.jamstec.go.jp/frcgc/jcope/htdocs/e/fukushima.html (Accessed August 30, 2013).

The area of contaminated soil depends on the threshold contamination level; for example, at a threshold level of 600 kilobecquerels per square meter, the area contaminated by the Chernobyl accident was more than 250 times larger than the area contaminated by the Fukushima accident. Of course, simply because the Fukushima accident contaminated a smaller area of soil does not mean that it did not cause serious problems. The International Nuclear and Radiological Event Scale rated both accidents to be Level 7 events. But the accidents themselves are not strongly comparable: The Chernobyl accident, which involved an uncontrolled nuclear reaction, contaminated a huge geographic area and cannot be easily compared to the Fukushima Daiichi Nuclear Power Station accident, which involved a meltdown of the reactor core due to a loss of cooling water.

That said, it is important to compare the land usage in the vicinity of the two nuclear power stations for a clear understanding of the impact. In the vicinity of the Fukushima Daiichi Nuclear Power Station, less than 5 percent of land is occupied by cities, less than 10 percent of land is occupied by rice paddies, less than 10 percent of land is used for other agricultural purposes, and more than 75 percent of the land is forested or mountainous. In contrast, in the area around Chernobyl (actually in the entire Belorussian Soviet Socialist Republic), 43 percent of the land was used for agricultural purposes, 39 percent of land was forested, and 2 percent of the land contained lakes, rivers, or other bodies of water.

Addressing food and tap water

Food contamination and shipping restrictions

On March 17, Japan's Ministry of Health, Labor, and Welfare (MHLW), noting that the Fukushima nuclear accident had released radioactive material into the environment, issued orders to regional governments that food products in which radioactive contamination had been detected in excess of regulated values were not to be provided to the public.[6] The next day, on March 18, monitoring agencies detected radioactive iodine in spinach in Ibaraki Prefecture. Then, on March 19, Chief Cabinet Secretary Yukio Edano held a press conference and reported that levels of iodine 131—in excess of the provisional restriction limits—was found in milk from Fukushima Prefecture and in spinach from Ibaraki Prefecture.

The government first imposed restrictions on food shipments on March 21, affecting milk from Fukushima Prefecture, as well as spinach and *kakina* (a green Japanese vegetable) from Fukushima, Ibaraki, Tochigi, and Gunma

---

[6] These orders were based on the Food Sanitation Act and took the values specified by the Nuclear Safety Commission's guidelines for restricting food and water consumption as provisional limits for the restrictions in question.

prefectures.[7] By March 23, shipping restrictions expanded to include some other foods produced in Fukushima and Ibaraki prefectures—and, simultaneously, consumption restrictions were also imposed on some foods produced in Fukushima Prefecture. On April 4, the government restricted the shipment of foods produced to specific cities and towns in Chiba Prefecture.[8]

On April 5, following reports that radioactive iodine had been detected in seafood and shellfish, MHLW made a public announcement, setting provisional limits for the first time; thus, regional governments across the country were instructed not to sell or provide to the public any food items that exceeded provisional limits.

Once the restrictions were lifted from all of the prefectures and after radiation measurements fell within provisional limits for three consecutive weeks, the government stipulated that food products would then be tested weekly and shipping restrictions on food items would be lifted for regions. On April 8, the government lifted shipping restrictions for milk produced in some regions of Fukushima Prefecture and for spinach and *kakina* produced in all regions of Gunma Prefecture; still, more bans began to follow.

But these restrictions were not so short-lived in other areas: Even two years after the disaster, products produced in some regions—including game and fish liver, edible wild plants, and Japanese mushrooms—continued to exceed radioactivity limits, and shipping restrictions remained in effect for these products.

Water contamination and drinking restrictions

On March 15, the MHLW—noting "the necessity of imposing restrictions on the consumption of water, including drinking water"—announced to all regional governments that the Nuclear Emergency Response Headquarters would make all decisions on the matter.[9] The ministry noted that the guidelines for restricting food and water consumption would serve as input for the headquarters' decisions. On March 18, MEXT requested that all regional governments test the tap water in their localities and report the results to MHLW, so that it could assess the progress in various regions. MEXT then advised local governments that tap water exceeded the guidelines. MEXT also recommended that tap water consumption should be avoided, but that it could be used safely for other purposes, like bathing or washing clothing; the ministry added that the water could be consumed if no other alternative water source was available.

---

[7] In accordance with Article 20, Item 3, of the Act on Special Measures Concerning Nuclear Emergency Preparedness (the Nuclear Emergency Act), Prime Minister Kan, the chairman of the Nuclear Emergency Response Headquarters, communicated these restrictions to the governors of the affected prefectures.

[8] At the request of regional governments—under the condition that radiation tests are performed every week, and radioactivity measurements for the food products in question are found to fall within provisional limits for three consecutive weeks—restrictions were lifted.

[9] This was in accordance with the Nuclear Emergency Act.

Local water bureaus in city and regional governments were left to announce actual consumption restrictions. In Tokyo, for example, water restrictions were imposed between March 23 and 24, and in Iitatemura, a town in Fukushima Prefecture, water restrictions were in place between March 21 and April 1 (and until May 9 for nursing infants).

Provisional restrictions

At the onset of the Fukushima accident, the quantitative values for restrictions on food shipments and water were based on "provisional restrictions," the legal basis of which is the Food Sanitation Act. Written in 1947, this legislation set out to "protect the health of the Japanese people" by "establishing restrictions and other measures necessary to ensure the safety of food products from the standpoint of public sanitation." However, prior to the Fukushima accident, the legislation did not include guidelines for establishing the decision-making criteria needed to enact regulations and other measures relevant to the sorts of radioactive contamination problems that Japan instantly faced in March 2011. Thus, within days, Japan's Nuclear Safety Commission created guidelines for restricting food and water consumption; used as decision-making criteria, these guidelines counted among the commission's disaster-prevention policies. Though they provided benchmark values for safe ingestion levels in emergency situations, these guidelines were not intended to be used as concentration thresholds for determining whether radioactive materials contained in food products could have harmful health effects. In lockstep with the International Commission on Radiological Protection (ICRP), the authority on setting guidelines for preventative measures, as well as the International Atomic Energy Agency (IAEA) and other bodies, the commission found the safe effective radiation exposure to be 5 millisieverts per year (50 millisieverts per year for the equivalent dose of radioactive iodine to the thyroid); further, this number also took into account the realities of Japanese diet and lifestyle.

In practice, decisions based on these guidelines inspired the shipping restrictions; however, because these guidelines were not officially included within the Food Sanitation Act, they were referred to as provisional restrictions under the Act. On April 1, 2012, new ministry ordinances, enacted by the Minister of Health, Labor, and Welfare, went into force; an amendment to the Food Sanitation Act allowed the ministry to set regulations on the maximum annual permissible radioactive materials found in food—a dose of 1 millisievert.

Trade-offs between protections for producers and consumers

In some cases, restrictions placed on food products and water consumption can encourage trade-offs between producers protecting their profits and producers protecting the public by minimizing health risks. Fukushima brought all of these cases to the fore.

Lettuce ships, despite moratorium

This first case illustrates the need to balance two distinct types of citizen protection: the need to minimize the health risks to consumers on the one hand, and the need to protect the profits of producers on the other hand. The case also describes that multiple levels of administrative bodies—civic, prefectural, and national—worked at cross-purposes.

On March 20, Tokyo announced that iodine 131 levels in excess of provisional limits had been detected in a March 18 harvest of *shungiku* (garland chrysanthemum) grown in Asahi City in Chiba Prefecture. Upon receiving this announcement, the Chiba-Midori Agricultural Cooperative of Asahi City instituted a self-imposed moratorium on shipments of agricultural products including *shungiku*. On March 25, Chiba Prefecture announced the results of radioactivity tests on fourteen agricultural products grown in Asahi City and harvested on March 22; the results revealed levels of iodine 131 in excess of provisional restriction limits in five of the fourteen products, *sanchu*, a type of lettuce, among them. The following day, on March 26, Asahi City announced test results on twenty-seven agricultural products harvested on March 21; once again, levels of iodine 131 in excess of provisional restrictions had been detected in eleven of the twenty-seven products, again including *sanchu*. On March 25, Chiba Prefecture announced that its radioactivity tests on the March 22 *sanchu* harvest revealed iodine 131 levels of 2,800 becquerel per kilogram, whereas Asahi City announced a day later, on March 26, that its tests on the March 21 *sanchu* harvest revealed levels of 4,800 becquerel per kilogram. The provisional restrictions specify the limit is 2,000 becquerel per kilogram.

The five products that Chiba Prefecture detected to have radioactivity levels in excess of the provisional restrictions were already among the products covered by the shipping moratorium imposed by Asahi City's Chiba-Midori Agricultural Cooperative on March 20. Nonetheless, on March 29, the Chiba Prefecture government notified Asahi City that "we request that you reduce shipments for some time, until safety can be reliably confirmed."[10] Later that day, Asahi City announced test results on ten agricultural products grown in that city and harvested on March 25; these tests revealed that six products, including *sanchu*, had radioactive levels that fell *within* the provisional restrictions.[11]

On April 4, the national government notified Chiba Prefecture that it was imposing shipping restrictions[12] on spinach grown in Katori and Tako in Chiba

---

[10] This may be taken as evidence of a slight sluggishness of response on the part of Chiba Prefecture. However, in view of the principles underlying the conditions for lifting the shipping restrictions imposed by the national government on April 4, the request for a moratorium on March 29 seems appropriate, and, moreover, inasmuch as there would have been no legally binding authority under which the government could have issued instructions, the sluggishness should not be considered a significant failure of government.

[11] *Sanchu* had iodine 131 levels of 1,700 becquerel per kilogram.

[12] This was based on Article 20, Item 3 of the Nuclear Emergency Act.

Prefecture, as well as on six agricultural products (including *sanchu*) produced in Asahi City; these shipping restrictions were lifted on April 22.

In mid-April, a large Japanese supermarket chain announced that, between March 30 and April 7, it had sold 2,200 packages of Asahi City-grown *sanchu* at fifty-seven of its stores in the Kanto area. Stating that the *sanchu* was within provisional restriction limits, the chain clarified that no new shipments had been made since shipping was restricted on April 4; thus, the chain later clarified, the *sanchu* that had been sold was actually shipped between March 29 and April 4. Despite this, the distribution company in Asahi City continued shipping the product even after the moratorium was announced mid-transit. "I decided on my own that there was no problem," said the distribution company's president.

On April 22, Chiba Prefecture requested that shipping restrictions be lifted, and they were lifted that same day. The basis for this decision was the fact that, since April 4, the radioactivity level had fallen within provisional limits during three rounds of tests. The limits imposed by the provisional restrictions are designed to ensure an effective annual radiation dose of 5 millisieverts per year, assuming that food products are consumed continuously throughout the year. Consuming foods in excess of those limits for a short period of time poses minimal health risk.

For the purposes of our investigation, we analyzed the behavior of the primary players: the large supermarket chain that acted as both the retailer and the distribution company, the governments of Asahi City and Chiba Prefecture, and the national government.

Supermarket chain and distribution company

Since the supermarket shipped the problematic *sanchu* before the government's legally binding shipping restrictions went into effect on April 4, it did not violate any law. Nonetheless, the supermarket chain blatantly disrespected the requests of Asahi City and Chiba Prefecture for a self-imposed shipping moratorium. The radiation tests conducted during this time period, however, found levels within the limits specified by the provisional restrictions.

City governments

Asahi City received Tokyo's test results—which revealed radioactive levels in excess of provisional restrictions—on March 20 and rapidly moved to request a shipping moratorium. (The Tokyo Metropolitan Government only tested the *shungiku*, but the moratorium extended to other agricultural products as well.) Tokyo's announcement was intended to prevent highly contaminated agricultural products from proceeding to market. In contrast, Chiba Prefecture, despite receiving test results on March 25 that revealed levels in excess of the provisional restriction limits, did not request a shipping moratorium until

March 29—the same day that Asahi City announced that its test results had fallen *within* the limits of the provisional restrictions. On the one hand, this may be taken as evidence of a slight sluggishness of response on the part of Chiba Prefecture. On the other hand, in view of the principles underlying the conditions for lifting the shipping restrictions imposed by the national government on April 4, the moratorium request on March 29 seems appropriate.

National government

Shipping was not restricted until April 4, more than two weeks after Tokyo announced its test results. Though concentrations of radioactive elements in vegetables continued to fall after March 29—when levels fell within the limits of the provisional restrictions—the government's imposition of shipping restrictions seems tardy. If, for example, the government had imposed shipping restrictions on March 20, the day that Asahi City decided on a self-imposed moratorium, then, based on the government's conditions for lifting shipping restrictions, those restrictions would have been lifted on April 12 or 15. Moreover, during the period in which radioactivity concentrations were highest, the government should have been able to restrict shipments in a legally binding fashion.

The national government was too slow to establish and announce the conditions under which shipping restrictions could be lifted. Consequently, distributors resumed shipments after Asahi City tested once and indicated that levels were within the limits of provisional restrictions; this did not comport with the government's rules, which required three consecutive rounds of successful testing. We cannot exclude the possibility that the self-imposed moratorium would have continued had the national government announced its conditions earlier.

Beef proceeds to market, despite high radioactivity levels

On July 8, Tokyo reported that, out of eleven head of cattle that had been transported from an emergency evacuation zone in Minamisoma City in Fukushima Prefecture to the Shibaura Meat Market, the meat of a single cow had been detected to contain levels of radioactive cesium (cesium 134 and 137) in excess of the provisional limits of the Food Sanitation Law. (The level detected was 2,300 becquerel per kilogram, while the provisional restriction limit was 500 becquerel per kilogram.) The Health Ministry commissioned this test on July 6, and, when the results were announced two days later, on July 8, the ministry requested that Fukushima Prefecture and six neighboring prefectures strengthen efforts to monitor and inspect beef. The following day, the ministry announced the test results on the remaining ten head of cattle; radioactivity levels for all ten cattle exceeded the limits of provisional restrictions, ranging from 1,530 to 3,200 becquerel per kilogram.

On July 11, Fukushima Prefecture announced plans to conduct urgent on-site inspections of all farms located within the planned evacuation zones and emergency evacuation preparedness zones in which food cattle were raised. In the same announcement, the prefecture announced the results of radioactivity tests, conducted the day prior, on livestock farms in Minamisoma City. Inspecting five types of feed—rice straw, oats, grass, mixed feed, and livestock drinking water—the test detected radioactive cesium in the first three, with the highest radioactivity concentrations in rice straw, at 75,000 becquerel per kilogram (or 17,045 becquerel per kilogram after correcting for water content).

On July 14, it became clear that a farm in the town of Asakawa in Fukushima Prefecture had shipped to market some beef from cattle that had been fed rice straw containing a high concentration of radioactive cesium. This prompted the prefectural government to ask all farmers within the prefecture to observe a voluntary moratorium on shipments until the urgent on-site inspections could be completed on July 18. That day, the results revealed that, between March 28 and July 13, beef from a total of 554 cattle—all of which had been fed contaminated rice straw—had been shipped to market. Upon receiving this news, the national government, on July 19, imposed restrictions on beef cattle from Fukushima Prefecture, prohibiting the transport of these cattle outside the prefecture and their shipment to meat markets. Subsequently, on-site investigations of beef-cattle farms in all regions across the prefecture were conducted, with results announced on August 6; during the period between March 28 and July 15, out of 143 farms, thirty farms shipped a total of 867 cattle that had been fed—or likely had been fed—rice straw contaminated with radioactive materials.

The causes of this incident were, first, that rice straw contaminated with radioactive material was used for cattle feed, and, second, that cattle that had consumed contaminated rice straw were not detected by inspections.

On March 19, the Ministry of Agriculture, Forestry, and Fisheries (MAFF), which oversees the use of rice straw and other feed, issued clear instructions to livestock farmers: They were to use feed harvested before the accident or feed that had been stored indoors, and they were to protect feed and drinking water to prevent the collection of radioactive material. In addition, on April 14, MAFF announced provisional tolerances for the content of radioactive material in coarse-grain feed as a temporary measure to ensure that milk and beef would not exceed the limits of the provisional restrictions. Although MAFF took measures to address the possibility that radioactive material might collect on stored feed, the ministry did not consider the possibility that grass might be harvested for use as feed.

For the shipment or transport of cattle from planned evacuation zones and emergency evacuation preparedness zones, the Livestock Hygiene Service Center used Geiger counters to inspect cattle, with any test subject measuring more than 100,000 counts per minute (cpm) marked for decontamination. (Out of 11,140 head of cattle screened between April 23 and July 11, some 85 percent exhibited readings below 1,000 cpm; the maximum reading was 16,000 cpm, and no cases required decontamination.) However, these tests are only capable of measuring surface radiation, and do not detect internal accumulations of radioactive material. Thus, screening alone is unable to prevent the shipment of contaminated beef cattle; however, if adequate inspections had been performed at meat markets, it seems likely that beef exceeding the limits of provisional restrictions would never have proceeded to market.

Consumption of tap water, in excess of restriction limits

On March 23, the Tokyo Metropolitan Government Water Bureau announced that it detected iodine 131 at the Kanamachi Water Purification Plant the morning before. The measurements—210 becquerel per kilogram—far exceeded that which is safe for infants to consume, 100 becquerel per kilogram; that limit is 300 becquerel per kilogram for adults. The bureau instructed that infant consumption of tap water should be reduced throughout Tokyo's twenty-three districts and in parts of the Tama area. The following day, on March 24, measured concentrations fell to 79 becquerel per kilogram, within provisional restriction limits, and the bureau announced that tap water was safe to drink, including for infants.

During this time, mineral water became scarce in many regions, and many citizens were faced with the difficult decision of whether to consume water containing radioactive material or to reduce consumption of water overall, while others faced the decision of whether to continue breastfeeding, in view of the risk of passing radioactive material from mother to child. On March 24 and 25, the Tokyo Metropolitan Government distributed 240,000 plastic bottles of water to twenty-three flagged districts, particularly to households with infants.

Both government and academic institutions offered expert advice and opinions concerning the consumption of tap water by infants and pregnant women. On March 21, the Health Ministry announced that though tap water outside the provisional restriction limits should be reduced, such water could be consumed in situations when no alternative water could be secured. Later, on March 24, three academic societies[13] offered joint advice based on a comparison of radiation dangers, feeding hard water to infants, and reducing water consumption altogether. The same day, the Japan Society of Obstetrics and Gynecology weighed in on the dangers of reduced water consumption for pregnant women.

---

[13] The Japan Pediatric Society, the Japan Society of Perinatal and Neonatal Medicine, and the Japan Society for Premature and Newborn Medicine.

After March 25, the radioactivity concentrations measured at water treatment facilities, including Kanamachi, began to decrease gradually; by the beginning of April, levels at all facilities had fallen below detection thresholds.[14]

But this narrative was not unique to Tokyo: In Iitatemura in Fukushima Prefecture, the local government announced on March 21 that tap water contained iodine 131 at a level of 965 becquerel per kilogram[15] and that tap-water consumption must be reduced. Thereafter, measured levels decreased, and test results announced on April 2[16] indicated levels below 100 becquerel per kilogram at three water treatment facilities—Hanatsuka, Takishita, and Tajiri. Iitatemura lifted its water-consumption restrictions for adults and children on April 1 and for infants on May 9.

In the case of tap water, neither the national government nor regional governments have the authority to impose forceful restrictions on water supply or consumption. Instead, each individual must independently evaluate the trade-off between the risk of consuming water containing tiny quantities of radioactive material and the risk of foregoing water consumption entirely. In this case, the government's responsibility is to provide appropriate information and thoroughly communicate risks so that the public can make informed decisions.

Further study is needed to answer several questions arising from the restriction cases of Iitatemura (where infants were restricted from tap water for six weeks) and Tokyo (where the same restriction lasted for ten days). That is, did the government need to implement measures—such as distributing safe drinking water (especially to households with infants)—in addition to communicating risks, as described above? Was the length of time that elapsed before the consumption restrictions were lifted appropriate? In view of the various risks posed by extended restrictions on water consumption, it seems possible that the period during which consumption restrictions remained in effect should have been shortened.

Protecting children

On April 19, MEXT, after receiving advice and guidance from the ICRP, turned its attention to reducing radiation exposure in school buildings and schoolyards in Fukushima Prefecture. The ministry announced that any schools at which air radiation levels in schoolyards or kindergarten yards exceeded 3.8 microsieverts per hour were to be used no more than one hour per day. MEXT periodically monitors air radiation levels in schoolyards and other locations and found that, on April 14, for example, thirteen out of fifty-two nursery and elementary schools in Fukushima City were subject to these restrictions.

---

[14] Detection thresholds vary from site to site, but are generally on the order of less than 10 becquerel per kilogram.
[15] Water samples were taken on March 20.
[16] Water samples were taken on March 28.

Eventually in August 2011, MEXT set the base radiation dose for children at 1 millisievert or less per year; it further established a benchmark of under 1 microsievert per hour for air radiation levels in schoolyards and kindergarten yards.[17]

The Japanese government enacted a special-measures law, which it announced to the public on August 30,[18] that authorizes the environment minister to establish criteria for the treatment of contaminated waste and soil and to conduct monitoring and measurement activities. The cost is to be borne by the relevant nuclear power companies, in accordance with the Act on Compensation for Nuclear Damage.

According to the Ministry of the Environment, this legislation calls for cities and towns designated as "contamination focus and testing zones" to be subject to decontamination efforts designed to bring annual levels of additional radiation exposure to within 1 millisievert. As a specific roadmap for decontamination work, model operations were implemented in November 2011, and actual decontamination work started in January 2012. Material will be stored at temporary storage sites for around three years. In special decontamination zones—warning zones and planned evacuation zones—the Ministry of the Environment, with help from cities and towns, secures the storage; in other areas, storage is secured by cities and towns with financial and technical responsibility borne by the national government. Thereafter, material will be transported to intermediate-term storage facilities.

The cost of decontamination efforts

The Ministry of the Environment has estimated the volume of removed soil, and the volume of waste, arising from decontamination efforts. Low estimates suggest volumes of approximately 15 million cubic meters in Fukushima Prefecture and approximately 1.4 million cubic meters in all other regions, while high estimates suggest volumes of approximately 28 million cubic meters in Fukushima Prefecture and approximately 13 million cubic meters in all other regions.

Two years after the disaster, officials still had not made public the total cost of decontamination efforts associated with the accident. In Iitatemura, designated as a planned evacuation zone, local officials[19] estimate the decontamination costs to come in around 324 billion yen ($4.2 billion).

---

[17] These calculations were based on the International Commission on Radiological Protection's recommendation of 1–20 millisieverts per year as a reference level in the aftermath of a state of emergency.

[18] The law is titled: Special Measures Law Concerning Treatment of Environmental Contamination due to Radioactive Material Released by the Nuclear Plant Accident Following the Earthquake in the Pacific Ocean off the Coast of the Tohoku Region.

[19] Iitatemura, Fukushima (2011) The decontamination plan of Iitatemura. Available at: www.vill. iitate.fukushima.jp/saigai/wp-content/uploads/2011/10/b2eb22467554edc1286c0f22672344be (accessed September 4, 2013). (In Japanese).

Because the costs of decontamination remain uncertain, it will be "a significant length of time" before specific estimates of damage costs can be made.[20,21] In addition, according to the Dispute Reconciliation Committee for Nuclear Damage Compensation, any costs in excess of the value of the affected assets are considered to lie beyond the scope of damage compensation, with the exception of some cultural assets and other items.[22]

The cost burden of decontamination efforts is a significant challenge for the national government. On December 9, 2011, the Cabinet Office issued its first announcement on these costs in Fukushima Prefecture and announced that the national government would assist in bearing the costs of decontaminating any region in which the additional annual radiation dose was 1 millisievert or greater due to the accident. For example, decontaminating a stand-alone house could cost around 700,000 yen ($9,000).

Decontamination work by citizens

On July 29, 2011, the Japan Society of Radiation Safety Management publicly released a manual containing instructions on finding and decontaminating hot spots. In addition, on November 22, a Cabinet team published safety tips and effective strategies for decontaminating households, roads, schools, parks, farmlands, and other locations.[23] The Ministry of the Environment also spearheaded volunteer-based decontamination initiatives.

Even as decontamination work proceeds, measurements taken by independent citizens have revealed high-radiation locations even outside Fukushima Prefecture. For example, in October 2011, residents of Adachi, a Tokyo ward, found a hotspot at an elementary school: Radiation levels clocked in at 3.99 microsieverts per hour, and the hotspot was cordoned off.

---

[20] Tokyo Electric Power Company, Inc. (2011) *Report of the TEPCO Management and Finance Investigation Committee*. Report, Management and Finance Investigation Committee. Tokyo: TEPCO. Available at: www.cas.go.jp/jp/seisaku/keieizaimutyousa/dai10/siryou1.pdf (accessed August 30, 2013). (In Japanese.)

[21] The Cabinet's Council established this committee on May 24, 2011, under the framework of the Act on Compensation for Nuclear Damage.

[22] This statement is based on the "Interim guidelines for determining the scale of nuclear damage compensation and other matters related to the accidents at TEPCO's Fukushima Daiichi Nuclear Power Station and Fukushima Daini Nuclear Power Station," which TEPCO released on August 5, 2011.

[23] Japan Atomic Energy Agency (2012) *Report of the results of the decontamination model projects—decontamination technologies*. Report, Cabinet Office's Team in Charge of Assisting the Lives of Disaster Victims, March 26. Available at: www.jaea.go.jp/fukushima/decon04/english/2-2-2%20 Decontamination%20Technologies.pdf

Japanese policies and legal framework on waste treatment

Procedures for treating radioactive waste had traditionally been codified by laws and regulations such as the Nuclear Regulation Act, but these referred exclusively to the treatment and disposal of waste arising *within* nuclear facilities, while there was no existing legal framework that could be applied to waste arising *outside* those facilities.

On August 26, 2011, Japan's Diet enacted a special-measures law that enabled the treatment and disposal of radioactive material contaminants. This legislation directed the Ministry of the Environment to formulate "basic policy regarding the treatment and disposal of environmental contamination caused by radioactive material released by the accident." Enacted by the Cabinet on November 11, 2011, the policy provided guidance on monitoring and measuring environmental contamination, treating contaminated waste, and decontaminating soil and other areas. Special priority was given to waste treatment, and the policy divided contaminated waste into three categories: waste contained within contaminated waste treatment zones,[24] designated waste,[25] and waste contaminated by radioactive material[26] released by the accident.

Treatment and disposal of incinerator ash

Incinerator ash at waste treatment plants contains high concentrations of radioactive materials, raising health concerns for citizens in nearby residential areas. The Japanese government monitors and measures the radioactivity of this ash: Ash with levels of 8,000 becquerel per kilogram or less is disposed of in the same way as ordinary waste products; ash with levels between 8,000 and 100,000 becquerel per kilogram is buried in ways that ensure no contamination of groundwater or public bodies of water; ash with levels in excess of 100,000 becquerel per kilogram is stored in shielded facilities.

Acceptance of waste

The destruction of buildings and other structures, primarily due to the tsunami after the Tohoku earthquake, generated massive quantities of waste and debris. The treatment and disposal of this waste was one major problem; in some instances, the possibility of radioactive contamination made it difficult to find places willing to accept waste products. For example, the Tokyo municipal government was designated to receive 500,000 tons of disaster waste over three

---

[24] The Ministry of the Environment will handle this type of waste.
[25] Various agencies will handle this, depending on the type of waste.
[26] This waste is to be handled in accordance with specified measures for monitoring, measurement, and dispersion control.

years. This, however, provoked public ire: According to press accounts, between the end of September and the beginning of November 2011, the city government received a total of 3,328 letters, calls, and e-mails, of which 2,874 were expressions of opposition or complaints. When, in August 2011, organizers of Koyoto's annual *Gozan no Okuribi* bonfire festival[27] planned to burn pine trees, the city of Kyoto, and the cultural heritage organizations that organize the event, received numerous complaints expressing concerns about radioactive-material contamination; indeed, the organizers had planned to use timber grown in the Takada Matsubara pine forest in the city of Rikuzentakata in Iwate Prefecture, an area heavily damaged by the recent natural disasters. Ultimately, the bonfire was canceled, despite the fact that tests showed that the firewood was not contaminated.

Low-level radiation exposure

Radiation—the invisible, odorless substance—must surely rank among the major causes of societal anxiety in the aftermath of Fukushima. This anxiety remains unassuaged to this day; as members of the general public continue to gather their own information, the increasing quantities of information released by the relevant authorities seems only to worsen the situation. One topic that has remained a particular focus of interest is low-level radiation exposure: not the large quantities of radiation to which the victims of the Hiroshima and Nagasaki bombings were subject, but the small quantities of radiation to which the general public is continually exposed.[28]

The impact of radiation on the human body[29]

Exposure to radioactivity—specifically, the absorption of radiation—has two major types of effects on living organisms. The first type of effect is deterministic and predictable; these effects are associated with certain radiation thresholds, with symptoms appearing once radiation exposure exceeds those thresholds. The second type of effect is probabilistic; in this case, symptoms may or may not arise, with some probability that depends on the radiation exposure. Because these effects may only be demonstrated in a probabilistic sense, it is not

---

[27] The bonfire is a ceremonial event to usher out the spirits of ancestors.
[28] Radiation limits for individuals involved in radiation-related work activities are capped at 50 millisieverts in any single year and at 100 millisieverts in any five-year period; consequently, the average exposure over five years is 20 millisieverts per year, corresponding to low-level exposure.
[29] See, for instance, the following textbooks (in Japanese):
Grodzinsky, D. (translated by Sato, M.) (1966) *Houshasen Seibutsugaku Nyuumon (Introduction to Radiobiology)*. Tokyo: Kagaku fukyu shinsho, Tokyo Tosho.
Kondo, S. (1990) *Genshiryoku no Anzensei (Nuclear Power Safety)*. Tokyo: Doubun Shoin.
Murray, R. (1965) *Genshikaku Kougaku Nyuumon (Introduction to Nuclear Engineering)*. Tokyo: Pearson Education.

possible to specify a certain radiation level below which symptoms are guaranteed not to appear. This raises the question of whether it is possible to assign a "threshold value"—that is, a certain level above which the effects of radiation begin to be observed—for low-level radiation exposure; by international agreement, there is no such threshold.

There are bodily effects, which are those experienced by a person exposed to radiation but not passed on to future generations; these include both immediate symptoms, which appear immediately after exposure, and delayed symptoms, which may not appear for years. In contrast, genetic effects refer to consequences passed on to future generations. In general, immediate symptoms are essentially deterministic, while delayed symptoms may be either deterministic or probabilistic. In addition, there are two types of exposure: external exposure, which occurs when radiation is absorbed from outside the body, and internal exposure, which results from the ingestion of radioactive substances within the body.

What is low-level exposure?

Whereas high-level exposure has deterministic effects that are relatively well understood, there is not even a clear international consensus on precisely what levels define "low-level radiation exposure." The Committee on the Biological Effects of Ionizing Radiations (BEIR) within the US National Academy of Sciences classifies anything under 100 millisieverts as low-level exposure,[30] while others suggest defining low-level exposure to be anything in the range of ten to 100 times the background radiation levels (or approximately 100 millisieverts).[31]

Present understanding has not reached consensus on the question of whether low-level exposure has definite effects on the human body. "Although the scientific community is doing what it can to estimate these risks, the reality is that we really don't know," wrote distinguished physician David Brenner in the April 2011 issue of *Nature*.[32] Referring to research conducted in 2003, Brenner continued, "More specifically, the uncertainties associated with our best estimates of the health effects of low-doses of radiation are large."

The greatest difficulty in assessing the health risks of radiation is their non-specificity; the symptoms of radiation damage are not exclusively associated with radiation, but may also be caused by other factors. For example, in the case of patients who contract cancer, there is no way to specify whether the cancer was caused by radiation exposure or by lifestyle habits.

---

[30] US National Research Council (2006) *Health Risks from Exposure to Low Levels of Ionizing Radiation: BEIR VII—Phase 2*. Washington, DC: The National Academies Press.
[31] See, for example, Grodzinsky, D. (translated by Sato, M.) (1966) *Houshasen Seibutsugaku Nyuumon (Introduction to Radiobiology)*. Tokyo: Kagaku fukyu shinsho, Tokyo Tosho.
[32] Brenner, D. (2011) We don't know enough about low-dose radiation risk. *Nature*, April 5. DOI: 10.1038/news.2011.206. Available at: www.nature.com/news/2011/110405/full/news.2011.206.html (accessed June 21, 2013).

The International Commission on Radiological Protection (ICRP) has taken a stance against defining threshold values for low-level radiation exposure.[33] The reason for this is that the paucity of data on the effects of low-level radiation on the human body requires that such effects be estimated by linear extrapolation from high-level radiation data. In other words, the effects of high-level radiation are extrapolated to the low-level exposure regime by drawing straight lines to the point of zero exposure, the linear no-threshold model. Aside from this, BEIR and the United Nations Scientific Committee on the Effects of Atomic Radiation (UNSCEAR) have issued reports, but both reports rely on the linear no-threshold model.

At the same time, there is no absence of countervailing opinions and research. If low-level exposure has no effect on the body, then health risks for low-radiation zones may be overstated, and some point out the need to consider biologically self-restoring system effects. Moreover, there is even some research to support the possibility that low-level exposure can have positive health effects (known as hormesis effects),[34] although the BEIR VII report essentially ruled out this possibility.

In November 2011, a committee to investigate prolonged low-level radiation exposure was convened by the national government's Cabinet Secretariat; their report was completed in December of that same year.[35] Their conclusion was that, in order to err on the side of caution, all radiation exposure should be assumed to have a direct connection to increased health risks, even in cases of low-level exposure at levels of 100 millisieverts or below.

The BEIR report demonstrated that the possibility of a threshold for low-level radiation exposure cannot be rigorously excluded. Notwithstanding this fact, the report takes the linear no-threshold model to be the most appropriate risk-assessment model.[36] In addition, the US Health Physics Society notes that, for the purposes of quantitative risk assessment, the health risks of low-level

---

[33] In 1980, Japan's Nuclear Safety Commission adopted a set of disaster-prevention policies, which, similar to those prepared by many governments around the world, are based on the conventional wisdom of the ICRP. Japan's policies have been revised many times, and terms have been adopted accordingly.

[34] For example, Japan's Radiation Safety Research Center, after studying the effects of low-level radiation exposure, claims to have "clearly demonstrated the existence of hormesis effects from radiation." (Started in 2001 as the "Low-Level Radiation Research Center," it has since been renamed the "Radiation Safety Research Center.") The center now states that, "[a]t present, this center adopts a neutral standpoint." See: Radiation Safety Research Center (2013) Aiming at further understanding of the biological effects of low-dose radiation. Central Research Institute of Electric Power Industry, Nuclear Technology Research Laboratory, Radiation Safety Research Center. Available at: http://criepi.denken.or.jp/jp/ldrc/index.html (accessed August 30, 2013).

[35] Cabinet Office, Government of Japan (2011) *Report of the working group on risk management of low-dose radiation exposure*. Report, December 22. Available at: www.cas.go.jp/jp/genpatsujiko/info/twg/111222a.pdf (accessed June 21, 2013). (In Japanese.)

[36] Lamarsh, J. (2003) *Introduction to Nuclear Engineering Vol. 2*. Tokyo: Pearson Education.

exposure are either un-observably small or nonexistent, and suggests threshold values of 50 millisieverts per year, or 100 millisieverts over an entire lifespan.[37]

The European Committee on Radiation Risk (ECRR), however, investigated the limits of the ICRP's exposure model and concluded that the commission underestimates the effects of low-level radiation exposure; the ECRR recommends that exposure limits for the general population be set lower than 0.1 millisieverts per year.[38]

The nuclear accident at Chernobyl may be thought of as providing an ongoing research experiment into the effects of prolonged low-level exposure similar to that resulting from the accident at the Fukushima nuclear station.[39] A 100-millisievert total dose of radiation may have different effects depending on whether the full dose is received in a short period of time or accumulated over time via prolonged low-level exposure; further comparison of these different dose profiles is needed.[40]

Societal anxiety exacerbated by government policy

Within a relatively short time of the accident, the government released comparisons stating that the radiation exposure resulting from the accident was on the same order of magnitude as the dose received from medical x-rays, CAT scans, or from international travel on an airplane. During the Fukushima crisis, MEXT frequently used comparisons that failed to take into account the distinctions between radiation exposure resulting from individual decisions and exposure resulting from the accident; this fact, together with the insufficiency of information provided on radiation levels, most likely had the effect of exacerbating a sense of anxiety among the public.

In addition, at a press conference at the time of the accident, then-Chief Cabinet Secretary Yukio Edano said, "There are no immediate risks to health."

---

[37] Health Physics Society (1996) *Radiation risk in perspective*. Report, a position statement of the Health Physics Society XXIV(3). Available at: www.hps.org/documents/radiationrisk.pdf (accessed June 21, 2013).

[38] European Committee on Radiation Risk (2010) *2010 recommendations of the European Committee on Radiation Risk: The health effects of exposure to low doses of ionizing radiation*. Report, regulators' edition. Brussels: ECRR. Available at: www.euradcom.org/2011/ecrr2010.pdf (accessed June 21, 2013).

[39] See, for example: Alexey, Y., Vassily, N., and Alexey, N. (2009) *Chernobyl: Consequences of The Catastrophe for People and the Environment*. (Annals of the New York Academy of Science Vol. 1181, December.) Boston: Blackwell Publishing, the New York Academy of Sciences.

[40] According to a report from the Cabinet Secretariat, epidemiological studies have found that a 500-millisievert dose does not increase cancer risks among residents of the Kerala region of India—a region in which natural radiation levels are high—but that the same dose *did* increase cancer risks among people exposed to radiation from an accident at a nuclear weapons facility in the Southern Ural region of the former Soviet Union. See: Cabinet Office, Government of Japan (2011) *Report of the working group on risk management of low-dose radiation exposure*. Report, December 22. Available at: www.cas.go.jp/jp/genpatsujiko/info/twg/111222a.pdf (accessed June 21, 2013). (In Japanese.)

This created general suspicions, at the time, that there would be *eventual* risks to health.[41]

Subsequently, MEXT stated:

> With regard to the decrease in radiation exposure levels after the accident, we refer to the ICRP's reference levels of 1–20 millisieverts per year, and in the long run we hope to return to ordinary annual exposure limits of under 1 millisievert per year for the general public.[42]

However, because this statement came in July, long after the accident, and because it failed to provide adequate explanation, it was widely criticized and merely increased public anxiety. The April 29 resignation of Cabinet adviser Toshiso Kosako, and his criticism of the government's decision to allow exposure levels of 20 millisieverts per year in schools, was yet another factor that stoked unease among the population.

Within Japan, organizations such as the ICRP and UNSCEAR are frequently cited as "institutions with international authority." However, the choice of the Japanese government and TEPCO to cite the opinions of these institutions in response to the Fukushima nuclear accident carries a certain undeniable taint of authoritarianism. In reality, there are critical assessments of both ICRP and BEIR's findings. For example, the ICRP report suggests annual radiation exposure limits of 50 millisieverts for workers and 1 millisievert for the general public, but in 1934, the former limit was 500 millisieverts, while the latter limit had not yet been established; it was not until 1958 that annual limits of 5 millisieverts were established.[43] Thus, the fact that these assessments are based on a body of present knowledge that has been established through extensive debate and discussion is not properly acknowledged. In addition, some experts feel that methodological problems (for example, situations in which analytical procedures are not properly disclosed) have contributed to a tendency to underestimate dangers.[44]

---

[41] According to a press conference that began at 4 p.m. on March 25, 2011, the statement that "there are no effects on the human body," was subsequently clarified to state, "From the various situations that existed at that time, at the present time there are no effects." For this reason, it is difficult to believe that the possibility of low-level exposure or genetic damage was taken into consideration. See, for example: MSN Economics News (2011) Press conference with Chief Cabinet Secretary Edano regarding radiation leakage: "I never said everything was fine." March 25. Available at: http://sankei.jp.msn.com/politics/news/110325/plc11032518580033-n1.htm (accessed February 28, 2012). (In Japanese.)

[42] *Sankei Shimbun* (2011) Statement from the Ministry of Education, Culture, Sports, and Technology: "On the May 27 statement that 'We target a level of under 1 mSv per year for schools,' which reflected 'our thinking at that time.'" July 20.

[43] Nakagawa, Y. (1991) *A History of Radiation Exposure*. Tokyo: Akashi Shoten, p. 185.

[44] Goffman, J. (1991) *Ningen to Hoshasen*. Tokyo: Shakai Shisosha. (Translated to Japanese.) Original text is: Goffman, J. (1981) *Radiation and Human Health*. San Francisco: Sierra Club Books.

In conclusion

Under such ambiguous scientific knowledge, the questions of how the Japanese government and TEPCO have shouldered their responsibilities to the Japanese people on matters such as evacuation orders and personal decisions to evacuate have not been questioned sufficiently. Although investigations into the appropriateness of government decisions are ongoing, to date there has been too little study of the scientific and societal problems posed by non-deterministic probabilistic phenomena. Many questions remain regarding the decisions of various policy makers who were forced to make decisive policy choices in the midst of uncertain risks.

In many foreign countries, discussions and debates tend to include individuals from a variety of different walks of life. For example, the ECRR includes not only experts in public sanitation and epidemiology, but also legislators, social scientists, and other members from outside the natural sciences. Many believe that, in the future, Japan must move to adopt similar methods.

It is also possible that Japan's unique situation of having experienced the bombings of Hiroshima and Nagasaki has contributed to societal anxiety. Domestically, Japan is often said to be characterized by a "nuclear allergy," but in reality the Fukushima response has provoked unrest and dissent even among experts on the medical impact of the atomic bomb.[45] And yet, we must acknowledge the reality that the Japanese possess a mentality where they bathe in radon hot springs for their positive health effects; in places such as the United States, for example, radon is a regulated substance.[46]

How do we manage the impact of radiation exposure on the people of Japan in the medium- to long-term? In Fukushima Prefecture, for example, a public survey was conducted with an eye toward understanding the future health needs of prefecture residents amid the impact of radiation released by the nuclear disaster.[47] According to the survey committee's report released in December,[48] out of 1,589 survey respondents, the accumulated radiation exposure statistics broke down as follows: 998 people (62.8 percent) measured less than 1 millisievert; 1,547 people (97.4 percent) measured less than 5 millisieverts; 1,585 people (99.7 percent) measured less than 10 millisieverts; and four people measured more than 10 millisieverts. The maximum was 14.5 millisieverts, reported by a single respondent. However, these were "estimates

---

[45] *Chugoku Shimbun* (2011) Dissenting voices heard on Fukushima response. July 11. (In Japanese.)
[46] US Environmental Protection Agency (2013) Radon. Available at: www.epa.gov/radiation/radionuclides/radon.html (accessed June 21, 2013).
[47] Fukushima Prefecture (2011) Health management survey for prefecture residents. Available at: www.cms.pref.fukushima.jp/pcp_portal/PortalServlet?DISPLAY_ID=DIRECT&NEXT_DISPLAY_ID=U000004&CONTENTS_ID=24287 (accessed June 21, 2013).
[48] Fukushima Prefecture (2011) *An overview of the basic tests (total exposure to external radiation) and thyroid tests*. Report, Health Management Survey for Prefecture Residents Survey Committee, December 13. Available at: www.pref.fukushima.jp/imu/kenkoukanri/231213gaiyo.pdf (accessed June 21, 2013).

based on activity records submitted on questionnaires." In addition, out of two million residents targeted by the survey, to date, only 370,000 have responded, a response rate of 18 percent. In the future, the national government will need to go beyond this survey to conduct fast, accurate, and long-term specialized surveys.

Chapter 8

Communicating the Fukushima Disaster

During the initial weeks after the earthquake, the shared fear among Cabinet and the Tokyo Electric Power Company (TEPCO) officials centered around what Yukio Edano, then Japan's top government spokesman, called a "devil's chain reaction." That is, they were concerned that, if they lost control of one reactor at the Fukushima Daiichi Nuclear Power Station and a hydrogen explosion occurred, high radiation dose levels would make it impossible for human intervention for the necessary repair; this handicap would lead to an overall loss of control—of all the reactors at the plant, spent fuel pools, and consequent dry heating—and then, ultimately, to a chain of explosions. At that point, it would become impossible to approach the plant, quite possibly necessitating the evacuation of people from Tokyo, nearly 124 miles away.

During this time, the sense of anxiety and dismay among the public in Japan was deepened by the government's vague descriptions of risk and by the confusion that arose from the inconsistencies published in communications by both the government and TEPCO. Further, the public distrust worsened by the government's delays in disclosing information provided daily by SPEEDI (System for Prediction of Environmental Emergency Dose Information) and other sources; the public found that, with each eventual disclosure, the situation was only worse than expressed in a previous statement. In response to international concerns about the spread of radioactive contamination and the evacuation of residents, the government could only transmit information that was even less firm in nature.

From the onset of the disaster, the government struggled with an underlying concern: How would the Kan administration communicate risk to assuage public anxiety? But just as the public did not have experience in these matters, neither did the government—which had never communicated the state of a nuclear reactor or the risk of exposure to low-dose radiation. Performing prompt and accurate information disclosure was, in large part, a matter of trial and error. Further, with respect to communicating such risks to international stakeholders and media, the Prime Minister's Office had long lacked the capacity for extensive public relations. Consequently, it adopted an ad hoc messaging approach to meet a pressing need.

Unlike the times of Chernobyl or Three Mile Island, when reporters still phoned in stories to their newsrooms from location, the Fukushima disaster unfolded during the age of social media—giving the government more tools at its disposal, but also more at society's disposal.

With the flood of information through social media, Japan's public became more confused about the state of the nuclear reactors and their release of radioactive materials. Consequently, the public became more anxious. In newspaper polls conducted in April, one month after the disaster, public opinion showed that nearly 70 percent of the Japanese public considered the government's provision of information and explanations to be inadequate.[1] A series of additional surveys conducted by the Japan Broadcasting Corporation, a public broadcasting corporation that is similar in nature and function to that of Britain's BBC, showed that between 60 and 70 percent of Japan's public disapproved of the government's response in the early stages of the disaster; it became even more critical in the subsequent months, climbing to 75 percent by June, just a few weeks after TEPCO publicly disclosed that a meltdown had indeed begun in the early stage of the accident.[2]

Within the government, Hiroshi Tasaka, special adviser to the Cabinet, continually touched upon psychosocial risk factors and the importance of public disclosure of information on the accident and its aftermath. In his retrospective on the series of government responses, he agreed with the polls—"the government," he said, "must first of all be deeply aware that it has lost the trust of the public in regard to atomic energy administration."[3]

Government crisis communication

In the initial weeks of the crisis, Prime Minister Kan shied away from direct public communication and media interaction. Instead, Chief Cabinet Secretary Edano took on the difficult task of accurately communicating the delicate nuances of health and safety risk to the public, while avoiding inciting public panic.

---

[1] In a public-opinion survey given by the *Asahi Shimbun* newspaper on April 16 and 17, 73 percent of the respondents stated that the government provided inadequate information on the Fukushima Daiichi Nuclear Power Station accident, and only 16 percent believed it to be adequate. In a survey (April 1–3) by the *Yomiuri Shimbun* newspaper, similarly, 66 percent said they did not believe it to be adequate, while 24 percent said it was adequate.
[2] Disapproval versus approval (in a NHK public opinion poll) of government's handling of the Fukushima Daiichi Nuclear Power Station accident: 68 percent versus 28 percent in April; 65 percent versus 31 percent in May; and 75 percent versus 19 percent in June.
[3] Tasaka, H. (2012) *An Inside Look at the Reality of the Nuclear Accident from the Prime Minister's Office.* S.l.: Kobunsha. (In Japanese.)

The prime minister's media interaction

Prime Minister Kan held his first press conference on the same day as the earthquake, but it was merely a brief administrative update that lasted under two and a half minutes. At 8:32 p.m. on March 12, he held another press conference—lasting just over nine minutes. Broadcast live by television and the Internet, the Japanese people—and the world—watched and listened for clarity and substantial information on the explosion. Kan simply read the prepared text without explaining the circumstances surrounding the explosion at unit 1, and he did not respond to reporters' questions.[4]

Immediately following the earthquake, some members of the Cabinet core team held the opinion that the prime minister should hold a press conference at least once every day. Within a matter of days, however, the team members reached consensus that Kan's press conferences should be held less often. As a result, his press conferences rapidly decreased in frequency. He only gave six more press conferences in total: on March 13, 15, 18, and 25 and on April 1 and 12.

Though before the earthquake Kan participated in daily impromptu interviews with the Cabinet *Kisha* Club (Press Club), this all but stopped after the Fukushima accident. Not only did Kan have a personal aversion to mass media, but his administration feared that the prime minister would not be adequately prepared to respond directly to the media. What also became problematic was Kan's own sense of pride in his scientific background. During

---

[4] It should be noted that the texts for the prime minister's press conferences were generally produced by administrative bureaucrats who, unlike the specialized speechwriters for heads of state frequently employed in the United States and Europe, were not specialized enough to suggest or draft speeches for the prime minister. In the wake of the earthquake, two of Prime Minister Kan's secretaries prepared the text for his press conferences: Shiro Yamasaki, formerly at the Ministry of Health, Labor, and Welfare, who served in a supervisory capacity for the secretaries to the prime minister, and Keisuke Sadamori, formerly at the Ministry of Land, Infrastructure, Transport, and Tourism, who was responsible for public relations. They summarized and drafted items relating to administrative announcements, and they produced a first draft with additional comments by Deputy Director General Kenichi Shimomura of the public relations office. Approximately twenty staff members work under the Cabinet public affairs officer and are engaged in the work of the Cabinet's public affairs, but they are seldom involved in preparing texts for the prime minister's press conferences. Implementation of Prime Minister Kan's full and impromptu press conferences inherently placed a heavy workload on the secretaries, in addition to their other duties. Ordinarily, the texts for Kan's press conferences are composed by the prime minister's five secretaries, who are administrative officers for finance, foreign affairs, police, economics and industry, and welfare and labor—each responsible for several of the ten ministries and agencies. They draw the required information for each subject to be covered in the press conference from the related ministries by administrative procedures and prepare the draft speech on that basis. In cases where the area of responsibility is unclear, the secretary for public relations plays the central role in drafting the communications. In preparations for impromptu press conferences, however, the usual procedure is for the Cabinet staff to query reporters assigned to the prime minister and for the Prime Minister's Office to compose the text based on the collected information.

the Cabinet's deliberations on accident response, Kan often inserted his own theories into technical discussions relating to nuclear power plants. The core team of his Cabinet and the prime minister's secretaries were concerned with the risk of misstatement or confusion that might arise if he were prompted to speak on either technical or specialized aspects of nuclear power or on his own personal interpretations. Rather, Chief Cabinet Secretary Edano imparted a sense of stability in describing and explaining such matters.

Edano, himself, admitted the difficulty of communicating such risk: "Frankly, in proceeding with information assessment and communication, it is a major struggle to determine how best to report the information and at what stage to do so. Particularly in regard to the present matter, I believe both that it is essential to present accurate information incisively and appropriately while avoiding the presentation of unreliable information. I will therefore only present prompt, sound, and reliable information."[5]

The shuffling of government, and the shuffling of public trust

The Japanese public came to recognize the faces of nuclear authority during the disaster; thus, when these faces changed, the public only grew more suspicious that the government was withholding important information on the state of the nuclear reactors.

NISA replacement

On the second day of the disaster, the Nuclear and Industrial Safety Agency's (NISA) Deputy Director General Koichiro Nakamura led a press conference, during which he mentioned, "the possibility of a meltdown exists" for unit 1. He also remarked that "a meltdown may be in progress" and "as a case of a meltdown occurrence, it may ... be in the same group [as the Three Mile Island nuclear power plant accident]." These phrases, of course, attracted wide and international attention.

The evening before, the agency's emergency-response support system indicated that there was a possibility that the reactor core of unit 2 could become exposed and that fuel melting might begin within a few hours. In actuality, however, it was unit 1 that first reached this crisis condition. By the next morning, according to multiple witnesses, the Cabinet core team had become well aware of the possibility of a reactor meltdown.

Nevertheless, when the Cabinet core team heard Nakamura's statements, they thought that it was inappropriate to communicate uncertain developments

---

[5] Edano, Y. (2011) Press conference, Prime Minister's Office. March 13. Available at: www.youtube.com/watch?v=3orovQa2K3w.

to the public.⁶ Edano reportedly yelled at a NISA representative at the Cabinet: "How could you make such public statements without notifying the Cabinet first?"⁷ He instructed both the agency and TEPCO to "at the very least contact the Cabinet about the content of any important press conference ... without fail."⁸

On the evening of March 12, the agency replaced Nakamura with NISA Public Relations Examination Officer Tetsuo Noguchi, who, at the press conference later that evening, responded to the meltdown question by simply saying: "It is my understanding that [the use of the term "core melt"] was not based on a clear understanding of the actual [reactor] state."

The next morning, NISA Deputy Director General Hisanori Nei announced Nakamura's replacement, saying only: "The replacement was made at the direction of the leadership." Skirting the use of the word "meltdown," he went on to say that "the possibility of damage to the fuel rods is undeniable." By the next afternoon, however, NISA replaced Nei with Deputy Director General Hidehiko Nishiyama, who stated that, "the appropriate description is 'damage to the outer covering of the fuel rods.'" And with that, he denied Nakamura's comment on a meltdown.

On May 15—two months after Nakamura's press conference—TEPCO announced that core melt had indeed occurred in unit 1 immediately after the onset of the accident, effectively indicating that Nakamura's comments were correct. This event heightened the public suspicion that the government was not communicating properly and that the state of the nuclear reactors was far worse than what was being disclosed.

A matter of semantics: replacement, dismissal of Nakamura

During the press conference, held on August 10, on Nakamura's resignation, NISA Director General Nobuaki Terasaka stated: "Nakamura was originally responsible for international affairs, and his duties had long been in that area. It then became necessary to make various personnel and organizational adjustments as the developments unfolded, and the change in the position for public affairs was made accordingly." He emphasized that "it is not his 'meltdown' statement that caused the replacement." The interim report of the government's investigation committee on the accident described that Nakamura himself volunteered to step down; however, it is easy to infer that Nakamura's "voluntary" resignation was encouraged by a very dissatisfied Edano and NISA.

---

⁶ Personal communication with related Cabinet personnel, n.d.
⁷ Personal communication with NSC Chairman Haruki Madarame, December 17, 2011.
⁸ Personal communication with former Chief Cabinet Secretary Edano. (Chief Cabinet Secretary Edano did not state the date of this instruction, but the context implies it was most probably March 12.)

Explanations to the public concerning low-dose radiation exposure

The government—and in particular Chief Cabinet Secretary Edano—repeatedly stated that the radioactive elements released into the environment after the accident had "no immediate effect on human health." The ambiguity of this statement fostered continuing public debate concerning the possibility of adverse effects on health due to long-term exposure to radiation.

Edano provided the following explanation at the Budget Committee meeting of the House of Representatives in November 2011:

> During the first two weeks following March 11, I held thirty-nine press conferences. At seven, I stated that "there is no immediate effect on the human body or human health." At five of these seven, my statement was made in the context of questions concerning food and drink. I never made any general statement that there would be no adverse effect to health. [For example,] I repeatedly stated that there is a specified annual limit for radioactivity contamination in milk. What I meant was that even if you drink such milk [with low-level contamination] once or twice, there would be no immediate problem.

Here, Edano claimed that the health assurance was limited to one or two ingestions. A review of his past statements, however, casts some doubt on this argument. For example, at the March 20 press conference, Edano said: "[E]ven if a person *continues* to ingest spinach containing radioactive materials in the recently detected concentration level, it is inconceivable that it would immediately affect health."[9] Likewise, at a press conference the day before, when responding to a question on the risks related to milk, he said: "If, hypothetically, a person were to *continue* to ingest milk containing radioactive materials at the recently detected concentration level for one year at the average rate of ingestion in Japan, the radioactive dose exposure would be about the same as one CAT scan."[10] The phrase "about the same as one CAT scan" may naturally be taken to imply that the dose was not large even after the one-year-long ingestions. If we assume that his response given at the Diet was true to his intent, then we must conclude that his statements at these press conferences were misleading.

With the benefit of hindsight, Edano has reflected on those press conferences, saying, "If I had directly delivered the content of the reports that I received, very few people could have understood them. ... For me, the most difficult problem was how to make my statement comprehensible to many."[11] Under severe time constraints, Edano had to translate the difficult scientific terminology

---

[9] Emphasis inserted by authors.
[10] Emphasis inserted by authors.
[11] Personal communication with former Chief Cabinet Secretary Edano, December 10, 2011.

to lay language on a largely ad lib basis. With this experience as an important lesson, the government should develop methods to communicate such information to the public, and formulate and adopt a standard risk scale, which would provide a more objective and comprehensible measure of the level of low-dose radiation risk to health.

Global information sharing

The Fukushima nuclear accident attracted the attention of a concerned public, governments, and media from around the globe. The Cabinet, ill equipped in its English-communication capability, faced the challenge of providing timely updates in English to an international audience.

The Cabinet's English-language information

When Motohisa Furukawa became deputy chief Cabinet secretary in June 2010, he found that the government, as a whole—outside the Ministry of Foreign Affairs, which works with the foreign press—did not have any international public relations capability. Within two months of his tenure, he established the Office of Global Communications within the Cabinet to increase the government's English-language capability and chose Noriyuki Shikata, who had formerly served as director of the International Press Division of the Ministry of Foreign Affairs and as press secretary at the Embassy of Japan in the United States, as the first director.[12]

Initially, the office's primary objective was to promote foreign knowledge and understanding of Japan's technology and economic policy, as well as Japan's new strategy for economic growth. On March 11, 2011, however, its central task became communicating the earthquake and the nuclear power plant accident. Immediately after the earthquake, the office translated and published the English-language transcripts from the prime minister's press conferences and other breaking news via Shikata's personal Twitter account, which he had only opened the month prior. Since the government's English-language resources were very limited, his followers quickly grew from about 100 to more than 10,000. On March 14, the Cabinet began to Tweet in Japanese and two days later, it opened an English-language Twitter account, which attracted more than 22,000 followers in two weeks; it rolled out its English-language Facebook page on March 23.[13] In addition, the office offered English-language news on its website devoted to information on its earthquake-response measures; the section included the prime minister's messages, as well as Edano's

---

[12] Edano, Y. (2011) Press conference, Prime Minister's Office. 3:30 p.m., March 13. Available at: www.youtube.com/watch?v=3orovQa2K3w.
[13] Prime Minister's Office, Government of Japan (2011) Great East Japan earthquake. Global communication activities of Prime Minister's Office. Press release, September 23.

press conferences. The office also provided hyperlinks to the ministries' English-language websites.[14]

Soon, the Office of Global Communications—with the approval of Chief Cabinet Secretary Edano and Deputy Chief Cabinet Secretary Fukuyama—became the official government voice for the international press. The office arranged interviews with Prime Minister Kan, Chief Cabinet Secretary Edano, Deputy Chief Cabinet Secretary Fukuyama, Special Advisor to the Prime Minister Hosono, and other officials; by functioning as the centralized media hub for the foreign press corps, the office was able to consider which political leaders would be most appropriate for what interview and strategically schedule interviews.

Assessment of the Cabinet's global communications structure

Immediately following the Tohoku earthquake, the foreign media were generally sympathetic to the people of Japan in the midst of the unprecedented devastation, with many reports expressing admiration for their orderly discipline and fortitude. A large number reported favorably on the Japanese earthquake-resistant architecture that had withstood a 9.0-magnitude earthquake, the superb Shinkansen railway (bullet train) technology that had no train accidents, and the voluntary support activities of young people in the stricken areas. Some reports on the nuclear power plant accident, however, were marked by exaggeration. German reporters, in particular, quickly fled Japan and moved bases to Hong Kong or Seoul, where they began issuing reports that stirred fear. *Die Welt*, for example, ran the headline "Tokyo in Deadly Fear" above an article that stated, "40,000,000 people are threatened."[15] CNN continually broadcasted images that stressed the danger of nuclear power plants. Media are generally characterized by a tendency to report, often and prominently, the negative and sensational. This was true of many of the reports on the earthquake and its aftermath, which tended to strongly emphasize nuclear accidents, radiation effects, and other negative aspects while slighting aid and rescue efforts, international cooperation, and other positive aspects.

Foreign students living in the Tohoku region immediately returned to their home countries or took refuge in the Kansai region, which they regarded as comparatively safe. Forty days after the earthquake, 65 percent of these students had not yet returned to the Japanese universities in which they had been enrolled. One major reason for this, in addition to the overall unease felt by the students, was undoubtedly the urging of their families who had seen the

---

[14] Prime Minister's Office, Government of Japan (2011) Great East Japan earthquake. Global communication activities of Prime Minister's Office. Press release, September 23.
[15] Heine, M. (2011) Tokio in todesangst. *Die Welt*, March 16. Available at: www.welt.de/print/die_welt/politik/article12841234/Tokio-in-Todesangst.html (accessed August 31, 2013). (In German.)

broadcasts produced by CNN and other international media groups. Many travelers canceled visits to Japan, and the number of tourists to Japan fell to half the normal level that existed before the earthquake. In one foreign-owned corporation in the financial sector, approximately 270 of the employees working in Tokyo sought refuge in Hong Kong.

In the business world, the flow of investment to Japan essentially froze. Businesses suspended plans to establish branch offices. Scheduled visits to Japan for medical treatment shifted to Thailand or Singapore. Many foreign-owned companies contemplated reducing their investments.

In April 2011, the Office of Global Communications implemented a media strategy to counter these effects. On April 17, the office contributed an article to *The Washington Post* signed by Prime Minister Kan and titled "Japan's Road to Recovery and Rebirth," which was also reprinted by the *International Herald Tribune* and thus reached much of the English-speaking world. Director Shikata requested through the Ministry of Foreign Affairs that Japan's embassies all over the world approach the foreign media with this message, and, consequently, 128 media outlets in sixty-two countries around the world reprinted it. The Cabinet's Public Relations Office, together with the Ministry of Foreign Affairs, composed a special message of gratitude signed by Prime Minister Kan for the cooperation and assistance provided by many countries. It was placed as a paid announcement in the *International Herald Tribune*, *The Wall Street Journal*, *Financial Times*, and in other newspapers in China, South Korea, France, and Russia. The message was subsequently carried as an unpaid announcement in 216 newspapers in sixty-three countries.[16]

The relentless stream of unfavorable articles nonetheless continued in the foreign media, criticizing in particular the Japanese safety management configuration and its processes of public information disclosure. This included opinion pieces asserting collusion between regulatory agencies and the government, and mounting accusations that the government was concealing information.

Use of social media

The 2011 Tohoku earthquake occurred in the new era of pervasive Internet use—that year, the number of users in Japan reached more than 94.6 million people, representing 78.2 percent of the population.[17] The Digital Age has greatly expanded the opportunities not only for information acquisition, but also for easy information dissemination through the proliferation of social

[16] Heine, M. (2011) Tokio in todesangst. *Die Welt*, March 16. Available at: www.welt.de/print/die_welt/politik/article12841234/Tokio-in-Todesangst.html (accessed August 31, 2013). (In German.)

[17] Ministry of Internal Affairs and Communications (2011) White paper information and communication in Japan. Available at: www.soumu.go.jp/johotsusintokei/statistics/statistics05a.html (accessed September 4, 2013). (In Japanese.)

media such as blogs, Twitter, Facebook, Mixi (a Japanese social-networking system), YouTube, and Nico Nico Douga (a Japanese video-sharing website), among many other services.

In Japan, the public used social media platforms after the earthquake to communicate and record damage, confirm personal safety, gather information, as well as to express thoughts and opinions. Likewise, those who were affected by the Fukushima Daiichi Nuclear Power Station accident turned to social media for information on safety, refuge, and radiation.

Through analysis of Google keywords, Tweets, and other social platforms, we can see where the public's interests were. But for an even more intimate view, we can look at the explosive effects of some Twitter pages and blogs. One of those personalities was Ryugo Hayano, a professor specializing in nuclear physics at the University of Tokyo. Hayano, who watched the Nuclear and Industrial Safety Agency's press conferences and synthesized the information into accessible, useful information—in 140 characters or fewer—became something of a celebrity as he went from 2,255 followers on March 7 to more than 151,000 by March 21. In his Tweets, he used observational data, graphs, and tables on radiation that were publicly available on the Internet.

Over time, Hayano's efforts transformed into something of a movement. He called on researchers to provide radiation data and drew attention to various other Twitter accounts of substance, including the "Radiation Summary Page" maintained by Ryo Ichimiya, a researcher at the High Energy Accelerator Research Organization,[18] as well as associate professor Keichi Nakagawa and his radiological team at Tokyo University Hospital. In less than a week, Team Nakagawa, as they were called, had approximately 200,000 followers.

Outside of this celebrity team, however, others were just as motivated to use and translate radiation dose data publicly disclosed by TEPCO. Regional organizations and movements emerged in which private individuals made their own measurements of radiation doses and organized the results. Hiroshi Ishikawa—a former engineer at Nippon Telegraph and Telephone Corporation and president of one of the corporation's subsidiaries—detected a sharp spike in the airborne radioactivity ratio on March 15 at his Tokyo home, where, since North Korea's missile test launch in 2005, he had dutifully measured

[18] The start of the "Radiation Summary Page" grew out of a Twitter conversation among physicist Ryo Ichimiya, theoretical physics professor Mihoko Nojiri (@Mihoko_Nojiri), and High Energy Accelerator Research Organization trustee Nobukazu Toge (@bunogeto). "The recording of time-series radiation-dose data is being performed by TEPCO, local bodies, institutes, and universities, but I think the need exists for a system that can organize and integrate the data and present it graphically. What do you think?" With this, the website was launched. Some seventy people ultimately registered, including researchers in Earth sciences, information technologies, as well as in physics. The group produced successive visualizations of nationwide radiation dose data. In addition to this researcher-led effort, on March 17, an individual Internet engineer created a website that listed nationwide radiation activity, using Google Maps to produce visualizations of open-source data provided by the Ministry of Education, Culture, Sports, Science, and Technology.

radiation doses and published his findings on the Internet. Until March 11, Ishikawa had ten visitors per day to his website; afterward, 60,000 people checked his data on a daily basis.

Ishikawa received several e-mails celebrating his citizen action, one of which read: "It is amazing that you can provide substantiated quantitative data, while the government continues to disseminate vague information. The values you provide are our only recourse, as we cannot rely on those announced by the government. We are deeply grateful." Another stated: "The people are not fools. They are cool-headed, and I do not believe that public disclosure of this type of data will lead to a sense of panic. It is rather an absence of public disclosure that arouses extreme unease."

For people who came to distrust the government information transmissions, the voluntary data measurement networks became an entity in which they could place their trust.[19] Private individuals translated international weather reports to Japanese,[20] administered Geiger counters to produce live productions of radiation counts,[21] and created online citizens' groups for monitoring radiation dose levels.[22] It is apparent, from the high level of concern about radiation and radioactivity—and from citizens' online radiation measurement activities throughout the country—that most people considered the situation less as an event occurring in Fukushima and more of one in their own local regions and affecting their own welfare. With very limited information provided by the government or the mass media on hotspots of high radiation doses, those concerned began to take matters into their own hands. This movement arose in part because of government failure to use social media for interactive information exchange and acquisition, and its consequent inability to transmit information of public concern.

An underlying problem was the weak state of the government's information transmission system: It lacked staff with close knowledge and understanding of social media. Rather, it employed people with a background in portal sites and used volunteers from advertising companies; but portal sites and advertising

---

[19] It must also be noted, however, that confusion occurred at an elementary level because of mistakes in operating the measurement instruments. Some people continued to post incorrect measured values due to an insufficient understanding of the instrument operating procedure, leading to dissemination of unreliable information in some cases.

[20] Kan Yamamoto, team staff member at Tosayama Academy, was living in Germany at the time of the Fukushima disaster. Noticing that valuable meteorological information was not in the Japanese language, he gathered other translators and published translations of weather predictions from various countries.

[21] From March 13 to 19, Ustream Asia Inc., a major online video-streaming company, reported that seven of the top twenty real-time video sites were Geiger counter live-stream sites, with the highest among these ranking fourth. From March 20 to 26, such sites were in six of the top twenty rankings.

[22] The "5cm50cmAboveGroundMeasurementNET" was a group that collected radiation measurements around children's parks and other places in Tokyo. Similarly, the National Network of Parents to Protect Children from Radiation was also founded.

largely use one-way communication models, and not real-time two-way communication. The Cabinet's Public Relations Office and IT public relations advisers would not welcome bloggers into their midst until September.

As social media flourished, mass media waned. A survey by the Nomura Research Institute found that, after the earthquake, the public did not have a heightened trust of any mass media outlet, other than NHK. Unsurprisingly, the broadcaster actively used Twitter and blogs. On Twitter, @nhk_kabun—the handle of NHK's Science and Culture Department—had just 3,000 followers before the earthquake, but by March 21, the community had grown to 223,903. On many days, the number of Tweets exceeded 100, and the staff members later related that they had strongly felt the concerns and feelings of those people and relayed them as feedback to the broadcast commentators.[23] This plainly demonstrated the need for mass media to receive, as well as transmit, information via two-way communication.

One problem that clearly exists for social media is its inability to reach all generations and levels in society. We fear that this digital divide, and the resulting disparity in information access between those who use social media and those who do not, may widen further.

Information transmission by government

As mentioned, @Kantei_Saigai, the Cabinet's handle, went live on March 13. Operated around the clock by three rotating young members of the Cabinet's Public Relations Office, the Tweets were typed, confirmed by a public relations officer, and then posted; in some cases, the public relations officer wrote the posts personally. By the end of the month, the account had 300,000 followers.

The Cabinet's Twitter account was more of a unilateral bulletin board of official events and announcements by the government. No replies were given to queries, even when requested. In short: There was no interactive communication. Despite the heightened interest in low-dose radiation, the Cabinet did not use the Twitter platform to communicate messages on such health risks—nor did it try to clarify messages that were wrongfully being perpetuated on other social media platforms.

For example, on March 13, the Cabinet published forty-eight Tweets, mostly consisting of quotes from Prime Minister Kan's press statements, but nothing on radiation doses, aside from a link to Deputy Chief Cabinet Secretary Fukuyama's definition of "microsievert" and a message noting that no abnormalities were found in monitoring results near the main gate of the nuclear power plant.

Two days later, the Cabinet published sixty-two Tweets, largely quoting from press conferences and announcements by Prime Minister Kan and Chief Cabinet Secretary Edano. Despite heightened concern about radioactivity in

---

[23] *Shimbun Kenkyu* (2011) Untitled. September. (In Japanese.)

mid-March, the Cabinet's number of Tweets stayed minimal—only twenty-three on March 23, and nineteen the following day—and vague. Such Tweets related to the detection of radioactive iodine in the water at the Kanamachi Purification Plant. While the Cabinet instructed parents to avoid giving tap water to infants in and around Tokyo, it also Tweeted comments like "It is safe to drink tap water" (March 23) and "For anyone other than infants, there would be no physiological effect (from drinking such tap water)" (March 24). Despite the contradictory messages, the government still did not interact with its Twitter followers.

TEPCO data transmission

Immediately following the Fukushima accident, TEPCO issued nearly ten daily press releases in PDF format. During the planned power outages on March 14, however, the PDF format made it especially difficult for users to search for the affected areas; this was only exacerbated by the fact that the company did not offer a complete list of the areas.

TEPCO opened its Twitter account, @OfficialTEPCO, on March 17. Within twenty-four hours, the company had 160,000 followers, which grew to 280,000 by the end of the month. The company's first postings merely provided guidance on planned power outages—and, as users criticized, did not provide any information on radiation. Users also criticized the company for not personally apologizing for the disaster. TEPCO continued to Tweet about planned power outages but, using the same approach as the Cabinet, did not interact with its followers.

Missed opportunities

Although the government used social media early into the disaster at Fukushima, it did not take advantage of the interactive potential of these platforms. With social media, it is often much more important to receive and analyze information, and then communicate to followers, fans, or readers than simply to transmit information. The government needs to construct techniques and systems to survey and analyze social media postings. With existing technologies, it is already possible to gauge Internet users' search terms; such information can help governments know how to respond to the public.

Chapter 9

US–Japan Relationship

Seventy years after they were enemies at war, the United States and Japan are now the strongest of allies. On foreign policy, the two nations work together on a range of security priorities in the Asia-Pacific, including efforts to shape China's growing regional influence and measures to contain the nuclear threat of North Korea. In military matters, the United States relies heavily on its bases in Okinawa and elsewhere to stage operations in East Asia; about 40,000 American troops are still stationed in the country. When it comes to the economy, the two nations are crucial trading partners. American consumers buy Japanese cars and electronics, Japanese consumers eat US soybeans and watch US movies, and the economic give-and-take continues. Japanese firms are also some of the biggest investors in US companies and US treasury bonds.

It is no exaggeration to say that the US–Japan relationship is key to both nations' prosperity and growth. But during the Fukushima Daiichi nuclear accident, that relationship was severely tested. While existing bonds between the governments were strong, particularly between the countries' militaries, a breakdown in communication caused misunderstandings and public disagreements between the two governments.

Many countries provided aid to Japan during the Fukushima nuclear accident, with nations such as France and Russia offering not only material resources like radiation monitors and protective suits, but also expert advice on how to stabilize nuclear reactors and spent fuel pools. But due to the special relationship between the United States and Japan, the Obama administration provided a particularly massive quantity of supplies, and also dispatched a huge number of personnel.

These personnel came not only from the US Embassy in Tokyo and the US forces stationed in Japan, but also military organizations like the Chemical Biological Incident Response Force (CBIRF, a unit of the US Marine Corps) and the US Pacific Command in Hawaii. On the nuclear power side, a contingent of 160 people, including staff from the Nuclear Regulatory Commission (NRC) and the US Department of Energy, was stationed in the embassy in Tokyo to assist Japan in crisis management.

Needless to say, the US government prioritized protection of its own citizens, including military personnel based in Japan—but multiple government officials have emphasized that the United States placed equal stress on helping to resolve the nuclear plant crisis.[1]

Yet despite the Americans' strong desire to be of assistance, the initial collaboration between the United States and Japan was so disorganized and ineffectual that it brought about a crisis in the two nations' historic alliance. For more than a week after the nuclear crisis began, poor communication fostered mistrust between the two governments, and officials struggled to share information and to coordinate the provision and reception of aid. The lack of coordination and communication prevented the governments from jointly identifying and implementing the most effective countermeasures at the plant. It also caused the US and Japanese governments to issue conflicting statements and advice to their respective citizens during the crucial and chaotic first week of the disaster, which gave rise to confusion among the public and resulted in mounting suspicion of the Japanese government.

In the most public display of the strife between the two nations, the United States and Japan ended up with different evacuation recommendations, with the US government urging more conservative measures. Although the US government initially instructed its citizens to follow the guidance of the Japanese government, it later counseled its citizens to protect themselves by leaving the area within 80 kilometers of the Fukushima plant, and advised them to consider departing Japan altogether. These American statements had a broad effect. Other nations expected the US government to have a clear understanding of the situation, given its in-country infrastructure of military bases and its close alliance with Japan; thus, many foreign embassies in Tokyo followed the US guidelines in their advice to their own staff and citizens.

Diplomatically, at least, there was a positive ending to the Fukushima crisis. The discord between the United States and Japan was eventually resolved by the development of "coordination conferences" in which key officials from the two governments met daily to discuss and agree on strategic priorities. Establishing this structured framework resulted in a close working relationship, and prevented a rift between the two countries. The ultimate resolution of the crisis shows the imperative of establishing strong, flexible communication structures between government agencies during times of peace and tranquility, which can be activated promptly and efficiently in times of a disaster.

Taking stock of what went wrong between the two nations can also be seen as a prudent investment for the future. The Fukushima catastrophe involved not only a nuclear accident but also a double-blow of natural disasters that made every task more difficult for the response teams. The extent and complexity of the calamity, as well as the rapid progression of events, can be compared to an attack of nuclear terrorism. As such, an evaluation of the US

---

[1] Personal communication with US government officials, December 1 and 2, 2011.

and Japanese collaborative response to the disaster may provide lessons for coping with future emergencies.

★★★

The US National Security Council learned of the earthquake off the Tohoku Pacific coast soon after it struck, before dawn reached the US east coast. The full team was assembled, and President Barack Obama was awoken and informed of the situation.[2]

At 7 a.m. government officials convened a meeting that included the NRC, the Energy Department, as well as agencies involved in security matters, including the National Security Council, the Defense Department, and the State Department. The room was completely packed. This first meeting focused on earthquake and tsunami relief, but the meeting also included a report from John Holdren, assistant to the president for science and technology, on the possibility of a nuclear disaster in Japan. National security personnel confirmed that they anticipated the worst-case scenario and were considering policy responses accordingly.[3]

That morning within its headquarters, the NRC initiated a "monitoring and response mode" to launch international actions. In addition, the US Agency for International Development activated its Disaster Assistance Response Team (DART) to provide emergency disaster relief. DART funded the first NRC staff members, who were dispatched to Japan in response to the Fukushima nuclear plant accident; these staffers left for Tokyo within twenty-four hours of the earthquake.

In Japan, the initial phase of collaborations between the two governments occurred within their respective defense authorities; both had mechanisms in place for disaster relief. The first US–Japan military coordination center, intended to synchronize earthquake and tsunami relief efforts, was established within the Ministry of Defense in the Ichigaya district of Tokyo as early as March 11. This was a procedure taken straight from the emergency-response playbook for defense officials. Military coordination centers were also established at Yokota (on March 11) and at Camp Sendai (on March 15). Liaisons from both Japan's Self Defense Forces (SDF) and the US military were stationed at these centers, and the US Embassy was brought into the meetings via videoconferencing links. As the nature of the earthquake and tsunami damage gradually became clear, the US and Japanese forces clarified their response measures and the distribution of labor. Coordinated relief efforts began to run smoothly, and the military forces started to implement their plans.

In the meantime, the situation at the Fukushima Daiichi Nuclear Power Station was worsening rapidly in those first few days, and defense officials were forced to shift their attention to the developing nuclear crisis. While the SDF

---

[2] Personal communication with the NSC, December 2, 2011.
[3] Personal communication with the NSC, December 2, 2011.

and the US military worked together very effectively on the earthquake and tsunami response, their limited mandate prevented full cooperation during the nuclear disaster. Although the SDF duties prescribed in the government's response manuals included *off-site* responses in the area surrounding the Fukushima Daiichi plant, such as attending to the needs of evacuees, they did not include *on-site* responses at the plant facility itself, such as measures to help resolve the nuclear accident.

As the crisis worsened and *on-site* activities became necessary, it was critical to develop a strategic outlook of the situation and to clarify the precise role and response policies of both the SDF and the US military. Because efforts had been made by the two militaries to gather information from agencies in the Japanese government and from the Tokyo Electric Power Company (TEPCO), some mechanisms existed for collecting data and sharing it with US agencies. However, these mechanisms were extremely constricted due to the limited mandates given to both militaries. Much broader and more open communication between the two governments was necessary as the nuclear catastrophe expanded, but it took many days, and much frustration and confusion, before such dialogue was achieved.

Communication breakdown

Initially, the US government respected Japan's leadership in responding to the events at Fukushima Daiichi. The Americans awaited formal Japanese requests for assistance and, in the meantime, restricted their own activities to gathering information about the nuclear reactors.[4] However, the Japanese government wasn't forthcoming with that information. Even the American nuclear experts who arrived just one day after the accident, the so-called NRC Tokyo Task Force, weren't able to establish regular contact with their counterparts in the Nuclear and Industrial Safety Agency (NISA) or the other Japanese agencies that had oversight of nuclear matters. With little input from the agencies that presumably had direct knowledge of conditions at the Fukushima Daiichi plant, the Americans did their best to cobble together an understanding of the worsening accident. This poor communication had serious repercussions: The two governments differed dramatically in their assessments of the situation in the nuclear reactors, and had conflicting opinions regarding the necessary evacuations.

In addition to the defense authorities' cooperative efforts during the first days of the accident, there were some ad hoc meetings and informal conversations between officials in the Japanese and US governments. But throughout the most critical days of the crisis there was no systematic coordination or exchange of information between the relevant agencies in the two governments, and the US–Japan relationship was in peril. As days went on

---

[4] Personal communication with US government officials, November 3 and 4, 2011.

and the situation at the Fukushima Daiichi plant seemed to spiral further and further out of control, the US government made increasingly desperate efforts to collect information that was urgently required.

As one of the US government's highest priorities was to protect the safety of Americans within Japan, it needed accurate data in order to deliver evacuation advice and instructions to these citizens. In the first days of the accident, the US Embassy in Tokyo, as well as the NRC and Energy Department personnel stationed there, attempted to establish contact with the Ministry of Economy, Trade, and Industry (METI), NISA, TEPCO, and other organizations. At the same time, embassy staff tried to secure information from channels such as the US–Japan military coordination centers. However, none of these efforts resulted in an adequate flow of information. The leader of the NRC Tokyo Task Force was busy with a flurry of high-level meetings with the ambassador and other officials, so contacts between the NRC and NISA were made by lower-level officials. In retrospect, it is clear that from the onset of the disaster the leaders of the NRC and NISA should have collaborated and established solid partnerships.[5] This oversight was one cause of the communication problems that ensued.

As information-gathering efforts proceeded in Tokyo, officials in Washington were placing direct telephone calls to any Japanese government officials with whom they had existing relationships. During this process, for example, US Deputy Energy Secretary Dan Poneman contacted Japan Atomic Energy Commission (JAEC) Chairman Kondo via telephone and e-mail;[6] in the initial days of the accident, contact with Chairman Kondo became an important communications channel for the Energy Department.

This disorganized, ad hoc system for information exchange fostered a sense of wariness: With information arriving in bits and pieces, the Americans didn't always trust the accuracy of the reports that they did receive. As the US government tried to piece together a complete picture of the scenario at the Fukushima Daiichi plant, it worried not only about the paucity of data provided by the Japanese, but also wondered whether sensors and other measurement equipment were functioning properly after the tsunami.[7]

While the US government awaited a formal request for aid, the Japanese assumed a posture of prudence toward initial US offers of assistance. In an e-mail that he sent on the morning of March 12 to NRC Chairman Gregory Jaczko, Japan Nuclear Energy Safety Organization (JNES) Chairman Katsuhiro Sogabe expressed polite thanks for the NRC's offer to help, while simultaneously turning down the offer.[8] It is unclear whether JNES had full access to information

[5] Personal communication with an official of the NRC Tokyo Task Force, October 28, 2011.
[6] Personal communication with a US Energy Department official, November 3, 2011.
[7] Personal communication with an official of the NRC Tokyo Task Force, December 17, 2011.
[8] US Nuclear Regulatory Commission (2011) Document ML11257A101. Obtained by a Freedom of Information Act request. Available at: www.nuclearfreeplanet.org/nrc-foia-documents.html.

and appreciated the full severity of the situation at Fukushima Daiichi at that time, and it has not been confirmed that JNES consulted with NISA, TEPCO, or the Cabinet in drafting this response. However, it seems apparent in the immediate aftermath of the disaster that Japan did not plan to accept international assistance; it intended to resolve the situation entirely on its own.

Regardless of Japan's preferences, the United States was not content to wait quietly in the wings for the duration of the crisis. Frustrations of American officials began to boil over several days into the accident, when they felt they were still receiving insufficient information despite the increasingly dire situation at the nuclear plant. The extent of the Americans' desperation for information is reflected by a request, made by US Ambassador John Roos in a late-night phone call on March 14 to Chief Cabinet Secretary Yukio Edano, to allow a NRC staff member to be stationed in the Cabinet—a request that Japan refused.

At a Cabinet meeting on March 15, some Cabinet members wondered if the Americans wanted nothing more than information on the nuclear accident, a statement which seemed to cast aspersions on the sincerity of US requests for true cooperation. In response, Defense Minister Kitazawa stated that it would be best to accept US offers of cooperation and assistance.[9]

The differences in perspective and the erosion of trust that had been growing between the United States and Japan since the beginning of the accident reached a crisis point between March 15 and 16. By then, explosions had already torn the roofs off of units 1, 3, and 4, and damage had occurred in unit 2. The sequence of explosions and the billowing clouds of smoke issuing from the wrecked reactor buildings were shown repeatedly on television and republished constantly on the Internet, and there was a general sense that the situation at the Fukushima Daiichi plant was getting more dangerous and unstable day by day.

On March 15, Ambassador Roos informed Defense Minister Kitazawa that the US government was unhappy with the quality of information sharing between the two nations. Prime Minister Kan also notified Kitazawa that the formal communication channels between the Americans and the Cabinet were not functioning properly,[10] and officially requested that the situation be improved.[11] Upon receiving these communications, Kitazawa contacted the NRC, which explained that it had been unable to establish contact with TEPCO or NISA. For this reason, on the morning of March 16, the Ministry of Defense convened a meeting at its headquarters, which brought together the Americans and representatives of the Ministry of Foreign Affairs, NISA, and

---

[9] Personal communication with Defense Minister Kitazawa, January 17, 2012.
[10] However, on the evening of March 14, and again the following morning, American nuclear power experts met with Deputy Chief Cabinet Secretary Fukuyama, Cabinet Advisor Yasui, and other officials.
[11] Personal communication with Defense Minister Kitazawa, January 17, 2012.

TEPCO. This gathering of agencies was subsequently repeated three times, and at one meeting, the US Pacific Command furnished a list of areas in which it could be of assistance.[12] In a further indication of the communication problems within the Japanese government, it seems that some staffers in the Cabinet were unaware of these meetings.

In Washington, there was confusion and fear following the explosion in the building of unit 4. Before dawn on March 16, officials of various US government agencies conferred by telephone. During this conference, based on data gathered by the Global Hawk unmanned surveillance aerial vehicle, officials concluded that the nuclear reactors were melting down, and, in view of concerns over the state of unit 4's spent fuel pool, discussed fears that the situation could grow even more dire.[13] After this conference, the White House shifted its posture from awaiting formal Japanese requests to active involvement. This shift reflected the intentions of the president himself.[14] On March 16, Assistant Secretary of State Kurt Campbell made an impassioned request to Japanese Ambassador Ichiro Fujisaki for greater information sharing. This request heightened concerns within the Japanese government (especially in the Ministry of Defense and the Ministry of Foreign Affairs) over the worsening US–Japan relationship.[15]

Crisis points

The communication problems between the United States and Japan became particularly apparent in several statements made by the American officials on the days of March 16 and 17. Namely, the opinions between the two countries clearly diverged regarding the danger posed by the spent fuel pool in unit 4, the proper extent of the evacuation zone, and the efficacy of dropping water on the reactors from helicopters.

On the early morning of March 15, an explosion ripped through the unit 4 reactor building. That reactor had been out of service at the time of the earthquake and tsunami, and all of its fuel had been removed from the reactor vessel and stored in the spent fuel pool. The situation at unit 4 had been considered stable. But when the building shattered, experts scrambled to determine the cause.

On March 16, NRC Chairman Jaczko publicly stated that most of the water in the unit 4 spent fuel pool had likely evaporated, which would allow for large releases of radioactive material. Speculations as to how Chairman Jaczko reached this conclusion have centered on two possibilities. The first is that his

---

[12] *Asahi Shimbun* (2011) US military prepares comprehensive aid list, anticipates large-scale dispersion in response to the nuclear accident. Available at: www.asahi.com/international/update/0521/TKY201105210528.html (accessed February 28, 2012).
[13] Maher, K. (2011) *The Japan that Can't Decide*. Tokyo: Bunshun Shinsho. (In Japanese.)
[14] Personal communication with US government officials, November 3 and 4, 2011.
[15] Personal communication with Ministry of Foreign Affairs and Ministry of Defense officials.

statement followed from a discussion in Washington about the various possible scenarios for the explosion in the unit 4 building. Among the situations envisioned, the scenario that seemed most logical was that the water in the spent fuel pool had evaporated or leaked away through cracks, causing the exposed, zirconium-coated fuel rods to warm up and melt, producing hydrogen gas that fueled the explosion.[16]

A second possibility is that a TEPCO engineer, in a private conversation with a visiting NRC staffer in Tokyo, expressed his personal opinion that the pool water might have dried up, and this statement was relayed to Chairman Jaczko. The NRC staffer who heard this opinion mistook it for the official position of TEPCO and communicated it as such to Chairman Jaczko's staff.[17]

Chairman Jaczko's statement received wide publicity and generated considerable fear. In response, TEPCO announced on March 17 that photographic images taken from a helicopter on the evening of March 16 showed reflections from the water surface in the pool, proving that it wasn't dry. The Japanese government hastened to provide that information to the Americans; for example, the leader of the NRC Tokyo Task Force visited the Cabinet and was shown photographic images of the pool taken by the SDF. While the photographic images of the pool were not clear, the photographs and other data given to American officials restored their confidence that there was indeed water remaining in the spent fuel pool.[18] Weeks later, Chairman Jaczko's conclusion was publicly refuted by Energy Secretary Steven Chu, who stated on April 1, "We have been able to measure the temperature in all pools, and these measurements indicate that there is water remaining."

But the miscommunication about unit 4's pool had significant repercussions. While the status of that spent fuel pool was still uncertain, the United States began to draw up new evacuation plans for its own citizens. And given the NRC's apprehensions about the pool, it was natural for the government to adopt a conservative approach.

Immediately after the accident, the US government had directed all American citizens in Japan to follow the evacuation instructions of the Japanese government.[19] With regard to the Japanese government's orders that citizens within a 12-mile (20-kilometer) radius evacuate and citizens within a 19-mile (30-kilometer) radius remain indoors, the NRC stated on March 15 that "the

---

[16] Personal communication with a White House Office of Science and Technology Policy (OSTP) official, November 4, 2011.

[17] Personal communication with a Nuclear Regulatory Commission Tokyo Task Force official. However, the likelihood that this information was erroneous was confirmed immediately afterward by consulting data and other resources; Tokyo conveyed this fact to Chairman Jaczko, but not until after he had made his public statements.

[18] Personal communication with Energy Department and OSTP officials, November 3 and 4, 2011.

[19] US Nuclear Regulatory Commission (2011) NRC sees no radiation at harmful levels reaching US from damaged Japanese nuclear power plants. Press release, 11-046, March 13.

steps recommended by Japanese authorities parallel those the United States would suggest in a similar situation."[20]

However, the NRC reversed course on the following day. Based on "the guidelines for public safety that would be used in the United States under similar circumstances" and referring to "the results of two sets of computer calculations used to support the NRC recommendations," on March 16 the NRC recommended the evacuation of all American citizens within a 50-mile (80-kilometer) radius of the Fukushima Daiichi Nuclear Power Station.[21] With the United States recommending a larger evacuation zone than the Japanese, residents of that disputed zone were afraid and confused, and were left wondering whether or not they were safe in their own homes.

The NRC had guidelines to consult as it worked to determine the extent of the recommended evacuation area. Its contingency plans for a serious nuclear accident include two types of "emergency planning zones," which would be the focal point of response efforts. The first zone contains the land within a 10-mile radius of a nuclear plant, where response efforts would seek to reduce the likelihood of direct contact with the plume of radioactive materials. In the second and larger zone, containing the area within 50 miles of the plant, efforts would focus on preventing people from consuming contaminated food and water. (In contrast, Japan's emergency plans include only one emergency planning zone extending 5 to 6 miles from the nuclear plant.) When the NRC was making its decision on the recommended evacuation zone around Fukushima Daiichi, it was primarily concerned with direct exposure to the plume—yet it didn't recommend a 10-mile zone, but opted instead for a more conservative 50-mile zone.

So what were the considerations underpinning the decision to recommend evacuation within 50 miles of the Fukushima Daiichi plant? An announcement[22] on March 16 states: "In the United States protective action recommendations are implemented when projected doses could be [more than or equal to] 1 rem to the body or [more than] 5 rem to the thyroid."[23] (Note that 1 rem corresponds to 10 millisieverts.) The report that forms the basis of this radiation threshold is the US Environmental Protection Agency's 1992 announcement of its "Protective Action Guides,"[24] which recommends evacuation in the early

---

[20] US Nuclear Regulatory Commission (2011) NRC analysis continues to support Japan's protective actions. Press release, 11-049, March 15.
[21] US Nuclear Regulatory Commission (2011) NRC provides protective action recommendations based on US guidelines. Press release, 11-050, March 16.
[22] US Environmental Protection Agency (1992) *Manual of protective action guides and protective actions for nuclear incidents*. Report, Office of Radiation Programs, 400-R-92-001. Washington, DC: EPA, pp. 2–6. Available at: www.epa.gov/radiation/docs/er/400-r-92-001.pdf
[23] US Nuclear Regulatory Commission (2011) Press release, March 15 and 16.
[24] US Environmental Protection Agency (1992) *Manual of protective action guides and protective actions for nuclear incidents*. Report, Office of Radiation Programs, 400-R-92-001. Washington, DC: EPA. Available at: www.epa.gov/radiation/docs/er/400-r-92-001.pdf

stages (within a few hours to a few days) of an accident in "cases in which total body exposure of 1-5 rem, thyroid exposure of 5-50 rem, and/or skin exposure of 50-250 rem may be expected."

According to the NRC's announcement,[25] the 50-mile (80-kilometer) evacuation order was based on numerical results obtained with the RASCAL (Radiological Assessment System for Consequence Analysis) program. This modeling program predicts radioactive releases from reactors and spent fuel pools during a nuclear emergency, takes into account weather systems, and estimates exposure for people in the fallout zone. However, the RASCAL program had not been designed to evaluate crises involving simultaneous accidents at multiple nuclear reactors within a power plant. Working quickly, the NRC constructed a model to aggregate the impact of three separate nuclear reactors and spent fuel pools, and RASCAL was used to simulate this model.

However, according to a letter later sent by NRC Chairman Jaczko to Senator Jim Webb, one of the two sets of calculations that were done to support the March 16 evacuation order assumed "a 100 percent damaged unit 4 spent fuel pool for approximately 15 hours."[26] Bill Borchardt, executive director for operations of the NRC Tokyo Task Force, later stated, "In the initial stages following the accident, the NRC staff feared that unit 4 would run completely dry."[27] Thus, in ordering a 50-mile evacuation zone on March 16, the US government appears to have relied on the findings of a model based on the assumption of the worst-case scenario: a total depletion of water in reactor 4's spent fuel pool.

Another account also exists of how the US government decided on the 50-mile evacuation zone. According to the NRC's Tokyo Task Force, the actual evacuation distance that emerged from RASCAL's simulation of radioactive material dispersion was around 19 miles. According to this account, the 50-mile evacuation zone was determined by first assuming that radioactive material could spread over a 25-mile region, and then doubling this number just to be absolutely safe.

It appears that the US government, driven by an imperative to protect its citizens, and faced with uncertain information, recommended an evacuation zone that was based on extremely conservative estimates. The letter from

---

[25] US Nuclear Regulatory Commission (2011) Document ML11257A101. Obtained by Freedom of Information Act request. Available at: www.nuclearfreeplanet.org/nrc-foia-documents.html.

[26] US Nuclear Regulatory Commission (2011) Letter from Chairman Gregory B. Jaczko to Senator Jim Webb providing information on assumptions used in recommending a 50-mile evacuation for US citizens following the Fukushima Daiichi nuclear facility events. Letter, June 17, p. 3. Available at: www.nrc.gov/reading-rm/doc-collections/congress-docs/correspondence/2011/webb-06-17-2011.pdf

[27] US Nuclear Regulatory Commission (2011) Briefing on the progress of the task force: Review of NRC processes and regulations following the events in Japan. Transcript, March 21. Washington, DC: NRC. Available at: www.nrc.gov/reading-rm/doc-collections/commission/tr/2011/20110321.pdf

Chairman Jaczko to Senator Webb spotlights the ambiguity of the situation: "Since communications with knowledgeable Japanese officials were limited and there was a large degree of uncertainty about plant conditions at the time, it was difficult to assess the potential radiological hazard."[28]

This point comes into sharper relief upon considering the actions of other US government agencies. For example, the US Defense Department distributed iodine tablets to Air Force personnel active within a 70-mile (approximately 110-kilometer) radius of the Fukushima Daiichi plant. Moreover, on March 17, defense officials announced a policy of allowing overseas leave on a voluntary basis for family members of military personnel living on the bases at Atsugi and Yokosuka, which lie about 150 miles and about 200 miles away, respectively, from Fukushima Daiichi.[29] Uncertainty caused by a lack of information has been cited as one reason for these actions.[30]

While the US government clearly felt the need to be conservative in its counsel, it also had fewer practical considerations than the Japanese government. The fact that Japan's recommended evacuation zone was smaller than that recommended by the United States can partly be attributed to the differences in how a nation responds to an accident *within* its borders versus an accident *outside* its borders. For Japan to enlarge the region covered by its evacuation order was no simple proposition and would not necessarily have been the most favorable option from a public safety standpoint. In contrast, the United States did not have to worry as much as Japan about securing destinations for evacuees.[31]

Even after issuing the 50-mile evacuation recommendation, the US government took further steps to ensure the safety of its citizens. On the evening of March 16, Ambassador Roos and Foreign Minister Matsumoto spoke by telephone. During this conversation, Roos discussed the need to make critical decisions regarding American citizens in Japan. Moreover, in the pre-dawn hours of March 17, Deputy Defense Secretary James Steinberg spoke by telephone with Chief Cabinet Secretary Edano to request that Japan provide the United States with more information. Later that morning, President Obama spoke with Prime Minister Kan by telephone for approximately thirty minutes; during this conversation, the president informed the prime minister that the

---

[28] US Nuclear Regulatory Commission (2011) Letter from Chairman Gregory B. Jaczko to Senator Jim Webb providing information on assumptions used in recommending a 50-mile evacuation for US citizens following the Fukushima Daiichi nuclear facility events. Letter, June 17, p. 3. Available at: www.nrc.gov/reading-rm/doc-collections/congress-docs/correspondence/2011/webb-06-17-2011.pdf

[29] Stewart, P. (2011) US readies to fly military families out of Japan. *Reuters*, March 17. Available at: www.reuters.com/article/2011/03/17/japan-quake-usa-military-idUSN1716166220110317 (accessed August 31, 2013).

[30] Personal communication with a DOD official, November 3, 2011.

[31] Chief Cabinet Secretary Yukio Edano has expressed similar opinions. (Personal communication with Chief Cabinet Secretary Edano.)

US State Department was planning to advise all American citizens in Japan to evacuate,[32] which it did that same day. The State Department also authorized voluntary relocations of family members of US government workers in Japan.[33]

The lack of communication had produced a public disagreement, with the US government issuing very conservative evacuation recommendations and appearing to doubt Japan's ability to manage its nuclear crisis. With relations between the two governments teetering on the edge of complete breakdown, the Japanese government scrambled to demonstrate that it would take strong and decisive measures to get the nuclear disaster under control. It was in that context that the Self Defense Forces were instructed to use helicopters to dump water into spent fuel pools, in an attempt to replenish the water supply and keep the fuel cool.[34]

Many officials in both the US and Japanese governments were skeptical that this operation would have any substantive benefit. The fact that water dumps were carried out nonetheless was largely due to political considerations, according to an interview with Defense Minister Kitazawa. The Japanese government had to do *something* to convince the Americans of its seriousness, and it also had to demonstrate to its own citizens that it was taking forceful action.

In a discussion about the helicopter operation, Ambassador Roos advised Kitazawa that the method used to cool fuel must be something "sustainable."[35] This was intended as a statement of opposition to the water dump. This conversation took place on the afternoon of March 17, as Roos and the NRC's Tokyo Task Force leader Charles Casto sat in Kitazawa's offices watching the helicopter maneuver on live television. Although Kitazawa indicated that he understood the US emphasis on using a sustainable cooling method, he emphasized that the most important thing was for the government to calm the people of Japan by demonstrating that it was responding to the crisis.[36]

Finding a better way

Meanwhile, according to Vice Minister of Defense Akihisa Nagashima, the increasingly tense state of US–Japan relations had gradually become more widely acknowledged within the Japanese government. After learning of the

---

[32] Cabinet Office, Government of Japan (2011) *Chronology report on Fukushima accident*. Report. Tokyo: Cabinet.
[33] In the United States, state governors issue evacuation orders based on NRC data as well as offer public advice and provide other announcements in the event of a nuclear crisis. Because the Fukushima accident occurred overseas, the US ambassador to Japan assumed these duties in place of state governors.
[34] Personal communication with multiple US and Japanese government officials, October 28 and November 3 and 4, 2011.
[35] Personal communication with a NRC official, December 17, 2011.
[36] Personal communication with former Defense Minister Kitazawa, January 17, 2012.

growing distrust between the allies, Vice Minister Nagashima (at the behest of Special Advisor to the Prime Minister Goshi Hosono) attended a meeting with the NRC convened by TEPCO on March 18. During this meeting, which was convened one week after the beginning of the crisis, the frustrated NRC representatives were still asking where they could go to receive accurate information.

Immediately afterward, Nagashima reported the meeting's conclusions to Deputy Chief Cabinet Secretary Yoshito Sengoku at the Cabinet, who noted that he had received similar messages from Ambassador Roos. At Deputy Secretary Sengoku's request, Nagashima contacted Roos at around 1 p.m. that day, and a US–Japan meeting was scheduled on the spot, to convene a few hours later, at 3:30 p.m. On the Japanese side, attendees included Hosono, Nagashima, JAEC Chairman Kondo, and Nuclear Safety Commission officials; and on the US side, attendees were Ambassador Roos, Political Minister Counselor Robert Luke, NRC's Casto, and Energy Department representatives.[37]

At the gathering, the participants agreed that a new protocol should be created to enable regular information sharing between Japan and the United States. According to a memo by Nagashima, the purpose of this new protocol was to facilitate "periodic discussions of the cooperative relationship between the United States and Japan regarding the crisis at the Fukushima nuclear plant, including matters such as information sharing, coordination of response activities, and the proper matching of US aid offers with Japanese needs (a particularly urgent priority)."[38] On March 19, Hosono and Nagashima visited the Cabinet and proposed meetings to coordinate the policies of the United States and Japan; the prime minister decided there and then to establish such meetings.

These US–Japan coordination conferences were established as an office within the Cabinet Office for National Security Affairs and Crisis Management. On the Japanese side, this office was headed by Deputy Chief Cabinet Secretary Tetsuro Fukuyama, with Special Assistant Hosono supervising day-to-day affairs; meetings were attended by representatives of the Ministry of Defense; the Ministry of Foreign Affairs; NISA; the Nuclear Safety Commission; the Ministry of Health, Labor and Welfare; the Ministry of the Environment; METI; and TEPCO. On the US side, meetings attracted US Embassy Deputy Chief of Mission James Zumwalt, US Embassy Political Minister Counselor Robert Luke, and representatives of NRC, the Energy Department, USPACOM, the US military in Japan, and the US Embassy. In this way, communications between the two countries were at last consolidated, with all relevant government agencies participating.

[37] Nagashima, A. (2011) Panel discussion. In: *Sasakawa Peace Foundation Third US—Japan Public Policy Forum*, November 2011. Tokyo: Sasakawa Peace Foundation.
[38] Nagashima, A. (2011) Proposal to reconvene US–Japan coordination capabilities in response to nuclear power generation. Memo, Ministry of Defense, March.

The first preparatory meeting for the US–Japan coordination conferences took place on the evening of March 21, and thereafter meetings were convened almost every night, for a total of around forty meetings. In the beginning, meetings lasted as long as two hours or more. Research and operation teams were established to discuss collaborative efforts in four distinct areas: (1) radioactive material shielding and urgent initiatives required to prevent the spread of radioactive materials; (2) nuclear fuel-rod processing to stabilize the nuclear plants in the medium-term; (3) the decommissioning of the reactors, a necessary long-term measure; and (4) medical and lifestyle assistance for citizens. The Americans referred to this bilateral consultative mechanism as the "Hosono Process" in honor of Goshi Hosono, who presided over the meetings and later served as Japan's nuclear disaster minister.

Though frustrated before establishing this structure, the Americans involved in the conferences praised the process as an extremely effective means of sharing information and coordinating policy.[39] Energy Department and NRC officials had attributed the problems to stonewalling by the Japanese agencies; now they realized that information from the site was not being accurately relayed even to the Joint Emergency Response Headquarters at TEPCO's Tokyo office, NISA, or even the Cabinet. The US–Japan coordination conferences served as a clearinghouse for all the data available about the reactors and spent fuel pools, and US suspicions began to be eased by the free flow of information.

Improvements in communication were evident not only between the United States and Japan, but also within each nation's government. The establishment of the coordination conferences forced both sides to recognize the importance of speaking with one voice. Both in the United States and Japan, prior to each evening's coordination conference, various government agencies met to share information and coordinate policy regarding the status of operations they were carrying out, future responses, the items needed to carry out assistance efforts, and other matters.[40]

This mechanism for US–Japan policy coordination established a process by which agreements were reached, strategies were crafted, and plans shifted into execution phase. The problems addressed included questions such as when to switch the water being injected into reactors from seawater back to freshwater,[41] when to inject nitrogen (since hydrogen collecting in the reactor containment vessels posed an increasing risk of explosion), and how to treat contaminated water.

---

[39] Personal communication with officials of the US government and the US Embassy in Tokyo, October 27 and 28, November 3 and 4, and December 1, 2011.
[40] Personal communication with Ministry of Foreign Affairs and Ministry of Defense officials, October 18 and 24, 2011.
[41] Moreover, immediately after the disasters struck, the United States advised Japan to inject freshwater, but the freshwater supply available for injection was limited, and the Japanese decided to inject seawater in an effort to prioritize cooling no matter what the liquid involved.

The Japanese were still the primary force behind any response measures.[42] The Americans recognized the Japanese leadership and clarified the nature of their own involvement: All short-term responses to the constantly changing situation were left to the Japanese, while the Americans would study and advise the Japanese regarding measures that would be necessary in the medium- and long-term to stabilize the reactors in the aftermath of the accident.[43] For example, at a meeting on March 22, TEPCO representatives shared the results of an analysis of salt that had collected on the floor of reactor pressure vessels.[44] The first piece of advice the Japanese accepted after the Americans' shift in posture was to switch the injected water from seawater to freshwater[45] on March 28.

The conferences also resolved the mismatch between the actual needs of the Japanese and the list of items the Americans offered in assistance. In the early stages of the crisis, the Japanese didn't expect that they would eventually need to accept foreign aid; Japan's nuclear community was also reluctant to accept offers in view of the resources that would have to be dedicated to setting up procedures for processing the aid received.[46] Moreover, questions arose over the appropriate relaxation of regulations in a time of crisis; for example, with regard to offers to provide iodine tablets, there was a question of how to accept these items under domestic pharmaceutical regulations. In practice, the domestic supply of iodine tablets turned out to be sufficient, so there was no need to accept foreign offers. In general, many of the foreign aid offers turned out to involve capabilities that Japan already had or equipment that was incompatible with Japanese technology. The US–Japan coordination conferences played an important role in rectifying such information gaps and mismatches.

Liability questions

In the course of the crisis, Japan switched from being a longtime provider of aid into a new role as a recipient of assistance from the international community. One question that emerged was whether liability exemptions applied; generally speaking, both government institutions and private-sector corporations expect

---

[42] Moreover, documents from the US NRC and other sources include warnings emphasizing the importance of playing only a supportive role and not appearing too far out in front.
[43] Personal communication with a US Energy Department official, November 3, 2011.
[44] "Preliminary Analysis on Salt Accumulation on RPV Bottom," a document prepared by TEPCO for a meeting on March 22, 2011; NRC Freedom of Information Act document ML11269A172. According to this document, both NISA and the Fukushima Daiichi Nuclear Power Station manager had expressed concerns on March 16 regarding the accumulation of salt on the reactor floors. The salt was predicted to reach the bottom of the fuel in units 1 and 2 on March 31, and in unit 3 on April 2; this contact would reduce the cooling capacity of the circulating water. See: US Nuclear Regulatory Commission (2011) Document ML11269A172. Obtained by Freedom of Information Act request. Available at: www.nuclearfreeplanet.org/nrc-foia-documents.html
[45] Personal communication with a US Energy Department official, November 3, 2011.
[46] Personal communication with a Ministry of Foreign Affairs official, October 18, 2011.

to be exempted from liability for any problems that may arise from the use of resources or equipment provided during an emergency such as the Fukushima Daiichi nuclear accident.

If the party that receives the assistance is a private-sector corporation, the provider's exemption from liability will typically be spelled out in a contract between the two parties. If the recipient is a national government, then the International Atomic Energy Agency's (IAEA) Convention on Assistance in the Case of Nuclear Accident or Radiological Emergency applies. This rule specifies that aid provided in the event of a nuclear power emergency must be registered and indicates that the providing country may ask the receiving country for a special exemption from liability based on regulations governing collaboration between the two nations. However, even under this treaty, if the two parties cannot agree on a liability exemption, then the provider may be forced to accept some risk of legal action.

On March 12, the IAEA's Incident and Emergency Center sent faxes to all national representatives anticipating aid requests from Japan and asking each nation to notify the emergency center of its capacity to assist.[47] However, the Japanese government did not apply the IAEA's assistance treaty in accepting aid. During the nuclear crisis, Japan received large quantities of aid, but all of it was provided under agreements between Japan and the donor countries. There is no indication that agreements on liability exemptions were drawn up between Japan and any other country, which would mean that, in the event of trouble, the donor country would have borne the risk of legal action.

Despite the absence of specific agreements on liability exemptions, to date, there have been no reported instances of conflict developing between governments or other parties. Even if such a conflict had arisen, the Japanese government would presumably have addressed it sincerely. However, in the case of liability exemptions within aid agreements drawn up between civilians (as well as liability waivers for aid workers), there is the possibility of incidents resulting in damages that couldn't be compensated by the Japanese government under domestic Japanese law.

How to supply international aid under existing treaties while minimizing liability risks to donor nations is a tricky question, which deserves more thorough study. This issue is particularly relevant in the context of a nuclear accident, where the monetary value of damage compensation could be huge.

Finding a framework for the future

The Fukushima nuclear accident tested the US–Japan alliance on its ability to manage a crisis, protect public safety, and operationalize a response. After the initial natural disasters of the earthquake and tsunami the situation at the

---

[47] US Nuclear Regulatory Commission (2011) Document ML1128A114. Obtained by Freedom of Information Act request. Available at: www.nuclearfreeplanet.org/nrc-foia-documents.html.

Fukushima Daiichi plant deteriorated rapidly, and required extremely rapid decision making. In many ways, the accident presented a threat that can be compared, without overstatement, to an act of nuclear terrorism, demanding simultaneous coordinated responses from many different government agencies. However, because no one had anticipated such a severe and complex disaster, the decision-making process regarding specific response measures was disorganized and groping.

By establishing a two-tiered system for exchanging information (consisting of daily US–Japan coordination conferences and separate meetings between lower-level personnel regarding technical operations), the United States and Japan developed a "whole-of-government" approach to their collaboration, which was effective on two levels: It allowed high-level political leaders in both governments to demonstrate their commitment to resolving the problem, while simultaneously achieving effective technical cooperation between experts on both sides. This process not only prevented a rift in US–Japan relations, which had been threatened by conflicts arising from poor communication, but also streamlined decision making within each of the two governments.

The creation of an operational framework that enables the US–Japan alliance to prepare for a nuclear accident or a nuclear terrorist attack is an important future challenge. The traditionally strong military alliance between the two countries played a critical role in the early days of the crisis, and the two military authorities facilitated communication between other government agencies to the best of their abilities. But the limited mandates of those military authorities meant that the militaries couldn't coordinate a response to the Fukushima crisis on their own. Ultimately, crucial roles were played by corporations and government organizations other than the defense and diplomatic agencies that have traditionally overseen the US–Japan alliance. The response demonstrated that the key to success is to design a systematic collaborative framework, which allows for the broad distribution of information. This framework must include not just the two national governments, but also organizations encompassing a full range of capabilities, both inside and outside the government. We must move beyond the "whole-of-government" perspective to institute a "whole-of-state" outlook or, indeed, even a "whole-of-alliance" approach.

Chapter 10

Lessons of the Fukushima Daiichi Nuclear Power Station Accident and the Quest for Resilience

In March 2011, the Tohoku earthquake and ensuing tsunami triggered a total loss of electrical power that caused the first Level 7 nuclear accident since Chernobyl, making it among the largest-scale nuclear accidents in history. At the Tokyo Electric Power Company's (TEPCO) Fukushima Daiichi Nuclear Power Station, reactor cores in units 1, 2, and 3 suffered meltdowns; unit 4's reactor building was largely destroyed; and the area around the spent fuel pool was damaged, while hydrogen explosions punctuated the disaster as it unfolded. Radioactive elements blanketed the area.

After the accident, environmental pollution wreaked both physical and economic havoc on the region: Chief among these problems was damage to crops, livestock, and marine products; a slump in food consumption; increased unemployment rates; and a decrease in real estate values. Though there were no reported deaths from severe radiation exposure, nearly three years after the disaster some 150,000 residents in the Fukushima Prefecture continued to live as refugees from the spread of radioactive material and the consequential environmental contamination. Forced to abandon the foundations of their lives and the towns in which they had lived for years, these refugees will forever grapple with anxiety over the long-term risks to their health and that of their children. The accident presented Japan with its worst crisis since World War II, and the dangers it posed have yet to be eradicated. These include not only the impact of radioactive contamination within Japan, but also the criticism from neighboring countries—and the major unease around the world—provoked by the release of contaminated water into the ocean.

The accident at the Fukushima Daiichi Nuclear Power Station and Japan's inadequate response to it have revealed flaws that characterize Japan's systems of governance and crisis management. And yet this accident was no one-off, no fluke, no exception: Instead, this was an accident that could have happened at any time, anywhere in the world.

Multiple natural disasters and a sequence of interrelated crises

The Fukushima accident was a compound disaster involving complex interactions among an earthquake, a tsunami, and a nuclear crisis—one that created extraordinary challenges. The magnitude-9 earthquake and a more than 10-meter-high tsunami damaged or cut off roads, communication, shipping, and logistics, paralyzing governments on both local and national levels and confounding initiatives to respond to the nuclear crisis. Worse, fears of radiation risks complicated responses to both the nuclear and natural disasters.

The nuclear accident itself was also a compound crisis, with meltdowns threatened in three separate nuclear reactors—units 1, 2, and 3—and four separate spent fuel pools. Unit 1's hydrogen explosion blocked work on unit 3, particularly cooling efforts, while the hydrogen explosion at unit 3 complicated efforts to vent and inject seawater into unit 2. Each individual accident obstructed efforts to respond to other accidents, and the result was a parallel chain-reaction disaster involving, for instance, hydrogen explosions that need not have occurred at other plants.

At the height of the crisis, Prime Minister Naoto Kan ordered Shunsuke Kondo, chairman of the Japan Atomic Energy Commission, to draw up plans anticipating the worst-case scenario. Kondo's conclusions, which we acquired for use in our investigation, were as follows:[1]

1 Even assuming that hydrogen explosions release additional radioactive material into the atmosphere and that radioactive emissions from other units continue in their present state, the results of radiation assessments indicate that there would be no need to change the existing 12-mile evacuation zone.
2 However, we anticipate that the damaged fuel in unit 4's spent fuel pool will begin to interact with core concrete, leading to another release of radioactive material. In this case, the shelter zone, the area in which all the residents are ordered to take shelter inside of the building and which is set outside the 12-mile evacuation zone, would no longer be sufficient. At a minimum, within fourteen days of the substantive start of this new radioactive release, the seven-day radiation dose should be measured, and, based on these results, a 31-mile region surrounding the evacuation zone, which is currently designated the area in which all residents are ordered to take shelter, needs to be rapidly evacuated.
3 The surrounding 44-mile region should, for the time being, be designated an indoor-confinement zone. However, because soil contamination levels are high even as far as 68 miles away, relocations should be mandatory for

---

[1] Kondo, S. (2011) *Outline of the Worst-Case Scenario for the Fukushima Daiichi Nuclear Power Station*. Report, Japanese government, March 25.

some regions. In addition, in regions within 124 miles, there will be citizens who will want to evacuate (due to annual radiation doses that significantly exceed natural radiation levels); the government should respect and recognize their wishes.

4 Next, fuel damage in other pools may continue, leading to a core–concrete interaction and causing the release of large quantities of radioactive material into the atmosphere. As a result, forced evacuations may be necessary in regions as far away as 105 miles or more, while voluntary evacuation zones (that is, regions in which citizens choose to relocate due to annual radiation doses that significantly exceed natural radiation levels) may develop up to 155 miles or farther.

5 The geographical area covered by evacuation zones and shelter zones will decrease in time; however, if matters are left solely to natural decay processes, the aforementioned 105-mile and 155-mile regions will remain in place for several decades.

This worst-case scenario, if it had come to pass, would have required planned evacuations of some thirty million people from the Tokyo region. Kondo's presentation suggested that, if hydrogen explosions in nuclear reactors forced work stoppages, or if the fuel rods stored in the unit 4 spent fuel pool were to melt and the melted fuel were to react with concrete to release massive quantities of radioactive material, the consequences would be even more devastating than had been anticipated up to that time.

In reality, the Fukushima accident illuminated the vulnerabilities of concentrating multiple reactors at a single site. These dangers include not only the risk of a compound accident leading to a negative spiral, but also the possibility that hydrogen explosions and the evaporation of water from fuel pools could leave fuel exposed: a danger even more severe than that posed by the reactors themselves.

In practice, Masao Yoshida, chief of the Fukushima power station, responded to the crises in the three nuclear reactors and the four fuel pools. In reality, Yoshida shouldn't have found himself in the situation in the first place.

Could the accident have been prevented?

What was the decisive moment, the point of no return, in the course of the accident? The reactor cores began to suffer damage with the total loss of electrical power on March 11, and the seeds of the crisis were probably sown in the first few hours leading up to that evening, when the need to vent became urgent and the injection of seawater became unavoidable.

There is no question that there were multiple human errors made at the nuclear plant site during the crisis. The failure to recognize the operational status of the isolation condenser system was perhaps the most serious of these errors.[2]

A man-made accident? The lack of protections against severe nuclear power accidents

But the human errors were not limited to misunderstandings on the part of any one individual: The plant's operational control manager, the plant's chief, and even the nuclear power department at TEPCO headquarters all equally failed to recognize the problem.

TEPCO's accident-procedure manual neither considers the possibility of a total loss of electrical power[3] nor incorporates severe-accident defenses. Prior to the accident, not one of the plant's operators had ever actually operated the isolation condenser system. They were pitched headlong into the crisis with no education or training in severe-accident response, with no instruction manual, and without even the ability to read their measurement equipment in the pitch-black environment of the powerless facility.

The fact that the loss of cooling capability—the facility's last hope—was not addressed until early the next morning, on March 12, vividly accentuates the man-made character of this disaster. But the incident was man-made in more than one way: TEPCO, as a company, was negligent in instituting defenses against severe accidents involving a total loss of electrical power, and this organizational negligence toward severe-accident protections is perhaps the most significant man-made cause of the disaster.

The International Atomic Energy Agency (IAEA) clearly states, "Responsibility must be borne by the organizations or individuals responsible for facilities and activities involving radiation risk."[4] In the case of the Fukushima disaster, TEPCO must bear primary responsibility. And yet, judging from that accident and the responses to it, it must be said that TEPCO, to a remarkable extent, seems to have shirked all sense of responsibility.

Behind the negligent approach to accident defenses lay flaws in the managerial style and corporate culture of TEPCO, which de-prioritized the culture of nuclear power safety. From the beginning of the disaster on March 11 to 10 a.m. on March 12, TEPCO's two top executives—its chairman and its

---

[2] Cabinet Office, Government of Japan (2011) *Interim report of the Government Investigation Committee on the Accident at TEPCO's Fukushima Nuclear Power Stations*. Report, Government Investigation Committee on the Accident at the Fukushima Nuclear Power Stations, December 26. Tokyo: Cabinet. Available at: http://jolisfukyu.tokai-sc.jaea.go.jp/ird/english/sanko/hokokusyo-jp-en.html.
[3] The Tokyo Electric Power Company, Inc. *Manual of Operational Procedures in Case of an Accident*. Tokyo: TEPCO.
[4] International Atomic Energy Agency (1991) *Basic safety principles for nuclear power plants*. Report, International Nuclear Safety Advisory Group. Vienna: IAEA. Available at: www-pub.iaea.org/MTCD/publications/PDF/P082_scr.pdf.

CEO—were both away from corporate headquarters. During this time, TEPCO was unable to engage in timely "organizational decision making." During the initial response to the accident, the Japanese government and TEPCO were unable to forge a collaborative crisis-response team. A major cause of this was TEPCO's failure to deliver a swift and effective organizational response.

However, the fact that TEPCO was allowed to operate despite having prepared only meager and inadequate accident response policies—which failed to anticipate severe accidents caused by prolonged loss of all electrical power—is not just the fault of the company. The responsibility for such deleterious actions are shared with Japan's Nuclear and Industrial Safety Agency (NISA), the body responsible for safety regulations, and the Nuclear Safety Commission (NSC), which serves as NISA's regulatory assessor.

The NSC's "Safety Evaluation Guidelines" for light-water reactors state:

> There is no need to consider the possibility of a prolonged loss of all AC power because, in such cases, it may be expected that power lines will be restored or that emergency power generators will be available. If the reliability of emergency AC power generators is sufficiently high (due to the structural or operational configuration of the system, for example, if emergency generators are kept in a continuously operating state), then total loss of electrical power is a possibility that need not be considered in plant designs.

This section of the guidelines does not address the possible loss of DC power.

In the case of the accident at the Fukushima Daiichi Nuclear Power Station, the possibilities that "power lines will be restored or emergency power generators will be available" did not materialize. Indeed, the accident was caused precisely by a prolonged loss of both DC and AC electrical power. The NSC clearly deserves a hefty share of the blame for this lax oversight.

The off-site center—an idea introduced after the 1999 Tokai-mura nuclear accident—offers yet another example of a failed defense. Such centers were created to establish a facility response headquarters to serve as a first-line command center immediately after any accident. But during the Fukushima accident, the off-site center proved utterly useless. Indeed, despite the ostensible role of the center as a disaster-response headquarters, it was not even equipped with filters to cleanse the air. NISA, which oversees the operation and control of the off-site center, clearly also deserves a significant share of the blame.

The most problematic incident of all was the case of SPEEDI (System for Prediction of Environment Emergency Dose Information). Before the accident at Fukushima, the system was widely acclaimed as a valuable defense technique allowing predictions of the spread of radioactive material; however, as the accident unfolded, the system was discarded and its impact was crippled by, what Ministry of Education, Culture, Science, and Technology (MEXT) and

NSC reported to be, the "uncertainties in obtaining data on radioactive sources."

Between March 11 and 15, the government (including the Fukushima prefectural government) implemented a five-stage citizen evacuation procedure. However, prior to the government's public release of SPEEDI's simulation results on March 23, Prime Minister Kan and his administration, despite being aware of the SPEEDI results, did not use those results as a basis on which to make decisions regarding citizen evacuations. Various agencies, including NISA, NSC, and especially MEXT, emphasized their suspicions of the SPEEDI results and the limits of the usefulness of those results; moreover, there was concern that publicly disclosing the results would provoke unnecessary chaos and confusion among the residents of the affected regions and the Japanese people as a whole.

In hindsight, the fateful day was March 15, when the spread of radioactive material jumped dramatically, and the guard-wall strategy that attempted to confine the spread of radiation was conclusively demonstrated to have failed. Which citizens, in which regions, at what time, should be evacuated—and to where? Which citizens should be ordered to remain indoors, and for how long? Weren't thirty years' worth of development efforts invested in SPEEDI to answer precisely these sorts of questions?

It may well be true that SPEEDI, which requires the Emergency Response Support System's input on the sources of radioactive emissions, could not have been expected to make accurate predictions without accurate input data. However, SPEEDI was still useful at times: If nothing else, its calculations on March 15 indicated that, on that afternoon, radioactive material would spread over a large region of northwestern Japan. Although government agencies and administrative bodies were aware of these results, no one—not a single person—said, "Northwestern Japan is in danger." Instead, all government agencies, and particularly MEXT, chose to pass the buck to their institutional counterparts, engaging in passive authority battles and shirking their own responsibilities as soon as this became a political issue.

In retrospect, it seems that both SPEEDI and the off-site centers were little more than shiny security baubles introduced to secure the docility of the public and obtain support for nuclear plant construction. The government should have disclosed the SPEEDI simulation results in a timely fashion.

Of course, the effectiveness of SPEEDI as a decision-making resource in ordering citizen evacuations cannot be overestimated. Indeed, citizen evacuations require the most difficult type of government policy deliberations and surely could never be reduced to simple yes-or-no decisions based on simulation results alone. Moreover, SPEEDI cannot serve as a substitute for difficult government decision making and must certainly not be used as an excuse for avoiding it. Instead, SPEEDI is best seen as a protective tool to be used in the early stages of a nuclear accident, where it can offer one important type of warning.

Myth of absolute safety and the trap it poses

But why were defenses against nuclear plant accidents so inadequate? The answer almost certainly lies in the pernicious myth of absolute safety that developed within the nuclear power community; the myth implied that the mere existence of severe-accident defenses would provoke public anxiety over the safety of nuclear power generation. The myth of absolute safety is an ideology comprising two supporting tiers: On the surface is a certain social psychology that considers any discussion of the risks of nuclear accidents to be taboo, while underneath lies the self-interest of the "nuclear power village" that seeks to promote nuclear power generation. The promotion of this myth was deemed a necessary step in securing societal consensus on the introduction of nuclear power, the creation of an atmosphere willing to tolerate plant construction on certain sites, and the entrenchment of a positive image. As a foundation to all of this was Japan's own experience as a target of nuclear bomb attacks over Hiroshima and Nagasaki during World War II and the entrenched antinuclear sentiment it engendered.

Since the 1970s, as nuclear plant construction proceeded in Japan, industrial, governmental, and academic institutions, as well as local governments in regions surrounding nuclear plant sites (which constitute the horizontal and vertical strata of the nuclear power village), downplayed the risks of nuclear accidents in order to mollify the antinuclear power movement. The nuclear power village feared that any honest exposure of the true risks would provoke calls to shut down nuclear plants until all risks could be removed. With regard to nuclear power safety, both power companies and regulatory institutions tended to disfavor open disclosure of safety policies and preventative measures for fear of provoking fear and misunderstanding among the populace. Japan's nuclear power politics and administration became steeped in a culture that sought to purchase the small-scale comforts of the Japanese people even as it put at risk the large-scale safety of the nation and its citizens. Any risks that might surpass minor comforts to threaten public fear were dismissed as outside the realm of expectation.

TEPCO has stated that, in terms of the Fukushima accident, "the tsunami significantly exceeded TEPCO's expectations."[5] Yet, as research progressed into the Jogan tsunami, which deluged much of Japan's Sanriku region in the year 869 AD, attention shifted to the significance of this natural disaster; that is, it became clear that the height of the tsunami that deluged the Fukushima plant indeed should have been anything but "outside the realm of expectation." In fact, at the Tokai Daini Nuclear Power Station, the anticipated tsunami height had already been raised, and tsunami defenses for seawater pumps had been strengthened. Similarly, at the Onagawa nuclear plant operated by Tohoku

---

[5] Tokyo Electric Power Company, Inc. (2011) *Fukushima nuclear accidents analysis report (interim report) supplement*. Report, December 2. Tokyo: TEPCO. Available at: www.tepco.co.jp/en/press/corp-com/release/betu11_e/images/111202e14.pdf (accessed August 31, 2013).

Electric Power Company, the anticipated tsunami height was raised above the expectations incorporated in the original plant construction, with additional leeway to account for the site's elevation. Moreover, in 2006, TEPCO's Nuclear Quality and Safety Management Department announced the results of simulations indicating the possibility of a tsunami higher than the "anticipated" tsunami height at the Fukushima plant, but this finding was rejected as "academic" by the upper-level management of TEPCO's Nuclear Power Division.

The tsunami that shook the Fukushima plant was clearly not "outside the realm of expectation." Voluminous research placed it squarely within that realm, but TEPCO turned a deaf ear to these findings. Quite simply, TEPCO's expectations were wrong. Indeed, the very utterance of the phrase "outside the realm of expectation" is itself an abdication of risk management responsibility. Meanwhile, regulatory bodies, despite recommending to nuclear plant operators that they take into consideration the results of new tsunami risk research, did not request any specific new provisions and considered the issue to lie outside their regulatory authority.

The myth of absolute safety blocked implementation of the so-called "backfit approach," in which new scientific knowledge and the latest technological developments are incorporated into existing nuclear power generation systems in order to improve security. In the case of Fukushima, Japan's power companies and regulatory bodies feared that any safety improvements would provoke criticism that the existing safety provisions and regulations were inadequate—and then such criticisms would have to be addressed. In addition, they feared that the public would demand that nuclear reactors be shut down until all such safety improvements had been fully implemented.

Flaws in safety regulation governance

A characteristic feature of Japan's nuclear safety regulatory system is its "double-check" structure, which assigns responsibility for promotion and regulation to two separate, independent bodies: NISA, a regulatory organization that exists within the Ministry of Economy, Trade, and Industry's (METI) Agency for Natural Resources and Energy, and the NSC, which exists within Japan's Cabinet Office and acts as NISA's "regulatory assessor." The two-tiered system of nuclear power administration dates back to two earlier institutions of Japan's government: the Ministry of International Trade and Industry (MITI) (today's METI) and the Science and Technology Agency, which was subsequently incorporated into MEXT.

Even before Fukushima, the international community noted the flaws in this structure. For example, the "Report to the Japanese Government" prepared by the IAEA's Integrated Regulatory Review Service (IRRS) in June 2007 recommended that the "role of the Nuclear and Industrial Safety Agency, as the regulatory body, and that of the Nuclear Safety Commission, especially in preparing safety guides, should be clarified."

The NSC responded to this recommendation by ignoring it. In March 2008, the NSC's "Chairman's Comment Regarding the Results of the IAEA/IRRS Assessment" stated: "Japan's regulations are unusually outstanding even in comparison to international standards, and we are delighted to receive this highly positive assessment of the functional efficacy of Japan's system for ensuring nuclear safety."

The extent to which this self-regard was utterly unwarranted is now, of course, tragically clear.

What this incident demonstrates is Japan's psychological tendency to think in terms of a unilateralist approach to nuclear safety. Japan somehow managed to convince itself that its nuclear power regulatory system and its culture of safety regulations, surpassed international standards—a sense of superiority accompanied by a "Galapagos effect" in safety regulations. This phenomenon is illustrated by the facts that mandatory defenses against severe accidents and the philosophy of international collaboration to prevent terrorist attacks on nuclear plants never took root in Japan.

It must also be said that, as a regulatory body, NISA suffered from inadequacies of mission, capabilities, and human resources. Ultimately, the agency failed to cultivate a generation of safety regulation professionals. When the Fukushima accident struck, top NISA officials were unable to provide direct answers to questions from core staffers in the Cabinet; the agency was unable to plan or even propose a sophisticated roadmap for responding to the accident—and, with regard to TEPCO, the agency was unable to do more than appear humbly before the company on bended knee, reporting on its activities, and forever playing catch-up on the latest developments in the accident response.

In the discomfiting words of a member of the prime minister's inner circle who had previously worked at METI, the agencies charged with regulating nuclear power safety "thought they were regulating TEPCO, but in fact were being used as tools." In terms of nuclear safety regulations: Within the relationship between the regulatory agencies and TEPCO, the latter always had the upper hand in practice, enjoying a position of superior technological, informational gathering and political skills. When the crisis struck, NISA had no choice but to rely on TEPCO's resources, capabilities, and information. And yet the crisis clearly exceeded the limits of TEPCO's own abilities. The Fukushima nuclear crisis was, above all, a crisis of governance in safety regulations.

In the future, Japan must find a way to overcome its bifurcated approach to nuclear power administration and the redundancies in its safety regulatory institutions; most importantly of all, it must establish nuclear regulatory bodies that are independent of government efforts to promote nuclear power.

After the Fukushima accident, the Japanese government established a new regulatory agency for nuclear power. It is designed to exist outside the Ministry of the Environment, but the question remains whether the agency's independence will be in its institutional DNA. Furthermore, the question remains whether this new agency will be capable of responding to severe accidents and nuclear terrorist attacks. There is ample cause for skepticism. The most critical priority must be to establish expert task forces capable of responding adequately to severe accidents.

However, the greatest challenge of all is posed not by organizations, but by people. Mere functionaries and empty suits are incapable of dealing with crises, a lesson the Fukushima crisis taught in more excruciating detail than we could ever have wanted. Why were there no professionals in the upper ranks of NISA? The answer lies within the flawed institutional culture and personnel policies of Japan's bureaucracy, in which the top-level positions at NISA were part of a rotating array of executive offices within the parent agency (METI's Agency for Natural Resources and Energy), occupied by a bureaucrat for two to three years before the next periodic personnel reshuffling. If regulations are ever to carry any force, then leaders of the regulatory bodies must be experts in their fields, with many years of supervisory experience. This is necessary because regulated parties typically do not play straight with the institutions that regulate them. Safety regulation is a field in which both politicians and bureaucrats find it difficult to accrue many perquisites or career advantages. It was expected that nothing would go wrong, and if it did, NISA or MEXT officials would be persecuted over its responsibilities. Within the culture of bureaucrats in Kasumigaseki (the Tokyo district that is home to many government agencies), the safety regulator position is considered anything but a plum career. And yet nuclear power safety regulators are charged with nothing less than protecting the nation's people, the most important government responsibility of all. Japan must cultivate a new generation of core experts: dedicated professionals, with a strong sense of mission, who have made safety regulation their life's work. Furthermore, Japan must be prepared to reward these professionals with the treatment they deserve.

The ambiguity of "privately administered government policy"

Nuclear power generation in Japan has traditionally operated within a framework known as "privately administered government policy" (*kokusaku minei*): The national government establishes government policy to promote the peaceful use of nuclear power, while the actual business of nuclear power generation is privately administered by private-sector corporations.

The Fukushima disaster exposed the reality that, even if the operation of nuclear power plants by private-sector corporations (power companies) creates no problems under ordinary circumstances, in the event of a nuclear crisis, only the government can shoulder the ultimate responsibility to respond. The crisis-response capabilities, the decision-making procedures, and the failures of governance exhibited by TEPCO throughout the Fukushima crisis have forced the Japanese people to grapple with the question of whether such private-sector corporations can be trusted to operate nuclear power plants.

However, these suspicions should not be directed solely at TEPCO. The Fukushima crisis also revealed an appalling lack of responsibility on the part of the national government. Thus, the Japanese people have been forced to consider another difficult question: Can we really entrust nuclear power administration to such a government? In any event, no matter what type of nuclear power system Japan adopts in the future, it is inescapable that the national government will bear the ultimate responsibility for regulating nuclear power safety, handling radioactive waste, and responding to severe accidents.

Within the parallel chain-reaction disaster at the Fukushima Daiichi Nuclear Power Station, it is suggestive that the unit 4 spent fuel pool was the weakest link in the chain. Due in part to delays in substantive operations at the Rokkasho Reprocessing Plant in the Aomori Prefecture, spent fuel at Japanese nuclear plants continues to be stored in pools inside buildings, but this obviously presents a major risk. The Fukushima accident has clearly identified fundamental problems in the nature of Japan's nuclear fuel cycle.

TEPCO's "abandonment" episode in the pre-dawn hours of March 15, and the action it provoked (namely, the dawn excursion of Prime Minister Kan and a team of his core staffers to TEPCO headquarters, where they set up shop inside TEPCO's Crisis Management Center), demonstrate that, in the end, the national government was forced to shoulder the final responsibility for resolving the accident. According to testimony provided to us by politicians in the core ranks of the Kan administration, during this excursion, Kan addressed more than 200 TEPCO employees in the operational room at TEPCO headquarters, delivering the following admonitions:

> "If you shirk these duties, within a few months every single nuclear reactor and spent fuel pool will fall apart, emitting radiation. That's ten or twenty units, each two or three times the size of Chernobyl."
>
> "If this comes to pass, there will be no more Japan. No matter what, above all else, you must put your lives on the line to get this situation under control."
>
> "I absolutely refuse to stand by silently while you abandon the plant. If you abandon the plant, do you think foreign countries, like the US and Russia, aren't going to do anything? They'll be saying, 'Step aside and let us take care of it!'"
>
> "You guys are the operators, for heaven's sake. Give over your lives. Even if TEPCO tries to run away from this, you will never escape it. I don't care how much money it costs. There can be no talk of abandonment when Japan is at risk of collapsing. If you abandon this plant, then TEPCO will absolutely 100 percent fall apart."

Here was Prime Minister Kan asking TEPCO employees to "put their lives on the line." At the accident site, a "suicide squad" of workers had been assembled to vent the reactors and perform other critical operations, all while suffering exposure to radiation.

And yet ultimately it was the Self Defense Forces (SDF) that came to the rescue. The SDF spearheaded efforts to inject water into the reactors and the spent fuel pools, even as radiation levels at the site increased. In the words of an official in the Ministry of Defense's Joint Staff Office, this was "an action without a plan, without a strategy, and without any information," and yet the members of the SDF unflinchingly carried out their duties.

Japan's legislative system must clearly define the national government's responsibilities and the roles played by task forces in responding to a severe nuclear accident. In the future, Japan should seek to establish a substantive task force organization for responding to severe disasters and accidents, comparable to the US Federal Emergency Management Agency (FEMA).

Safety without security

Another reason that it is necessary to clarify the government's ultimate responsibility is the increasing importance of security at nuclear plants. The Fukushima accident not only awakened the public to the threat of nuclear terrorism, but also revealed the utter inadequacy of existing defenses against this threat.

The accident exposed the fact that the absence of defenses in the nuclear security realm had served as a convenient excuse for the absence of defenses in the safety realm. The international community, and especially the United States, had repeatedly expressed concern over Japan's meager provisions for nuclear plant security. In particular, after the 9/11 terrorist attacks in 2001, the United States moved to strengthen counterterrorism measures at nuclear power facilities; as part of these efforts, in February 2002, the US Nuclear Regulatory Commission (NRC) released its B.5.b guidelines for nuclear plant operators, which sought to achieve "damage minimization" in the event of an attack.

The B.5.b guidelines require provisions to ensure that cooling capability for nuclear reactors and spent fuel pools is maintained even in the event of an explosion or fire provoked by a terrorist attack. More specifically, the guidelines require the following:

1 Preparation of equipment and personnel capable of responding to certain contingency situations.
2 Measures to maintain or restore the capabilities of spent fuel pools.
3 Measures to maintain or restore the capabilities of reactor cooling and containment vessels.

Among these requirements, item 2 in particular requires multiple redundant methods of supplying water and flexibility and dynamic independence in water supply equipment. In addition, item 3 mandates that emergency-response plans be established for twelve distinct contingency situations, including loss of all AC or DC electrical power, or both, needed to operate a plant.

In 2006, the NRC informed NISA of the B.5.b guidelines and encouraged Japan to strengthen its nuclear terrorism defenses, but NISA is said to have rejected even the mere suggestion. To this point, between August 2011 and September 2012, Hiroyuki Fukano, NISA's director general, suggested that the handling of sensitive information in relation to nuclear security was an obstacle to policy collaboration between the two governments. Fukano stated, "According to NISA officials at the time, they were unable to get documents and resources from the NRC, and they weren't permitted to take notes." On the other hand, a NRC official we interviewed for our investigation stated, "The NISA folks insisted from the beginning that the problems of terrorism were nonexistent in Japan," articulating the NRC's distrust of Japan's approach to counterterrorism and nuclear security initiatives.[6]

"In the '90s, we showed the Japanese our severe-accident guidelines," continued this NRC official, "but the Japanese, and particularly TEPCO, were totally uninterested. Then, after the 9/11 attacks, we started preparing defenses for nuclear terrorism scenarios involving airplanes crashing into nuclear plants, but this time it was NISA that was totally uninterested." If, as this official's testimony suggests, Japan's safety regulatory institutions failed to heed direct warnings from their American counterparts, this would amount to grave misconduct.

It seems likely that stronger security measures to prevent nuclear terrorism would have bolstered the defense-in-depth of nuclear plants; meanwhile, by strengthening the defense-in-depth implementation, it might have been possible to get past the myth of safety and instead transition to a more realistic risk management approach.

In the final analysis, there seems to have been only the faintest awareness that nuclear power might pose problems beyond the serious risks to human life and health presented by the possibility of accidents; in addition to these dangers, nuclear power presents national security risks that can threaten Japan's very existence as a sovereign nation, as well as international security risks that can threaten the safety and security of the entire planet. TEPCO's dumping of contaminated water into the ocean—without first notifying affected nations—is one example of this unilateralist approach to nuclear safety at nuclear plants.

A nuclear accident anywhere is an accident everywhere. No matter where the accident may occur, its impact will spread far and wide.

Japan must take this fundamental lesson to heart, and it must seek insight from the framework of international collaboration—a framework with a proven track record of effectiveness in nuclear power safety, nuclear security, and nonproliferation.

---

[6] Personal communication with NRC official, August 24, 2011.

Between March 14 and 16—the darkest hours of the crisis—the very core members of the Cabinet repeatedly said, "Fukushima has become a war zone."[7] If the worst-case scenario had come to pass, it would have been necessary to evacuate some thirty million residents from the Tokyo area. And in the face of such an existential threat, Japan's government, and indeed the nation of Japan itself, proved to be equipped with profoundly inadequate capabilities.

Fukushima forced the nation to come to grips with this painful reality. On March 16, the US government, in commenting on the Japanese government's nuclear accident response, insisted on the need for a whole-of-government approach, which was an unusual order to make.

Without consolidated control, the Japanese government simply cannot mobilize the nation's full supply of resources. Such mobilization is hampered by the barriers of a stovepiped bureaucracy in which information tends to sink to the bottom of each agency's file cabinets, never again to see the light of day. Where does the ultimate responsibility for protecting the people of Japan reside, and who decides? The Fukushima accident has presented the most profound challenge to Japan's core identity since the end of World War II.

Crisis management and leadership

The accident responses of TEPCO and the Japanese government raise fundamental questions on the nature of Japanese crisis management and leadership. When plant chief Masao Yoshida received orders from TEPCO headquarters to cease the injection of seawater into unit 1, he chose—of his own accord—to continue the seawater injection nonetheless. During a videoconference with TEPCO headquarters, Yoshida uttered, what colleagues at TEPCO headquarters said sounded like, "terminate." In reality, Yoshida had stated to his employees, sotto voce, so as not to be heard over the microphone: "Shortly there will be an order to terminate seawater injection; however, whatever you do, do *not* stop injecting seawater." This incident vividly illustrates the twofold and even threefold breakdown in the chain of trust that arose between TEPCO headquarters and the Fukushima Daiichi Nuclear Power Station, between administrators and the nuclear power village, and between the government and nuclear plant operators; the incident may also be seen as an example of how the leadership vacuum at the highest institutional levels forced on-site personnel to shoulder the burden of critical decision making.

From plant chief Yoshida on down, employees of TEPCO's Fukushima Daiichi Nuclear Power Station were putting their lives on the line to respond to the accident. Many observers have praised Yoshida's courage, sense of mission, and decisiveness throughout the crisis, and indeed such praise is unquestionably warranted. The incident of the seawater injection further

---

[7] Personal communication with an official from the Prime Minister's Office, January 19, 2012.

illustrates how Yoshida's independent-mindedness saved the day from the weaknesses of decision making and the paucity of governance that prevailed at TEPCO headquarters, which neither produced a clear policy or precise action plan for responding to events at the Fukushima site nor relayed feedback and information from the site to the Prime Minister's Office. When the crisis struck, TEPCO headquarters was reduced to doing little more than spinning its wheels in an unproductive tizzy.

Yet, having pointed out these serious flaws in governance at TEPCO headquarters, we must also acknowledge that the independent-minded decisions made at the site, whether or not they were ultimately correct, unquestionably created new problems. In particular, the prevailing attitude that there is real power in snap, on-site decisions is problematic from a crisis-management perspective. Relying on site supervisors to act on their own whenever a serious crisis or accident occurs can lead to unexpected complications and even an unchecked escalation in the scale of an event. If such an accident escalates into a large-scale crisis, the final responsibility must rest with the higher-level institutions, ultimately the national government, and cannot be replaced by the authority of plant managers.

One of the most critical decisions for the political leadership—and one that is parallel to that of reactor stabilization policy in its life and death importance—is that of citizen evacuation. Within a twenty-four-hour period beginning on the night of March 11, the government progressively broadened the citizen evacuation zone in four stages, proceeding from 1.2 miles (ordered by the Fukushima prefectural government), to 1.9 miles, to 6 miles, and finally to 12 miles. This was primarily a preventative measure, and it is praiseworthy for having protected many citizens from radiation exposure.

However, once the orders were issued, efforts to double-check them and assist in their implementation were inadequate, and disclosures of the data and assessments that formed the basis for the orders were even more inadequate. The government should have paid more attention to the possibility of hardships in daily life due to prolonged indoor confinement and should have anticipated measures to mitigate these hardships. Moreover, the voluntary evacuation order issued on March 25 invoked a concept that is not addressed by disaster-prevention policies; this was an ambiguous proclamation that seemed to assign a difficult decision to citizens, who lacked the proper information to make an informed judgment. In the future, it seems clear that the use of voluntary evacuations as an emergency protection measure should be avoided as much as possible.

Once the crisis began and unexpected events unfolded, the government was unable to grasp the full complexity of the situation and was unable to keep events from spiraling out of control. The government not only lacked scientific and technological expertise, but also failed to exhibit decisive political and administrative authority. The exchange of information and status evaluations between TEPCO and the government was particularly rocky, as was communication on their decisions and intentions.

Indeed, these communications were not even possible until March 15, when TEPCO and the government established a Joint Emergency Response Headquarters (JER-HQ) within TEPCO's headquarters. It was only after the establishment of this crisis center and the launch of the US–Japan coordination conferences on March 22 that Japan adopted a whole-of-government approach to the crisis response.

The true story of the incident that led to the establishment of the JER-HQ—namely, TEPCO's "abandonment" episode—remains shrouded in mystery. TEPCO CEO Masataka Shimizu told the Prime Minister's Office: "Because the plant is in a critical state, we wish to explore the possibility of temporarily withdrawing workers without a direct role in response efforts, a step that we think will probably be necessary one way or the other." As for the abandonment (*tettai*) of the plant by all workers, Shimizu has insisted that "I neither considered it nor mentioned it."[8]

However, in an interview we conducted with Minister of Economy, Trade, and Industry Banri Kaieda and Chief Cabinet Secretary Yukio Edano, Kaieda stated that Shimizu[9] said nothing about the nature, scale, or means of any "withdrawal," while Edano stated that Shimizu's words were interpreted as signifying an abandonment of the plant.[10]

Nonetheless, why did Shimizu take the unusual step of placing multiple phone calls, in the middle of the night, to politicians in the inner circle of the Prime Minister's Office? TEPCO has yet to offer a convincing explanation for this behavior. Even if we suppose that the interpretation of the Cabinet—that TEPCO was planning to abandon the plant—was only a misunderstanding, why would Shimizu choose words so susceptible to misunderstanding? By suggesting that TEPCO was planning a full-scale abandonment of the plant, was Shimizu attempting to force the government into a full-scale intervention, and thus to pin the responsibility on the government? Taking this line of inquiry one step further, did TEPCO intentionally delay the venting of unit 1 in the early morning hours of March 12 in an attempt to force the government to issue mandatory orders, and thus evade responsibility for the release of radioactive material? During our investigation, we did everything we could to understand this point and other aspects of TEPCO's judgment and decision making during the crisis response, but TEPCO refused to grant interviews to our investigators.

The perhaps excessive involvement and intervention of the Prime Minister's Office has been criticized as micromanagement. It is certainly undeniable that

---

[8] International Atomic Energy Agency (1991) *Basic safety principles for nuclear power plants*. Report, International Nuclear Safety Advisory Group. Vienna: IAEA. Available at: www-pub.iaea.org/MTCD/publications/PDF/P082_scr.pdf

[9] This was said during a telephone conversation Shimizu conducted with Kaieda and Edano.

[10] NISA Director General Nobuaki Terasaka, who received a call from CEO Shimizu on the night of March 14, prior to Shimizu's call to Kaieda and Edano, stated to us in an interview: "I am not aware that [Shimizu] informed me of 'total abandonment.'"

Prime Minister Kan, by immersing himself in individual tasks of accident management, failed to direct sufficient attention to overall crisis management. The prime minister tended to direct his focus toward technical details and on-site logistics; for example, Kan personally supervised, by cell phone, the arrangement of mobile power-generator vehicles and spoke directly, by telephone, with low-level NISA section chiefs after concluding that upper-level NISA officials were unreliable. In many cases, Kan's strong use of words and his aggressive interrogations of his counterparts intimidated the bureaucrats and advisers. In some instances (for example, after unit 3's hydrogen explosion on March 14), Kan gave the impression, in the words of NSC Chairman Haruki Madarame, of "a man at the end of his rope."

In hindsight, it appears that, within a day or two of this period, the inner circles of the Prime Minister's Office, especially Kan himself, began to be gripped from time to time by a particularly strong sense of fear and powerlessness. These officials were nervous that public information disclosures of radiation levels would lead to panic among area residents and the Japanese people as a whole. Deputy Chief Cabinet Secretary Tetsuro Fukuyama recalled, "We discussed whether panic would break out in Fukushima, or across Japan," over the relationship of venting, hydrogen explosions, and radioactive material dispersion to citizen evacuations. And yet, without question, this excessive fear of mass panic gave the Japanese people the impression that the prime minister's inner circle had itself descended into a form of elite panic.[11]

However, we must acknowledge that Prime Minister Kan and his staff faced a variety of constraints that severely limited their decision-making abilities. Nor can we deny that these limiting factors essentially forced Kan and his advisers to micromanage the situation, even to the point of excess. The political leadership suffered from an extremely weak scientific and technical advisory capacity. At the height of the crisis, the prime minister appointed six Cabinet advisers, one after another, with whom he communicated directly by cell phone. Fukushima demonstrated just how crucial it is, in the event of a mishap in a massive technological undertaking such as a nuclear power plant, to have expert advisers on hand to deliver technical guidance. Japan must establish a Science and Technology Assessment Agency (or similar institution) to strengthen the mechanisms of scientific and technical advice to the prime minister.

The government's crisis management capabilities were also substandard. The most serious problem was that the emergency-response headquarters did not function effectively. Responsibility for this lies with NISA, which was not equipped with adequate crisis-response capabilities. NISA's response headquarters, its on-site response center, and its off-site center were all incompetently organized and managed. It also must be said that the Nuclear

---

[11] Clarke, L. and Chess, C. (2008) Elites and panic: More to fear than fear itself. *Social Forces* 87(2): 993–1014. Available at: www.bupedu.com/lms/admin/uploded_article/eA.491.pdf.

Emergency Response Headquarters (and the crisis management oversight capability), established in the Cabinet, was ineffective. "Not a single member at the Crisis Management Center ever passed anything upward. No one ever tapped an upper-level person on the shoulder and said, 'Excuse me, but we have received this new information on the nuclear accident,'" recalled a member of the prime minister's inner circle who had previously worked as a bureaucrat. "This is usually described as a crisis of political leadership, but in fact things were even worse at the bureaucratic level. When you get right down to it, any prime minister would have had a tough time working with these people."[12] The Crisis Management Center in the basement of the Prime Minister's Office may have been a working environment in which officials from various government agencies could communicate with each other, but it was certainly not a "situation room" in which calm, calculated decisions were handed down based on information and options discussed thoroughly by political leaders, agency supervisors, and expert staffers.

The key challenge in government crisis management is how quickly a bureaucratic institution can switch from routine operational mode (and the values and behaviors that go along with it) to emergency-response mode. Under ordinary circumstances, bureaucratic institutions prize the values of fairness, efficiency, respect for the law, and a bottom-up approach. But times of crisis call for flexibility, dynamic response, clear priorities, redundancy, and a top-down approach to decision making. In such situations, the boundaries among vertically siloed organizations must be broken down, resources and authority must be consolidated, and skills and abilities must be augmented in unison. But bureaucratic institutions are generally incapable of adapting in these ways, and TEPCO is the epitome of a bureaucratic institution.

The Joint Emergency Response Headquarters established at TEPCO headquarters was really a shortcut to establishing crisis management capabilities inspired by just this type of situation. Although the center offers a classic case study in Japanese administrative leadership in a time of crisis, and although it arose in conjunction with an utter collapse of regulatory governance, nonetheless we must acknowledge that, given the realities of the situation, it was fairly effective in achieving the maximum possible sharing of information, resources, and abilities among TEPCO (both at headquarters and at the Fukushima Daiichi Nuclear Power Station) and government institutions (including the Cabinet, various government agencies—especially the Ministry of Defense, the Self Defense Forces, the police, the fire department—and even the US government).

We must also put a positive spin on the decision, made at the height of the crisis, to anticipate the worst-case scenario. There is undoubtedly room for debate on the way in which the scenario was publicly announced and the records of it that were left behind. Nonetheless, not only did the exercise serve

---

[12] Personal communication with a key Kan administration official, January 6, 2012.

the purpose of pinpointing precisely how bad things could get, but the process of envisioning this scenario also forced officials to face head-on the full impact of the dangers confronting Japan, awakened the government to its own blind spots and gaps in understanding, and provided the impetus to search for and institute dramatic and sustainable response measures. The Kan administration—based on Kondo's scenario—organized multiple project teams to address various tasks: radiation shielding and measures to reduce the emission of radioactive material (including reinforcements for unit 4's spent fuel pool), fuel recovery and relocation, remote-control capabilities, and more.

In the face of the crisis, the government could not have based its response on the nuclear disaster manual, a document riddled with holes and omissions of contingencies declared outside the realm of expectation. Prime Minister Kan's insistence that "there is no theory for accidents"[13] was probably right on the money. We must question the political responsibility borne by previous administrations that established no provisions for responding to accidents, anticipated no worst-case scenarios, and left behind only this utterly inadequate manual.

Perhaps the most troubling aspect of the Kan administration's crisis-response efforts was the way in which risks were communicated. Most citizens are unable to grasp the significance of numerical radiation levels; indeed, it is almost impossible for non-experts to know the relevant criteria and to evaluate which levels are dangerous and which are safe. Citizens need easy-to-understand benchmarks to place risks in context, but during the Fukushima accident, the population was merely bombarded with numbers—from food contamination levels to schoolyard radiation standards—that only served to promote chaos and confusion among the Japanese people.

Even in evacuation zones, remaining residents found men in white protective suits standing on their doorsteps; these men insisted that people evacuate, but refused to answer questions as to why or how. This sort of no-questions-asked communication style was all too prevalent throughout the present crisis.

At the heart of the Fukushima tragedy is the fact that, in the darkest hours of the crisis, the Japanese people lost faith in their government. Above all, emergency communication serves as the central mechanism for establishing trust between the government and the people, and crises cannot be overcome unless the government and the people combine their strengths. The government must be fully invested in safety initiatives and must do everything in its power to protect the nation's people. Experts can contribute to this effort. Indeed, only when citizens have faith in their government can they rely on the comfort of knowing that it will protect them. Of course, the nation's citizens, themselves, bear the ultimate responsibility for crisis management by participating in society and politics—not as consumers of small-scale comforts, but as the architects of large-scale safety.

---

[13] Kan, N. (2012) Panel discussion. In: Annual Meeting of the World Economic Forum, January 6. Davos: World Economic Forum.

The advent of social media has strengthened the nature of communication between government and citizens in times of crisis. Indeed, social media has demonstrated extraordinary effectiveness as a tool for emergency communication. Though we must not overlook the flashing SOS signals that citizens tried desperately to send their government through the noise and fog of the Fukushima disaster, we must be wary of unqualified praise and acceptance of social media: This new medium encouraged widespread rumor and innuendo based on imprecise information. In addition to sending messages to citizens, the government must pay more attention to receiving messages from citizens to avoid mass confusion.

Resilience

Not all aspects of the crisis response were failures. For example, the quake-resistant critical building at the Fukushima plant—the construction of which was inspired by lessons learned from the 2007 earthquake off the coast of the Chuetsu region of Niigata Prefecture—played a major role as a Local Nuclear Emergency Response Center throughout the accident. Although there is room for improvement in the capabilities of the quake-resistant critical building (for instance, it was not sufficiently airtight to protect against radioactive contamination), nonetheless, without this building, it would have been essentially impossible to direct crisis-response efforts from within the plant site.

There is no question that TEPCO was negligent in instituting tsunami defenses. However, at Tohoku Electric Power Company's Onagawa Nuclear Power Plant and Japan Atomic Power Company's Tokai Daini Nuclear Power Plant, tsunami defenses functioned as intended and succeeded in avoiding a total loss of electrical power. The Act on Special Measures Concerning Nuclear Emergency Preparedness, and the nuclear disaster manual, were not usable in their existing form, but nonetheless were not entirely devoid of meaning. The problems resided not in the manual itself, but in the managerial and political agendas—and the individuals executing those agendas—that defined which risks were considered "expected" and which were considered to be "outside the realm of expectation." Of course, no manual based on the experience of previous crises can be used verbatim to dictate responses to the next crisis. Instead, crisis-response policy must incorporate adequate leeway from the start. What is required in the event of a crisis is not so much a predetermined disaster-response plan as a capacity for planning that is ready at all times to respond to a crisis. After all, the same crisis never occurs twice.

Although the Fukushima Daiichi Nuclear Power Station accident was a massive event that rose to the status of a Level 7 disaster, and although it tragically forced the evacuation of some 110,000 citizens who, even after three years, remained unable to return to their homes, to date no victims of acute radiation exposure have emerged. A key official in the Prime Minister's Office surprised our team by confessing, "I keep thinking that the gods must have been looking out for Japan after all."

Since the accident, simulation-based analyses of the accident have progressed, and it is now hypothesized that the majority of the nuclear fuel that melted in the high-temperature environment penetrated the reactor pressure vessel and settled to the concrete floor of the reactor containment vessel.[14] It seems the situation could have come dangerously close to realizing Kondo's worst-case scenario.

There was unquestionably an element of good fortune here. But crisis management does not operate on the basis of good fortune. After all, the same good fortune never happens twice.

There can be no resolution of the Fukushima crisis until we study the causes of the accident and the havoc it wreaked, learn the lessons of the steps taken to respond to it, and forge a new national consensus. Those have been the key objectives of our investigation. Above all, we must strive to enhance Japan's resilience: the resilience of Japan's government, its institutions, and its people.

We are all too aware that our investigation leaves many questions unresolved. A partial list of these questions includes the nature of information sharing and communication among national, prefectural, and local governments on citizen evacuations; the role and capabilities of crisis management provisions, such as the use of SPEEDI results by prefectural governments; the rationale for delays in instituting radiation controls and performing tests for internal radiation exposure among citizens immediately after the accident—namely, why internal radiation tests were not performed on children until August, by which time the radioactive iodine had vanished; the causes and responsibility for the spread of food contamination; the role of the media in risk communication; and the reasons behind the Japanese government's refusal to accept the US government's request for stronger measures to prevent nuclear terrorism.

Efforts to investigate, and extract lessons from, the accident at TEPCO's Fukushima Daiichi Nuclear Power Station and the damage it caused must continue long into the future.

We would like to propose that March 11 be officially designated "Nuclear Power Safety Day."

Remember the lessons of the Fukushima Daiichi Nuclear Power Station accident. Understand the critical importance of nuclear power safety and security. Check and re-check defenses and protections against accidents. Train workers to prepare for all contingencies. Finally, etch permanently into the consciousness of political leaders the crucial importance of leadership and crisis management.

Fukushima must never be forgotten.

---

[14] Tokyo Electric Power Company, Inc. (2011) *On hypotheses regarding the status of damage to units 1–3 reactor cores*. Report. Tokyo: TEPCO.

Epilogue

Committee Member Tetsuya Endo: problems of nuclear power generation revealed by the Fukushima accident

Before Fukushima, nuclear power in Japan had enjoyed a relatively smooth evolution. As a former member of the Japan Atomic Energy Commission and a promoter of nuclear power, I, myself, played a role in this history. Nuclear energy has grown to account for some 30 percent of Japan's total power production, the technical standards of Japan's plants are high, the country respects the IAEA's protocols for ensuring the peaceful use of nuclear power, and in general Japan had come to see itself as an A-plus student among the world's class of nuclear power producers. However, behind this seductive picture lay a raft of serious problems that Japan's nuclear power community and government bureaucracies both chose to ignore, opting instead for an overly confident attitude that was entirely lacking in humility.

The Fukushima accident has performed the service of laying bare many of the problems of nuclear power production in Japan, at the expense of inflicting a horrendous tragedy upon the nation. Can Japan's nuclear power community overcome these voluminous problems to rise again in strength? Before we can answer this question, we must first acquire a thorough understanding of the causes of the Fukushima accident.

This investigative report, in contrast to the reports of the Japanese government and the Diet, is significant for being an independent civilian report with no official ties or constraints; unique characteristics of this report include the unflinching analysis it offers of the factors that lay in the background of the Fukushima accident and the light it sheds on the international context of the accident and its aftermath. In the most direct sense, the Fukushima accident was caused by an earthquake and tsunami—that is, by natural disasters—and natural disasters, representing the wrath of nature, lie outside the realm of human control. But the tragic large-scale nuclear accident that these natural disasters provoked was, in my view, largely a man-made disaster. Hindsight, of course, allows for "if onlys": If only the warning voices of outsiders had fallen on more humble Japanese ears; if only nuclear plants had incorporated more appropriate design constructions and had been equipped with adequate protections against severe accidents. If these had been addressed, I cannot help but think that it would have been possible, at the very least, to reduce the impact of the accident by a large margin. To some extent, man-made disasters can be surmounted by man-made wisdom and man-made skill; I believe that catastrophe can be turned into good fortune.

The problem of the immediate future is how the lessons of Fukushima will be learned, both within Japan and internationally, and how rapidly these lessons can be incorporated in practice. I hope that our investigation will contribute to this effort. I hope that dramatic reforms are in store for nuclear power safety. Although this investigation was a civilian effort, many officials and other involved parties generously agreed to lend us their cooperation; sadly, we were unable to secure the cooperation of the most directly involved party—that is, TEPCO itself.

Committee Member Keiichi Tadaki: Japan's national government must acknowledge its responsibility in the Fukushima accident

The media have focused on TEPCO. The company's chairman and CEO have apologized and the latter was forced to resign. But who has apologized for the government, and how? Not a single government official has taken responsibility for the Fukushima accident and resigned: not politicians, not the chairman of the Nuclear Safety Commission, not the leaders of METI or MEXT. Can there be any doubt that, perhaps due to the media's misguided focus—perhaps due to sheer outrage over the monstrosity of TEPCO's negligence, or for whatever reason—questions regarding the true responsibility for the accident are not being asked?

The national government bears responsibility for ensuring safety at nuclear power plants. Failures of understanding and awareness among relevant government officials must surely rank among the most important factors underlying the failure to prevent the present accident and the inability to minimize the ensuing damage.

Corporations can never fully get past their focus on costs and the bottom line. For example, TEPCO was well aware of advances in academic research into the Jogan tsunami in 869 AD, and the company also knew that it needed to anticipate a tsunami of significantly greater height than previous estimates had indicated.[1] And yet, despite knowing these things full well, the company focused on details such as the cost of installing tsunami guard walls (tens of billions of yen), the duration of construction, and the difficulties of equipment

---

[1] Their knowledge was based on calculations in accordance with the "Long-Term Evaluation of Seismic Activity between the Sanriku Coast and the Boso Coast" prepared by the national government's Headquarters for Earthquake Research Promotion.

installation, and it chose to kick the issue down the road by concluding that the findings of the national government's evaluations did not need to be reflected immediately in plant design considerations. For a corporation, the possibility of a massive once-per-millennium tsunami and a worst-case assessment of the potential damage are not sufficient basis for allocating a gargantuan sum of money and bracing for prolonged operational disruptions. This is why the ultimate responsibility for safety can never be entrusted to a corporation.

Every single aspect of nuclear power production—that is, the choice of site for a nuclear plant, the design, construction, and operation of nuclear reactors, the handling of spent nuclear fuel, disaster-prevention policy, emergency disaster responses, damage containment measures, and other areas—requires sophisticated input from experienced experts in each area; it simply is not desirable to have all of these resources under the control of a single private corporation. The people of Japan placed their trust not in any one private corporation, but in the safety assurances of the national government. This is precisely why the national government is entrusted with strong regulatory authority, covering not only obvious areas such as the choice of plant sites and the installation of earthquake defenses, but all other major and minor details as well, right down to the last solder joint on each individual nuclear reactor. The government must legislatively mandate fine-grained plant inspections in advance of operation, as well as screenings for operational defenses against accidents; the administrative bodies must continually update their safety policies, based on the latest research and knowledge, and must continue to conduct strict plant inspections.

And yet, within the government, the officials charged with ensuring safety suffered from a serious lack of awareness. The off-site center they established as a local emergency-response headquarters for the event of a nuclear disaster was equipped neither with shields to block exposure to radioactive material, nor even with emergency communications equipment. For years, the Nuclear Safety Commission underestimated the importance of tsunami risks; indeed, their subcommittee that was charged with updating seismic-resistance inspection policies did not bother to include a single tsunami expert. Meanwhile, the Nuclear and Industrial Safety Agency, after convening its own working group whose committee members heard warnings based on new research into the Jogan tsunami, readily agreed to TEPCO's proposal to kick those warnings down the road. And the national government never even delivered to citizens the most important information they possessed: the monitoring results of the region surrounding the Fukushima Daiichi Nuclear Power Station between March 11 and 15, as well as the results of SPEEDI (System for Prediction of Environment Emergency Dose Information) simulations.

The confidence of the Japanese people in the administration of nuclear power has plummeted, and the government's declarations of safety have not engendered sufficient support to restart nuclear power generation operations. The days in which the government could avoid a forthright acknowledgement of its responsibility are over. Whether or not Japan resumes nuclear power generation, the relationship between the nuclear power community and the Japanese people will remain in place for many years. In the future, there can be no administration of nuclear power without the understanding and consent of the Japanese people. To secure these, the government must first come forth and acknowledge its own responsibility for the present accident, and must then make clear its intention to assume full responsibility for nuclear plant safety in the future.

I am grateful for the opportunity to serve as an individual Japanese citizen on this civilian investigation committee, and if our work proves to be of even the slightest service to the future of mankind, then my most extravagant hopes will have been exceeded.

Committee Member Ikujiro Nonaka: the government's crisis management methods failed to face reality

Nuclear power generation should be understood as a highly sophisticated intellectual system that aggregates many areas of cutting-edge science. In our analysis and investigation of the Fukushima accident, we not only focused on the technical management and evidence related to the accident, but also on the relationships behind the accident. What this multidimensional approach revealed is that the deterioration of intelligence, resulting from a closed and isolated community, caused the man-made disaster at the Fukushima Daiichi Nuclear Power Station. It became clear from our investigation that both TEPCO and the Cabinet lacked leadership and preparedness at a time of crisis, and thus could not perform adequately.

Prime Minister Kan did not stimulate dialogue among the people involved, but instead focused on analysis of virtual data that was detached from what was happening at the actual Fukushima site. Any real sense of commitment and empathy was destroyed by the distance between the Crisis Management Center—an important connection to the Fukushima front line—based on the basement floor of the Cabinet building, and the Prime Minister's Office on the fifth floor. Information needed to be relayed through multiple layers, and consequently, the internal organizational collaboration was delayed, data was covered up, and trust was lost among the stakeholders. Because Prime Minister Kan had intervened in micro-level issues and relied on particular advisers, he failed to exhibit agility, which would have allowed him to base discussions on collective wisdom, grasp the whole process, and make timely judgments.

We may contrast Prime Minister Kan's leadership style with that of Winston Churchill, who organized a "war room" at a time of crisis. Working around the clock, Churchill shared time and space with the top leaders of industry, government, and military; he formed a task force with these leaders to discuss issues and to encourage multiple viewpoints; and he made judgments based on both macro- and micro-perspectives. In contrast, Japan's Democratic Party advocated "prime minister leadership," which merely led to a weak human network among the administration, the Self Defense Forces, fire departments, police, companies, nonprofit organizations, and community leaders. And even then, at the time of the crisis, the party neglected to establish relationships among all these players. The disaster at the Fukushima Daiichi Nuclear Power Station was similar to war. That is, a "dictator by discussion"—a wise person who grasps the total situation, makes flexible and timely judgments, and empowers the front line—was vital. However, Prime Minister Kan did none of these things; he was nothing like the more practical Churchill. Instead, the Self Defense Forces and the US Marine Corps took on the heavy lifting, and their successful collaboration triggered America's proactive support of Japan.

Masao Yoshida, who was in charge of the Fukushima Daiichi Nuclear Power Station, made a series of quick and thoughtful judgments, including continuing to cool down the fuel rods with seawater. But employees like Yoshida were not rewarded at TEPCO. The company instead promoted only its top management—people with professional backgrounds in planning or purchasing. Though these top-management types were often good at dealing with administrative matters in Kasumigaseki and cutting costs, they lacked the practical wisdom necessary to respond to this crisis. In the past, TEPCO did have a leader, named Kazutaka Kigawada, who aimed to "contribute to society" and promoted an integrated training system. However, at some point, TEPCO began to pursue efficiency merely through virtual knowledge; this led to an over-compliance, over-analysis, and over-planning syndrome.

TEPCO was supposed to foster alliances and establish an intelligent network among industry, government, and academia in order to promote the creation of high-quality thought. It is now ridiculed as just another member of the nuclear power village—where heterogeneous knowledge is rejected and organizational knowledge deteriorates. The media and Japan's citizens must bear some responsibility, too. They did not see the reality of TEPCO as it was, and they were manipulated by the ideology of the Japanese government.

The accident at the Fukushima Daiichi Nuclear Power Station left Japan with many issues to consider, especially the fact that its preparations were sorely lacking and that its judgments and actions after the incident were not nearly responsive enough. In particular, the government's weak crisis-management approach led to a neglect of national security and a failure to respond dynamically at a time of crisis. The fact that no meeting minutes were kept or recorded during the meetings about the Fukushima Daiichi Nuclear Power Station only makes it harder for the nation to learn from its past failures.

We must all face reality honestly, no matter how heavy it may be. We must foster wise leaders who take us toward the future and who lead the world in organic energy systems, national security, and crisis management. We must rebuild Japan to be a resilient and knowledge-based country. It is my sincere hope to contribute to this effort as much as possible.

Committee Member Mariko Fujii: major issues in information disclosure during the crisis

For daily life within the 20-kilometer warning zone, time has come to a standstill. Even today, more than two years later, approximately 150,000 regional residents have no choice but to continue to live as refugees without even the solace of clear projections for the future. Whether Japan abandons nuclear power or moves toward less nuclear power, future energy policy needs to be discussed in realistic and forward-thinking ways; and yet at the same time, we must acknowledge the reality that Japan still has many nuclear power plants, and the process of managing our nuclear powered society will be ongoing for many years, including the road to decommissioning reactors. The studies and investigations that will be required for this process—and the problems it will raise for all of us to solve—encompass the full range of Japan's society and economy.

Risks cannot be controlled, and preventions cannot be put in place, without a firm grasp of the full spectrum of possibilities. The potential problems that a tsunami might cause were well known, but still the preventative measures that were instituted were insufficient. We need to rethink the basic societal framework for overseeing the very process by which plant operators institute all necessary protections, a process in which the crucial stage between recognizing risks and protecting against them must be filled in with input and advice from a wide variety of perspectives.

Accidents and disasters can take many forms. It goes without saying that electric power companies (the operators of nuclear plants) must test and retest each individual stage of their preventative and response measures under a variety of hypothetical accident scenarios; but this alone is not enough. In the event of a compound disaster, the only way to ensure an effective response incorporating all available resources is for government to assume its responsibilities and play a commanding role. Moreover, the question of how best to utilize the contributions of experts must be clarified in advance by performing thought experiments. Indeed, some problems, such as determining the proper operational framework for conducting citizen evacuations, incorporate questions that cannot be answered by nuclear power experts alone. Nonetheless, there are many decisions that must be made on the basis of scientific knowledge, and there can be no rigorous foundation for such decisions without a platform for responsible experts to make their informed opinions clear.

The response to the Fukushima crisis raised many serious questions about the way in which information was shared with the Japanese people. It is now believed that experts and officials were well-informed from the start regarding the status of reactor core meltdowns and other aspects of the progress of the accident, but this information was not adequately disclosed to the Japanese people. Nor was information on the spread of radiation disclosed to the public in a timely manner. As a result, evacuations were chaotic, and trust in the government deteriorated. The inadequacy of disclosure has been explained as stemming from fears of inciting panic, but in the absence of information, there can be no informed decisions, and ultimately the citizenry is left to shoulder the burden. The relevant agencies and institutions need to take a long, hard look at their protocols for disclosing information in a time of crisis.

At the same time, each and every one of us, as citizens, needs to improve our understanding of the risks of radiation and of nuclear plant accidents within our daily lives. We need to absorb this information calmly, in times of normalcy, so that it is available to serve as a basis for our individual decisions in times of crisis. Of course, we have also learned that the decisions we must make and the choices that are available to us vary with time. We have to separate out the decisions that can be made immediately from the decisions that can be made over time, depending on the information that becomes available to us. To this end, the public announcements regarding safety considerations that have been standard in the past should be replaced with a direct sharing of data and information, and each and every citizen should be required to make the effort to stay well informed.

The lessons learned from any severe accident must be put to use in designing crisis management techniques in other areas as well. It is my sincere hope that the conclusions of our investigation will contribute to building a better-defended society capable of facing realistic risks head on, and I hope to continue these efforts within my particular field of expertise.

Committee Member Kenji Yamaji: the response to the accident provoked a new crisis—a breakdown of trust

During the course of our independent investigation into the accident at the Fukushima nuclear station, it became clear that the process of responding to the accident provoked a serial breakdown of trust. Underlying this phenomenon was the inadequate communication of information, which had the result of obstructing efforts to muster available resources and craft a response to the accident based on expert knowledge and opinion; even after the accident had been largely resolved, it left in its wake a tragic legacy of fear and anxiety among the public regarding the radiation it released.

Upon reviewing the results of our investigation into the detailed timeline of the accident, one cannot help but be filled with regret. How could the shutdown of the unit 1 emergency condenser go unnoticed for so long? Why was the reactor containment vessel not vented more quickly? What caused the delay in utilizing SPEEDI results to help plan evacuations?

What seems undeniable is that, in the past, nuclear reactor safety precautions focused primarily on preventing damage to reactor cores, while largely disregarding measures for responding to severe accidents and evacuating nearby regions: that is, levels 4 and 5 of the defense-in-depth hierarchy. However, provisions to this end were not entirely absent. The containment vessel vents and the backup reactor-core water injection systems that were used during the Fukushima accident were installed as secondary precautions against severe accidents. Even acknowledging that the accident occurred under the extremely trying conditions of a massive earthquake and tsunami, it seems clear that, if appropriate responses based on expert opinions had been put in place more quickly, the enormous release of radiation in which the present accident culminated could probably have been avoided.

The crisis management structure that emerged—in which Prime Minister Kan, himself, directly oversaw the accident response, and the core decision-making center was placed on the fifth floor of the Cabinet (an environment that suffered from insufficient sharing of information)—was utterly inappropriate. Although in truth we must acknowledge that this structure arose because the off-site center, which was supposed to have served as a command-and-control center for directing the accident response, turned out to be ill-equipped and lacking in functional capabilities. TEPCO should have been on the front line of the accident response, assisted by expert knowledge and opinion coordinated by the Nuclear Safety Commission and NISA, and the Cabinet's Nuclear Emergency Response Headquarters should have played the role of central coordinator and repository for information sharing.

A team of individuals with no preparation, who have never worked together under ordinary circumstances and who meet face-to-face for the first time in the aftermath of a crisis are simply not going to be effective at sharing information with one another. Prime Minister Kan's lack of trust in the Nuclear Safety Commission's chairman, who was unable to predict the unit 1 hydrogen explosion; the METI minister's lack of trust in TEPCO for its delay in venting; the Kan administration's lack of trust due to TEPCO's "abandonment" incident before dawn on March 15—all of these unfortunate incidents arose from misunderstandings caused by a lack of information and expert knowledge. In this way, a chain of distrust prevented any rapid mobilization of resources based on expert knowledge, and the decision makers were forced to bear an excessive psychological burden.

The Fukushima accident has heavily damaged the faith of the Japanese people in the safety of nuclear power, a fact with serious societal consequences. The population's lack of trust has had a profound impact on awareness of exposure risks due to the emission of radioactive material. When we discuss questions of nuclear power safety, what we are really asking is how much risk is an acceptable level to tolerate, and ultimately this is a decision that must be made by society as a whole. It is not a question that can be resolved on the basis of scientific understanding alone. The assurance of nuclear power safety requires a comprehensive approach based on science and technology, and it is never possible to eliminate risks entirely. However, on a societal level, we strive for peace of mind, but peace of mind is only obtained by appealing to human emotion. If people are unable to trust experts on matters of safety, then their imaginations—fueled by their emotions—run wild, and anxieties increase without bound. We must restore faith in the opinions of experts, including in areas such as the health risks of low-level radiation exposure, where scientific understanding is incomplete.

Expert Reviews

Frank von Hippel, Senior Research Physicist and Professor of Public and International Affairs Emeritus, Princeton University

A series of Japanese commissions have attempted to understand the human and institutional failures that contributed to the core meltdowns in three of the six reactors at the Fukushima Daiichi Nuclear Power Station in March 2011 and to the prolonged loss of cooling of four of the plant's spent fuel cooling ponds.

This independent report provides important new insights to help inform the debate over the future of nuclear power in Japan and the world.

Here I comment on the following issues raised by this report:

- myth of safety;
- need for nuclear crisis management experts;
- need for thyroid protection;
- evacuation radius;
- danger from spent fuel pools;
- uses and abuses of probabilistic safety assessments; and
- importance of risk communication.

Myth of safety

"Until the 1980s, we had the sort of spirit that motivated us to do things of our own volition, even if the government didn't tell us to," says former TEPCO Vice President Toshiaki Enomoto. ... "We studied safety-design philosophy for nuclear plants and learned the lessons of past accidents, and our people were trained in an atmosphere of constant initiatives to improve and upgrade things. But in the 1990s, the core focus of nuclear power work shifted to maintenance and operational issues, and the practical aspects of safe design started to recede into the background." (Chapter 4)

The lulling effect of the myth of nuclear safety also is a problem for the global nuclear enterprise—both civilian and military.

The nuclear energy community—aptly called the "nuclear village" in Japan—sees public fears of nuclear power as irrational and counterproductive. And, by the usual measure, they are right. Even taking into account those nuclear power plant accidents that have occurred around the world, the risk of death and injury per nuclear kilowatt hour generated is much less than, for example, the risk associated with a kilowatt hour generated by coal power plants.[1]

But as the Chernobyl and Fukushima accidents have demonstrated, nuclear accidents can be very disruptive socially. They can suddenly render large areas uninhabitable; they can cause governments to fall; they can result in costly shutdowns of large fractions of a nation's generating capacity; and they inspire fears of death from radiation.

The nuclear village's way of dealing with the public's "irrational" fear of nuclear accidents was to convince itself, in order to be more convincing to the public, that a major release of radioactivity just could not happen. This undercut regulatory vigilance.

The huge economic and political influence of Japan's electrical utilities and the intertwining of their managements with the government's bureaucracies exacerbated the loss of regulatory bite in Japan. To a considerable degree, however, this muzzling is a worldwide phenomenon. Major nuclear reactor accidents are so infrequent that the long periods between them provide ample time for "regulatory capture," in which the regulating agency comes to serve the interests of the regulated entities.

In the United States in 2001, for example, the leadership of the US Nuclear Regulatory Commission (NRC) surrendered to the resistance of a reactor operator and decided not to follow the recommendations of NRC inspectors to force a power reactor to shut down for an emergency inspection of its pressure vessel. When the pressure vessel was finally inspected during the next routine shutdown, it was found to be almost corroded through. After investigating the incident, the commission's inspector general concluded:

> NRC appears to have informally established an unreasonably high burden of requiring absolute proof of a safety problem, versus lack of a reasonable assurance of maintaining public health and safety.[2]

It is therefore necessary to institutionalize sustained pressures on nuclear regulators to balance industry pressure to eliminate "redundant" safety arrangements. In the United States, independent experts working for the Union of Concerned Scientists at the national level and for citizens' groups monitoring

---

[1] Epstein, P.R., Buonocore, J.J., Eckerle, K., et al. (2011) Full cost accounting for the life cycle of coal. *Annals of the New York Academy of Sciences* 1219 (February): 73–98.

[2] US Nuclear Regulatory Commission (2002) *NRC's regulation of Davis-Besse regarding damage to the reactor vessel head*. Report, Inspector General's Office, case no. 02-03S, December 30. Available at: www.nrc.gov/reading-rm/doc-collections/insp-gen/2003/02-03s.pdf (accessed January 4, 2013).

individual nuclear power plants at the local level have managed to raise the visibility of important issues. But, in 2004, the US NRC changed the adjudicatory process through which outside groups can participate in its licensing process to make it "more effective and efficient."[3] David Lochbaum of the Union of Concerned Scientists has characterized these changes as the NRC "bowing to industry pressure ... to virtually eliminate public participation, except in the role of casual observer."[4] A lengthier critique quotes the 1974 judgment of the NRC's Atomic Safety Licensing Appeal Board (which was abolished by the NRC in 1991) on the contributions of outside groups to the licensing process:[5]

> Public participation in licensing proceedings not only "can provide valuable assistance to the adjudicatory process," but on frequent occasions demonstrably has done so. It does no disservice to the diligence of either applicants generally or the regulatory staff to note that many of the substantial safety and environmental issues which have received the scrutiny of licensing boards and appeal boards were raised in the first instance by an intervenor.

Need for nuclear crisis management experts

One of this report's striking observations was that there were no experts in the Japanese government or on call who had thought about how to deal with the consequences of a loss-of-coolant accident. It may be unreasonable to expect that every nuclear power plant should have such a team, but a country should be prepared to quickly assemble and deploy one. The United States has experts at its national laboratories ready to assemble as a Nuclear Emergency Search Team, which could be dispatched to any part of the country on short notice in case of a credible report of the possible presence of an illicit nuclear or radiological device.

In the case of reactor safety it would make sense to have a full-time team to supervise the training of plant operators for the possibility of an imminent or actual core meltdown event. Such a team would be prepared to provide expert technical advice and support to the central and local governments should an actual event occur.

Need for thyroid protection

> Government officials issued the first directive for the public to take potassium iodide and restrictions on consumption of food and water on March 21 [ten days after the accident began and six days after the largest

---

[3] US Nuclear Regulatory Commission (2004) Changes to adjudicatory process. *Federal Register* 69(9): 2182.
[4] Lochbaum, D. (2004) *US Nuclear Plants in the 21st Century: The Risk of a Lifetime.* Cambridge, MA: Union of Concerned Scientists, p. 9.
[5] Roisman, A.Z., Honaker, E., and Spaner, E. (2009) Regulating nuclear power in the new millennium (the role of the public). *Pace Environmental Law Review* 26(317).

releases of radioiodine]. The directive for the mandatory issuance of potassium iodide included the inhabitants of several villages and towns within the affected area. Enough potassium iodide for 900,000 people was distributed within a 31-mile (50 kilometers) radius of the plant. Because the evacuations had already been completed, however, the potassium iodide was not issued to the population.[6]

An epidemic of thyroid cancer among people in Belarus, Ukraine, and nearby regions of Russia exposed to radioactive iodine from the 1986 Chernobyl accident as children is the largest statistically detectable cancer consequence of the accident. As of 2005, the UN Scientific Committee on the Effects of Atomic Radiation estimated that about 7,000 excess thyroid cancers had already occurred in this population.[7]

Potassium iodide was used in Poland by 10.5 million children under 16 and 7 million adults with no significant adverse health effects. In light of these experiences, the US Food and Drug Administration recommended in 2001 that, in case of a predicted thyroid radiation dose exceeding 50 millisieverts, children 18 years of age and younger should be given a protective dose of nonradioactive potassium iodide to saturate the thyroid and block uptake of any inhaled or ingested radioactive iodine.[8]

The question is how to make potassium iodide available quickly in case of need. In the United States, there are preparations to provide it within a 10-mile radius of a nuclear power plant. Whether to provide it beyond that limited radius has been a matter of controversy—with the Nuclear Regulatory Commission opposed.[9] Japan's official reports to the IAEA on the Fukushima accident cite surveys in which no children were found to have received thyroid doses above 50 millisieverts. However, there appears to have been at least the danger of such doses. According to the report of the Institute of Nuclear Power Operations:[10]

> The first air samples from the site boundary, on March 22 and 23, had iodine-131 concentrations that were equivalent to approximately 80 mrem (0.8 mSv) of thyroid dose each hour if inhaled by an unprotected individual. The concentration remained between 25 and 200 percent of this value until April 18, 2011.

---

[6] Institute of Nuclear Power Operations (2011) *Special report on the nuclear accident at the Fukushima Daiichi Nuclear Power Station.* Report. Atlanta: INPO.
[7] United Nations Scientific Committee on the Effects of Atomic Radiation (2008) *Sources and Effects of Ionizing Radiation.* Report, Vol. II, Annex D: Health effects due to radiation from the Chernobyl accident, p. 14.
[8] US Food and Drug Administration *Guidance: potassium iodide as a thyroid blocking agent in radiation emergencies.* Report, p. 6.
[9] Frommer, F. (2011) Nuke crisis reignites debate on protective pills. Associated Press, March 30.
[10] Institute of Nuclear Power Operations (2011) *Special report on the nuclear accident at the Fukushima Daiichi Nuclear Power Station.* Report. Atlanta: INPO, p. 42.

At a thyroid dose rate of 0.2–1.6 millisieverts per hour, 50 millisieverts would be accumulated within one to ten days of exposure. Note that these high dose rates were being measured a week after the major overland release on March 15 when airborne iodine from that release would have long since blown away.

During the US Three Mile Island accident in 1979, there also were delays in distributing potassium iodide—indeed, it was not distributed. Fortunately, it was not needed in that case:[11]

> [T]hose located in the NRC headquarters and the governor's office ... assumed that almost complete information about possible consequences had to be obtained before certain types of preventive and protective action are initiated. There seemed to be a rather persistent attitude that "somehow" evacuation would be costly; that affected populations would "panic"; that declaring an "emergency" would create an emergency; that taking potassium iodide might be physically safe, but psychologically damaging. It is perhaps fortunate that such deep-seated attitudes did not have more negative consequences in the response to [Three Mile Island].

In the case of the Chernobyl accident, the Soviet government did not distribute potassium iodide widely even though it would have prevented thousands of thyroid cancers.

It would be advisable in preparing for the next major nuclear reactor accident for nations to have well-thought-out policies with regard to making potassium iodide available to the public.

Evacuation radius

It is painful to read about the evacuation. On the evening of March 12, the day after the earthquake, the government ordered evacuations within a 12-mile radius of the Fukushima station—affecting about 100,000 people. This was unplanned and proved to be challenging. In the case of a nearby hospital with 130 difficult-to-move patients and a nursing home with ninety-eight residents, evacuation took three days with eight patients and six residents dying on the way.[12]

Given the dire situation created by the evacuation of 100,000 people, one can understand how unthinkable it was for the Japanese government to follow the advice of the United States and order evacuation of the two million people living

---

[11] President's Commission on the Accident at Three Mile Island (1979) *Report on the Emergency Preparedness and Response Task Force*. Staff Report to the President's Commission on the Accident at Three Mile Island. US Government Printing Office: Washington, DC, pp. 96–97.

[12] *Asahi Shimbun* (2012) Government probe: Many Fukushima hospital patients died during botched rescue operation. July 24. Available at: http://ajw.asahi.com/article/0311disaster/life_and_death/AJ201207240092

within 50 miles of the Fukushima Daiichi plant.[13] Larger populations are within this distance of other nuclear power plants in both Japan and the United States. A notable example in the United States is the Indian Point Nuclear Power Plant 40 miles up the Hudson River from New York City.

The obvious lesson is that planning for evacuations must encompass a larger radius than the immediate neighborhoods of the nuclear reactors. But the evacuation of large urban areas remains unthinkable. This is a reason to have potassium iodide available in urban areas even tens of miles from nuclear power plants.

Danger from spent fuel pools

We learn from this report that a team led by Japan Atomic Energy Commission Chairman Shunsuke Kondo, charged by Prime Minister Kan with conducting a worst-case assessment, delivered a calculation on March 25 that considered how far the long-term evacuation area might extend. The conclusion was that, if the fuel in the spent fuel pool in unit 4 were to become uncovered and catch fire, the long-term compulsory evacuation zone contaminated by thirty-year half-life cesium-137 might extend out to 68 miles and the area of voluntary evacuation even farther (Chapter 1).

Similar results of possible spent fuel pool fires had been obtained by some colleagues and myself in a 2003 paper. We calculated potential downwind evacuation areas if cooling were lost in a US spent fuel pool and the fuel heated up to the point where the zirconium cladding on the fuel caught fire.[14]

The Fukushima accident has provided additional evidence that, no matter what happens, the cooling of spent fuel must be maintained. It remains to be seen whether this can be assured for nuclear power plants such as Fukushima and similar boiling water reactors in the United States and elsewhere, in which the spent fuel pools are located high in the reactor buildings outside of the reactor containment structures.

In addition, the Fukushima accident has focused attention on the dense packing of spent fuel pools. In the United States, spent fuel is typically kept in pools for twenty-five years to delay as long as possible the cost of buying storage casks for older, cooler spent fuel. This results in the spent fuel being packed almost as densely as in a reactor core, with each fuel assembly tightly surrounded by a box with neutron absorbing walls to prevent criticality. But does this exacerbate the danger if cooling water should be lost? Should spent fuel be transferred to safer storage in air-cooled dry casks after only five years

---

[13] Cox, A., Ericson, M., and Tse, A. (2011) The evacuation zones around the Fukushima Daiichi Nuclear Plant. *The New York Times*, March 25. Available at: www.nytimes.com/interactive/2011/03/16/world/asia/japan-nuclear-evacuation-zone.html?_r=0.

[14] Alvarez, R., Beyea, J., Janberg, K., et al. (2003) Reducing the hazards from stored spent power-reactor fuel in the United States. *Science and Global Security* 11: 1. Available at: irss-usa.org/pages/documents/11_1Alvarez.pdf

cooling as the chairman of Japan's new Nuclear Regulation Authority urged during his first press conference?[15]

Establishing dry cask storage also would provide Japan with an alternative to its unnecessary, hugely costly, and dangerous reprocessing program, which thus far has resulted in a stockpile of 44 tons of separated plutonium as of the end of 2011, enough for more than 5,000 Nagasaki-type bombs.[16]

Uses and abuses of probabilistic safety assessments

The objectives of probabilistic safety assessments, according to the IAEA, are "to determine all significant contributing factors to the radiation risks arising from a facility or activity, and to evaluate the extent to which the overall design is well balanced and meets probabilistic safety criteria where these have been defined." These assessments, the agency states, "may provide insights into system performance, reliability, interactions and weaknesses in the design, the application of defense in depth, and risks, that it may not be possible to derive from a deterministic analysis." In the United States and other countries, probabilistic safety assessments have a history of serving precisely these roles. In Japan, however, probabilistic safety assessments were never promoted to the status of regulatory conditions. (Chapter 6).

This report recommends that probabilistic safety assessments be made a central tool of nuclear reactor safety regulation in Japan, as they already are in the United States. These assessments are indeed a useful way to systematically look for possible vulnerabilities in a complex system. But their limitations and abuses must be guarded against.

Indeed, probabilistic safety assessment was first used as a part of a US government effort to shore up the myth of safety in the United States at a time when it was under attack. The NRC's 1975 *Reactor Safety Study: An Assessment of Accident Risks in U.S. Commercial Nuclear Power Plants* was an attempt to prove that a reactor accident with a major release of radioactivity to the human environment was exceedingly unlikely. This study estimated that the probability of a core-melt accident in a light water reactor (LWR) was about one chance in 20,000 per LWR-year. The 1979 core-melt accident at Three Mile Island occurred about 600 LWR-years later.

The NRC study also estimated that the chance of a reactor accident releasing enough radioactivity to cause $14 billion of damage (the maximum damages imagined, about $50 billion in 2013 dollars) would be one chance in a million

---

[15] Nuclear Regulatory Commission (2012) Press conference. Available at: www.nsr.go.jp/kaiken/data/20120919sokkiroku.pdf (In Japanese.)

[16] *Asahi Shimbun* (2013) Experts urge Japan to break away from "failed" nuclear reprocessing program. August 28. Available at: http://ajw.asahi.com/article/0311disaster/fukushima/AJ201308010057

per LWR-year.[17] The 2011 Fukushima accident occurred about 10,000 LWR-years later.

On the basis of congressionally mandated outside review of the study in 1979, just before the Three Mile Island accident, the NRC concluded that probabilistic safety assessment was unreliable in predicting the overall risk of nuclear reactor accidents:[18]

> In light of the Review Group conclusions on accident probabilities, the Commission does not regard as reliable the Reactor Safety Study's numerical estimate of the overall risk of reactor accident.

The causes of the Three Mile Island and Fukushima accidents illustrate the limitations of these assessments. At Three Mile Island, because of a lack of indicators in the control room, operators mistook the symptoms of a stuck-open safety valve that was causing a steady decline of water level in the reactor pressure vessel for the indications of an overly full pressure vessel. They therefore turned off the high-pressure injection system, which had automatically been turned on to replace the water being lost.[19] That scenario had certainly not been predicted in advance.

In the case of the Fukushima Daiichi Nuclear Power Station accident, the risk analysts did not update their assessments to take into account new evidence that huge tsunamis had struck the coast of Japan's Tohoku region three times in the past 3,000 years.[20] For the Fukushima Daiichi reactors, therefore, the accident had a probability of one in a thousand reactor years, not one in a million reactor years as had been calculated in the NRC's reactor safety study.

Importance of risk communication

Any traumatic accident or event can cause the incidence of stress symptoms, depression, anxiety (including post-traumatic stress symptoms), and medically unexplained physical symptoms. Such effects have also been reported in Chernobyl-exposed populations. Three studies found that exposed populations had anxiety levels that were twice as high as controls, and they were 3-4 times more likely to report multiple unexplained

---

[17] US Nuclear Regulatory Commission (1975) *Reactor Safety Study: An Assessment of Accident Risks in U.S. Commercial Nuclear Power Plants*. Report, Executive Summary, p. 11.

[18] US Nuclear Regulatory Commission (1979) *NRC statement on risk assessment and the reactor safety study report (WASH-1400) in light of the risk assessment review group report*. Report, January 18.

[19] President's Commission on the Accident at Three Mile Island (1979) *Report of the President's Commission on the Accident at Three Mile Island: The need for change, the legacy of Three Mile Island*. Report, October. US Government Printing Office: Washington, DC, p. 28. Available at: www.threemileisland.org/downloads/188.pdf.

[20] Nöggerath, J., Geller, R.J., and Gusiakov, V.K. (2011) Fukushima: The myth of safety, the reality of geoscience. *Bulletin of the Atomic Scientists* 67(37). Available at: http://thebulletin.org/2011/september/fukushima-myth-safety-reality-geoscience.

physical symptoms and subjective poor health than were unaffected control groups. ...

[I]ndividuals in the affected population were officially categorized as "sufferers," and came to be known colloquially as "Chernobyl victims," a term that was soon adopted by the mass media. This label, along with the extensive government benefits earmarked for evacuees and residents of the contaminated territories, had the effect of encouraging individuals to think of themselves fatalistically as invalids. It is known that people's perceptions—even if false—can affect the way they feel and act. Thus, rather than perceiving themselves as "survivors," many of those people have come to think of themselves as helpless, weak and lacking control over their future.[21]

Risk communication is very important. The psychological effects of the Chernobyl accident on the population exposed to the radioactivity was huge. Japan, given its similar experience with the survivors of Hiroshima and Nagasaki, should be especially sensitive to this issue. Putting the added cancer risks into perspective by showing how small an addition they represent to the risk of cancer from other causes may help to some degree. For example, for a dose of 100 millisieverts, approximately the maximum first-year estimated by the government for the population between 12 and 19 miles of the Fukushima plant,[22] the added lifetime risk of cancer death would be about 0.5 percent on top of the 20 percent preexisting cancer death risk in Japan.[23]

---

[21] Chernobyl Forum (2003–2005) *Chernobyl's legacy: health, environmental and socio-economic impacts: Recommendations to the governments of Belarus, the Russian Federation and Ukraine (second revised version)*. Vienna: IAEA, pp. 20–21. Available at: www.iaea.org/Publications/Booklets/Chernobyl/chernobyl.pdf.

[22] Permanent Mission of Japan to IAEA, Government of Japan (2011) *Report of Japanese government to the IAEA Ministerial Conference on Nuclear Safety: The accident at TEPCO's Fukushima Nuclear Power Stations*. Report, June 7, Attachment V-13-2. Available at: www.iaea.org/newscenter/focus/fukushima/japan-report/.

[23] US National Research Council (2006) *Health Risks from Exposure to Low Levels of Ionizing Radiation: BEIR VII—Phase 2*. Washington, DC: The National Academies Press. Available at: http://ganjoho.jp/public/statistics/backnumber/2011_en.html.

Jessica Tuchman Matthews, President of the Carnegie Endowment for International Peace, and James M. Acton, Senior Associate of the Nuclear Policy Program at the Carnegie Endowment for International Peace

While physical aftershocks from the earthquake that struck Japan on March 11, 2011, have long ceased, societal aftershocks are still reverberating. Most obviously, 110,000 of the evacuees forced to flee from the worst nuclear accident since Chernobyl have yet to return to their homes. For many Japanese—not just the evacuees and their families—their dislocation is an open wound.

At the same time, Japanese society has gained a new interest in its country's governance. Until the accident, it had been willing to let bureaucrats make most key policy decisions with a minimum of interference. Now, a wave of public involvement, which was started by dissatisfaction over energy policy, is having effects in unrelated areas, such as pensions.

A robust and engaged civil society is surely a welcome development. Japan has been in a sustained economic slump for almost two decades, partly as a result of a political system devoid of imagination and unable to innovate. (Although, hopefully, Japan's new prime minister's signature economic policy of "Abenomics" will turn out to be an exception to that rule.) It may be painful for the Kasumigaseki bureaucrats, but an injection of energy from civil society could—perhaps—be a turning point.

There is no better example of constructive and impactful engagement by civil society than the Rebuild Japan Initiative's report on the Fukushima Daiichi nuclear accident. Before the accident, it was almost unthinkable that government officials, all the way up to the prime minister, would have cooperated with a private investigation such as this. Their involvement is a clear mark of the seismic changes under way within Japan. It also enabled the groundbreaking analysis presented in the report.

Perhaps unsurprisingly, therefore, the report's most significant contribution is its detailed dissection of the government's response to crisis (Chapter 1), which is discussed below. But it contains many other highlights, including analyses of the way that private citizens used social media to fill the informational voids left by the government (Chapter 8), the historical foundations of the weaknesses in Japan's regulatory system (Chapters 2 and 4) and the initial difficulties—and eventual success—of the US–Japan alliance in responding to the disaster (Chapter 9).

For all the progress the report has made, however, it is vitally important that it marks a start—not an end—to the involvement of Japanese civil society in nuclear energy policy. As the report presciently notes, the problems uncovered by the accident are deeply rooted and will not be rectified quickly. Japanese think tanks and research institutions have a critical role to play in scrutinizing the government as it works to implement much-needed regulatory reforms. Moreover, if, as the report argues, the Japanese people were complicit in the creation of "a myth of safety" that inhibited effective regulation, civil society has a positive obligation to involve itself in the debate over corrective measures.

Rethinking regulation in Japan

Paradoxically, in spite of the critical role that Japanese civil society must play in scrutinizing regulatory reform, the weakest sections in what is still an extremely impressive and important report are its forward-looking passages concerning where Japan should go from here.

Since the accident, Japanese society has sometimes treated the United States as some sort of Platonic regulatory ideal. As the disaster unfolded, an often-heard refrain, both from Japanese experts and the general public, was that an accident like Fukushima could not happen in the United States, and, even if it did, there would not have been such problems with disaster management. The Diet, for example, used the United States as a model in 2012 when it modified the bill to reform Japan's regulatory system by adding a new commission of four members and a chairman along the lines of the US Nuclear Regulatory Commission (NRC).

The report is not immune from a tendency to idealize the US model of nuclear regulation. Chapter 6, for instance, presents a trenchant and persuasive critique of the weaknesses in Japanese regulations that led to the accident. However, the report doesn't apply the same critical standards when invoking US regulations for the purposes of comparison. The report is complimentary about the US procedures to deal with severe accidents (the unsurprisingly named Severe Accident Management Guidelines). It appears to endorse the suitability of US rules to prevent station blackouts—the complete loss of AC power—caused by external events. The report also calls upon Japan to officially adopt quantitative goals for nuclear safety, citing the United States as a model. Other sections of the report show similar deference to the United States. Chapter 10, for instance, criticizes Japan for not putting in place antiterrorism measures equivalent to the so-called B.5.b standards developed in the United States after the terrorist attacks of September 11, 2001.

Yet, weaknesses in *all* these aspects of US regulation were highlighted by the Fukushima accident. In some cases, these weaknesses were identified by the Near-Term Task Force created by the NRC to understand the lessons that should be learned from Fukushima. The task force, for instance, called for a strengthening of provisions to manage a station blackout, since existing rules did not require consideration of events involving multiple reactor units and ignored the challenges of augmenting personnel during a major natural disaster.[24]

---

[24] Miller, C., Cubbage, A., Dorman, D., et al. (2011) *Recommendations for enhancing reactor safety in the 21st century: The Near-Term Task Force review of insights from the Fukushima Dai-ichi accident.* Report, Nuclear Regulatory Commission, July 12, pp. 53–56. Available at: http://pbadupws.nrc.gov/docs/ML1118/ML111861807.pdf

The task force was also critical about the existing Severe Accident Management Guidelines. US power companies were *not*, as the Rebuild Japan Initiative's report states, "forcibly required to comply with these guidelines." These guidelines were actually voluntary—just as similar guidelines were in Japan.[25] While all US plant operators did eventually implement them, the quality of implementation was variable. The task force also identified the need for a greater focus on quality assurance of measures put in place to deal with a terrorist attack (the guidelines developed pursuant to the B.5.b requirements).

Other potential problems with the US regulatory system should also be recognized in light of the Fukushima accident. As noted above, the Rebuild Japan Initiative's report—very sensibly—urges Japan to adopt quantitative safety goals and goes on to cite the US approach as a model. However, US goals are focused narrowly on preventing damage to human health from radiation. Depending on exactly how severe the health consequences of the Fukushima accident turn out to be, it is possible that Japan might actually meet these goals. This possibility illustrates that US goals fail to take into account the full range of effects associated with nuclear accidents, such as the economic effects of a long-term evacuation. The Fukushima accident should prompt the NRC to adopt broader goals; it should not prompt Japan to copy the United States' current approach.

This critique is not meant to imply that Japan can learn nothing from the United States. Of course it can. Rather, the point is that it is critically important that Japan avoids moving from isolation to emulation. In Chapter 10, the report argues that, prior to the accident, Japanese regulators were internationally isolated:

> [Japan's regulators had a] psychological tendency to think in terms of a unilateralist approach to nuclear safety. Japan somehow managed to convince itself that its nuclear power regulatory system and its culture of safety regulations, surpassed international standards—a sense of superiority accompanied by a "Galapagos effect" in safety regulations.

For two reasons, the other regulatory extreme—treating the United States as an infallible role model—is also misguided.

First, the United States is not the only source of best practice. In terms of protecting reactors against external hazards, for instance, Taiwan and many European countries probably have higher standards than the United States.[26] Moreover, different regulators have emphasized different responses to Fukushima. European regulators have generally stressed expensive large-scale

---

[25] Miller, C., Cubbage, A., Dorman, D., et al. (2011) *Recommendations for enhancing reactor safety in the 21st century: The Near-Term Task Force review of insights from the Fukushima Dai-ichi accident.* Report, Nuclear Regulatory Commission, July 12, p. 47. Available at: http://pbadupws.nrc.gov/docs/ML1118/ML111861807.pdf

[26] Acton, J.M. and Hibbs, M. (2012) *Why Fukushima was Preventable.* Report. Washington, DC: Carnegie Endowment for International Peace, pp. 19–24.

hardware improvements to plants to defend them against externally initiated events and hence prevent accidents; the NRC has argued that such changes are not cost effective and has emphasized improvements in procedures to respond to an accident should one be triggered by an external event.[27] Japan needs to perform its own risk-informed analysis of the best way forward. It should not *assume* that the United States has the best approach or that the right solution for the United States is the right solution for Japan. (After all, the external threats to nuclear power plants in each country are not identical.)

Second, if the Japanese approach to regulation is largely based on copying foreign standards, it will not be able to push forward the boundaries of nuclear safety. Given Japan's extraordinary expertise in nuclear energy, it should seek to be a leader, not a follower. It should aim for an exchange of best practices with the international community.

Unanswered questions about crisis management

The report's greatest strength is its forensic analysis of decision making within the Cabinet core team during the crisis (Chapter 1). It identifies serious problems that arose at the very highest levels of the Japanese government, including a lack of information reaching senior officials, a breakdown of trust between elected officials and their technical advisers, and the prime minister's attempts to micromanage the response. The specific lessons need to be heeded worldwide. To give one very practical example: Japan's top leaders were unwilling to use the official Crisis Management Center in the basement of the Cabinet building because they were not allowed to bring in their mobile phones; their relocation to the Prime Minister's Office complicated the flow of information. The desire of senior officials to stay connected during a crisis is a reality that, in the twenty-first century, disaster planners must take into account.

The report's critique is, by no means, entirely negative; it also identifies what the government did well. However, it doesn't discuss the question of what constitutes an "acceptable" response to a major disaster. No response to a disaster that hits without warning—let alone one as complex as the triple catastrophe that struck Japan—ever looks pretty. By its very nature, disaster management involves curtailing, not preventing, human suffering. Errors are inevitable. So, what is the relevant benchmark for judging a response to be substandard? This is an important question in an assessment of government performance.

Another critical issue that merits greater attention is the balance between procedure and improvisation, such as the establishment on March 15 of TEPCO's Joint Emergency Response Headquarters.

Given the complexities of disaster management, a decision such as this will inevitably have both positive and negative consequences. Without a doubt, the

---

[27] Acton, J.M. and Hibbs, M. (2012) Fukushima could have been prevented. *International Herald Tribune*, March 9.

balance between procedure and improvisation is hard to strike. This is an issue that merits further attention.

A final issue that deserves much greater discussion is the appropriate role for elected politicians in managing a disaster. The report identifies the problems created by the politicians leading the Fukushima response and argues for more training and exercises. But is this suggestion feasible? Can busy politicians really be persuaded to participate in training for unlikely disasters? If not, could the response be led by an appointed expert instead? Or would this system break down as elected politicians tried to take charge for fear of being seen as too passive? How effective is gaming ("table-top" simulations in which the role of the principals is generally played by high-level advisers) at uncovering potential pitfalls and developing solutions?

Underlying all these questions is a more fundamental and important one: Is there a tension between effectiveness and accountability in disaster response? Particularly in a democracy, politicians are under obvious pressure to lead the response to a disaster. Past experience of nuclear accidents gives contradictory evidence of how capable elected politicians are of performing this task effectively. The Three Mile Island accident in 1979 provides a positive example of how elected politicians can wield their authority—authority that bureaucrats lack—to marshal an effective response. On that occasion, US President Jimmy Carter and Pennsylvania Governor Richard L. Thornburgh were able to cut through different layers of government, work together, and help reassure a terrified public.

The administration of Naoto Kan failed to use its authority nearly so effectively in managing the Fukushima accident—although it must be acknowledged that the situation it faced was significantly more complex and serious than Three Mile Island. The potential tension between accountability and effectiveness in disaster response is not limited to nuclear accidents, but it is made much more acute by the extraordinary complexity of a response to one. All nations that possess, are building, or are planning to build nuclear reactors must face this conundrum, and it deserves further discussion.

★★★

In the final analysis, the Rebuild Japan Initiative's report adds enormously to our understanding of the causes underlying the Fukushima accident and the response to this catastrophe. While some of its analysis is specific to Japan, much is not. The structural problems it identifies with Japanese regulation almost certainly exist in many other states. They include—but extend much further than—the need for a truly independent regulator, the main focus of international discussions so far. The report's most important lessons, however, pertain to the challenges facing central and local governments in managing the consequences of a nuclear accident. Emergency planning and preparedness almost certainly deserves greater attention in most states that have or want nuclear reactors. These states ignore this report's lessons at their peril.

Paul 't Hart, Professor of Public Administration, Utrecht University, and Associate Dean at the Netherlands School of Public Administration

No easy answers

When the tsunami hit the northeast coast of Japan, the entire nation shuddered. The natural disaster exposed the limits of Japan's crisis preparedness, and the human and institutional failures contributed to the catastrophe at the Fukushima nuclear power plant. The story of that disaster cruelly highlights the key paradoxes of risk and crisis management in advanced societies, which pose challenges for all those trying to learn the lessons of Fukushima.

There are no easy solutions. The urgent desire in Japan to prevent future Fukushimas should be leveraged to engage in a broad societal learning process. Building safer societies that continue to rely on high-risk technologies won't just require technical work by risk regulators and emergency managers. First and foremost, it involves major adaptations in politics, industry, and citizenry. Hopefully this insight will not get lost in the spiral of anger–accusation–defensiveness–avoidance that we often see in the wake of a major and mishandled emergency.

Moreover, the potential lessons of Fukushima should extend well beyond the nuclear power debate. Not only Japan but all advanced societies should be fundamentally reexamining their ability to profitably harness technology while hedging against its unintended consequences.

Paradoxes

The first paradox we need to face is that there is no natural correspondence between objective and subjective risk. For any society, its level of concern about the risks it faces provides the major impetus for its investment in safety and security. Yet that level of concern may be out of kilter with what knowledgeable and dispassionate observers would describe as the actual nature of the risks the community faces.

Risk inoculation is one potential hazard brought on by the discrepancy between real and perceived danger. People may not appreciate the real risks they run, and therefore may not be motivated to invest appropriately in risk prevention and emergency preparedness. This blindness can be the result of many factors. On the institutional side, the risk-producing or risk-regulating entities can manipulate public perceptions through a lack of transparency or purposeful "framing" of the issue. On the community side, a collective illusion of invulnerability (the classic "it won't happen here" mentality) can take hold as years or decades go by without incident. Such tranquil periods can lead people to conclude that existing risk regimes must be sufficient, whereas in reality they have been largely untested or unobtrusively eroded.

A corollary of this type of misperception is what has been called the vulnerability paradox: The more invulnerable a society has seemingly become, the more vulnerable it will prove to be when a major disturbance occurs regardless. Put differently: The more a risk-regulation entity has banked exclusively on prevention, the less resilient system operators, communities, and governments will prove to be when an emergency actually arises.

The Fukushima disaster revealed the sundry causes of risk-denial to be at work in Japan. The cozy regulator–industry relations in the "nuclear village" reduced official scrutiny, and local populations who owed their livelihoods to the power plants didn't ask for more. Add in a cultural illusion of Japanese superiority in all matters technological and therefore nuclear, and you get a culture that had collectively willed away the reality that the national energy strategy was hostage to a fundamentally high-risk technology. Yes, preparedness structures seemed to be in place, and emergency-response responsibilities were distributed across a wide range of government authorities. But these response plans proved to be largely "fantasy documents," to use sociologist Lee Clarke's term: exercises in wishful thinking that proved to bear no correspondence to the difficult on-the-ground realities of organization, communication, and collaboration in the midst of a crisis.

Although the Fukushima disaster didn't reach the level of the government's worst-case scenario, which suggested the evacuation of Tokyo, the crisis will nevertheless be enshrined in collective memory as one of the worst in post-war history worldwide. Learning from such a low-probability but high-impact contingency is a tricky thing. There's a danger of policy makers "not learning" because status quo interests prevail over the groundswell of support for change.

That persistent danger is of particular concern in Japan, where the industrial and politico-bureaucratic power structures lack the pluralism, transparency, and accountability that make institutions agile in the face of critical disturbances and responsive to negative public feedback. Also, after a rare yet extreme event such as the 2011 tsunami, risk-denial can take new forms. Status-quo interests have argued that the likelihood of reoccurrence is exceedingly low. Making Japan strong enough to withstand a worst-case scenario disaster, the critics argue, would require massive investments that may not constitute reasonable and proper public policy.

However, risk denial isn't the only possible pitfall. The variance between real and perceived danger can also produce an opposite but equally undesirable consequence, which I call risk inflation. In this situation, societies obsess about danger and therefore engage in overregulation, at great cost. According to German sociologist Ulrich Beck, this risk inflation has taken hold throughout the Western world. In so-called "risk societies," there is a widespread preoccupation with vulnerability, including the threats posed by technologies designed to sustain our modern, urbanized lives. These perceived risks can become inflated based on questionable evidence, prompting regulations that do not discernibly enhance security yet impose great economic costs, or curtail rights and liberties. The latter occurred, the critics argue, in the wake of the September 11 terrorist attack, when most governments adopted far-reaching anti-terror legislation and greatly expanded their institutional capacities for counterterrorist intelligence and crisis management operations.

In the aftermath of the Fukushima Daiichi nuclear accident, policy makers have to guard against the possibility of undue risk amplification. In their determination to prevent another such accident, they could take the lessons of Fukushima too much to heart and extol the virtues of a no-regret, zero-risk approach. If this mentality becomes an overbearing ethical imperative, it could crowd out rational debate about the full societal cost–benefit ratio associated with such policies.

To avoid both these unhelpful extremes, Japan needs prudent management of the "crisis after the disaster." It needs a debate that involves not just industry and government elites, and does not place the twin burdens of blame for the past and reform for the future exclusively at their feet. In the end, Japanese society as a whole will have to interrogate and perhaps reshape its attitudes. This reexamination should encompass not just nuclear power, but a range of man-made risks that Japanese society has implicitly embraced on top of the ongoing threat posed by natural phenomena such as earthquakes, typhoons, and volcanic eruptions.

This insight rests partly on the second paradox: that many of our most sophisticated socio-technical systems are also our greatest risk amplifiers. The compound disaster of earthquake and tsunami was turned into a calamity of much bigger temporal, spatial, and political proportions by the unsuspected vulnerability of the Fukushima Daiichi Nuclear Power Station. That weak spot transformed a disaster for Japan into a global concern and changed a natural disaster into a man-made emergency. Nuclear power stations are not the only technological systems that have that potential. The list of risk amplifiers includes petrochemical installations, power grids, water systems, urban mass-transport systems, nuclear aircraft carriers and submarines, the Internet, satellite communications, and—as we have seen in recent years—the global financial system. These systems are crucial in sustaining our cities, countries, and economies.

Yet precisely because these systems are so crucial, it is to them that we must apply what we might call the Charles Perrow prophecy, after the eminent sociologist. For these technological systems, complete failure prevention is impossible because of their inherent complexity; when one part fails, real-time escalation processes are set in motion by the proliferation of tight couplings between constituent components. In this light, the Fukushima disaster was an inevitable and therefore "normal" accident, as Perrow has grimly concluded. The policy implications of Perrow's conclusions are radical: If we want to prevent incidents like Fukushima, we need to both reconsider our reliance on nuclear power systems and accept that mitigating the risk of catastrophic failures might involve costly regulatory interventions as well as drastic changes to spatial planning and land use.

The conundrum raised by Perrow's prophecy is stark. These intricate technological systems have become integral to successful capitalist societies.

Reducing their complexity and releasing the tightness of the couplings among their components entails major costs. Those costs affect the corporations that own and run these systems, the economies that will forego efficiencies and see growth diminished, and the citizens who will experience noticeable changes to the efficiency-driven urbanized and globalized lifestyles they have become accustomed to.

Perhaps because reducing our reliance of complex systems would bring consequences that are unpalatable to many, the search has been on for an alternative way of tackling the risk-amplification potential of high-tech industrial and infrastructural systems. Hence the recent quest for reliability and resilience, long-established notions in engineering that have in recent decades become buzzwords among students and practitioners of safety management. The two concepts are closely related. Reliable components remain safe even under extreme operating conditions; resilient systems are able to bounce back vigorously from major disturbances.

After three decades of intensive research on organizations that have managed high-risk systems in a nearly error-free fashion, we now know that achieving high levels of reliability and resilience is not just a matter of the technical design of industrial hardware, but also involves meticulous attention to the softer factors: staff training, work process design, group culture, and leadership styles. Creating a safety culture both reduces the chances that human errors will create accidents and improves the capacity of organizations to effectively mitigate and thus contain the escalation potential of disruptive incidents.

The Fukushima investigations have revealed, beyond any doubt, that such safety cultures were lacking at all levels in TEPCO prior to and following the incident. The reconstruction of risk management and incident response at the plant, TEPCO's head office, and key regulatory agencies suggest that a culture of complacency had taken hold, built on a mythical belief in perfect prevention. This mind-set is deeply at odds with the attitude of permanent weariness and dedication to fault-finding that is characteristic within high-reliability organizations.

Likewise, the current report testifies to existing Japanese proclivities for orderly and top-down modes of decision making among key institutional actors. Such beliefs run against the principles of resilient incident management, which emphasize decentralization, local knowledge, and deference to experts by management, strategies that enhance an organization's capacity for effective improvisation in real-time. This flexible approach is especially important in the face of surprising or fast-moving events, when centralized systems tend to delay action and give authority to people lacking an adequate understanding of operational realities.

Back in his 1984 case study on manned space exploration, Perrow sang the praises of authorizing local operators to respond to emerging incidents, demonstrating convincingly that this local empowerment made the difference between the death and survival of *Apollo 13*'s beleaguered crew. However, in the case of the Fukushima Daiichi accident, even this report's authors cannot fully escape what might well be a broader cultural predilection in Japan for top-down action in times of crisis when they question the desirability of the plant manager's acting in contravention of the head office's instructions with regard to venting.

In my view, the plant manager was one of the very few people on the day who "got it." He decided to rely upon his own local expertise rather than on the command structure of a hierarchy, whose top members he knew had little idea of what was exactly going on at the plant. The man was a lone hero. One enormous challenge for Japan is to transform its corporate, industrial, and governmental cultures into systems where improvised problem-solving behavior is actively nurtured and protected.

The third paradox to be discussed here is that community empowerment is the last resort rather than the first instinct in corporate and governmental crisis management. The Fukushima disaster provides many examples of the counterproductive paternalistic attitudes toward at-risk communities held by plenty of authorities around the world. This mentality may spring from the best of intentions, rooted in a belief that the public is panic-prone and unable to rationally handle information about danger. In reality, as the sociologist Clarke has shown, there is not a shred of scientific evidence that the public is likely to panic when given information about a threat—quite the contrary.

As others see it, the elites' lack of communication in times of crisis may spring from the self-serving desire to hide evidence of incompetence, collusion, or corruption. As author Naomi Klein has argued, it may even be part of an opportunistic political stratagem to "shock and awe" citizens with bad news at a time of the authorities' own choosing, in order to further their own policy aims. Whatever the driving motives behind this official reticence, it manifests itself in a reluctance to share information about an emergency in a timely and comprehensive fashion with citizens.

As a result, three things happen. First, citizens gather information from alternative sources that provide quicker and seemingly more meaningful intelligence. Citizens then base their behavior on their own instincts, information from mass media, and suggestions from trusted sources in their own local and virtual networks, which may make them seem to act irrationally to government officials who have a fuller understanding of the situation. Finally, citizens lose trust in the government's competence or integrity, further diminishing officials' ability to steer community behavior in desirable directions. We saw all three of these developments during the Fukushima crisis. And we will continue to see these patterns in major emergencies until governments and corporations switch toward more proactive, transparent, and simply smarter crisis communication strategies. These institutions seem to naturally tend toward uncommunicativeness, whether by design or default. This has to change.

Avoidable failures

Governments have a decent record of emergency preparedness and response when it comes to "expected" crises: disasters that have occurred before and are likely to periodically reoccur given a nation's geology, geography, industrial mix, and so on. But when it comes to unexpected mega-crises—such as Hurricane Katrina, veterinary as well as human pandemics, the Boxing Day tsunami, the Icelandic volcanic ash cloud, and now the Japanese nuclear catastrophe—the governmental response often falls short. In the Japanese response to Fukushima, we can clearly see a set of recurrent pathologies that often afflict corporate and governmental emergency preparedness and response.

Illusory planning

Disaster plans can be essential tools for office holders and organizations in times of crisis. The process of recording procedures, routines, actors, and venues in thick and detailed plans helps to prepare for contingencies. Such plans work especially well for routine disturbances, when uncertainty and time pressure are relatively low and the scale of threat is limited. A major crisis is qualitatively different. The pervasive uncertainty and overwhelming scale that characterize major disasters often shatters some of the key presumptions of existing plans.

This is not to say that all emergency planning is useless. On the contrary, when done properly, it serves important start-up and network-building functions, as I'll discuss below. But by attaching too much value to the plan as document, a false sense of security can emerge. "Fantasy documents" underrate the damage and chaos that disasters entail and overrate the capacity of organizations to quickly and effectively minimize their impact. They do not consider worst-case scenarios. They also suffer from risk-selection bias, in that they are unduly focused on a narrow set of seemingly most salient threats (for example, floods in Bangladesh, winter havoc in northern Ontario, or terrorist attacks in post-9/11 Washington, DC).

Many of the plans pertaining to potential accidents at Fukushima—and indeed other nuclear establishments—rested on a fantasy of safety. Cutting through all the word games, the plans' essential assumptions appeared to be: This is not the United States or Russia, and nothing bad will ever happen; if something bad does happen, we will contain the incident on-site; therefore, we need not plan for what might happen if we can't.

Communication breakdowns

Crisis management depends critically on smooth communication flows within and between organizations. Moreover, these entities need to communicate with the public, either directly or through the media. During most crises, however, communication breaks down for a variety of reasons, only the simplest of which involve technical problems with equipment. The most debilitating communication difficulties are often caused by cultural factors, such as a lack of preexisting communications channels, lack of trust between organizations, and a narrow definition of the crisis situation and what is important to divulge to others.

In the case of Fukushima, there is ample evidence of serious breakdowns of trust between the government and TEPCO, and between the Japanese Cabinet and regulatory agencies. These communication failures could have been avoided with proper planning. However, Japan approached emergency planning as a purely technical, routine, and highly hypothetical exercise (since nuclear disasters were held to be impossible in Japan anyway). If the government and utilities had instead viewed emergency planning as a vital endeavor, they might have addressed the underlying cultural factors and divergent interests between the regulators and those they regulate, which inhibit transparency and trust.

Undue insistence on top-down management

According to a persistent planning myth, any crisis management operation is best organized in a command and control mode. However, the first phase of a crisis will inevitably be marked by a lack of information and communication, making it impossible to control every move of first responders. The same goes for fast-moving emergencies like bushfires, when centrally organized response systems often break down and time is lost pushing requests up the line and waiting for orders to come down. Effective responses in such extreme circumstances are necessarily improvised, flexible, and networked (rather than planned, standardized, and centrally led). They are driven by the initiative of operational leaders and the strength of the preexisting ties between the teams and organizations they represent. Any attempt on the part of strategic decision makers to command every aspect of crisis response impedes flexibility and local initiative, and constitutes an avoidable failure.

This report documents the top-down, micromanaging leadership style of Prime Minister Naoto Kan. His dedication and stamina were admirable, but his tendency to involve himself in operational detail and his opinionated and vocal *modus operandi* had clear drawbacks. His attempts to control the disaster response did not enhance the speed or flexibility of governmental decision making, nor did it provide a route to obtain broad political support for the far-reaching and controversial decisions that were deemed necessary.

Moreover, Kan's imperious methods made it difficult for subordinates to "speak truth to power" and tell him things he did not want to hear, as the head of the Nuclear Regulatory Authority found out in no uncertain way. Such a style can easily result in a poorly informed leader. Also, this top-down leadership style made the prime minister literally irreplaceable, a major handicap in handling a protracted and all-consuming crisis such as the Fukushima nuclear disaster. After all, even a prime minister like Kan gets tired, sleeps, and occasionally needs to put some distance between himself and the disaster's vortex of challenges and emotions. But who was empowered to take urgent action in Kan's absence?

Mismanaging media

There is no doubt that media outlets provide crucial information to both the crisis-response network and the outside world. But they do more than that. They set the terms by which the performance of crisis managers will be evaluated. In addition, the Internet and its social networking sites have brought both opportunities and complications in organizing communication in times of crisis, which need to be addressed in the emergency plans. Reporters will not abandon their critical faculties or ignore commercial pressures to produce news that sells, just because the story of the day is one of disaster and tragedy. When the media turn critical, crisis managers' disappointment at this "betrayal" often creates an "us-versus-them" mentality with journalists, bloggers and Twitter users.

Clearly, media management during the Fukushima crisis was deeply problematic. The vacuum created by the prime minister's reluctance to engage with the press resulted in a hitherto little-known Cabinet official, Chief Cabinet Secretary Yukio Edano, becoming the public face of the government's response by default. Given the circumstances, he performed admirably in the role. But Edano could not stem the groundswell of criticism about the government's handling of its regulatory responsibilities before the accident and about its communication with the public during the emergency, when the government often delayed providing information and made statements that lacked clarity and consistency. As a result, the government lost the opportunity to shape the narrative of the disaster, and would gradually lose control of its nuclear energy policy in the weeks and months following the acute stage of the emergency.

Collaboration breakdowns

It is not easy to achieve full-blown collaboration during crisis management. Empirical studies have shown that pivotal actors accord collaboration a low priority, that tribal identities are strong (particularly among the uniformed services), and that there's a persistent divide between full-time and volunteer crisis management workers. The consequences of poor inter-organizational relations in crisis-response networks are clear (one only has to think of the botched response to Hurricane Katrina). If the response network falters, even the simplest tasks (such as bringing bottles of water to the New Orleans Superdome) may become unbelievably complicated. Networks fall apart when organizations act independently, producing disjointed response operations that are confusing to citizens. Time-consuming conflicts can emerge over the division of labor, the hierarchy of decision making, and the choice of methods of operation. What's more, the institutional architecture of crisis response can become too complex and involve too many actors, resulting in inevitable misunderstandings.

In Japan's response to the Fukushima catastrophe, interdepartmental and intergovernmental collaboration was severely impaired. This was due in part to the devastating infrastructural damage wreaked by the quake and the tsunami, but also to the sheer complexity of Japan's nuclear incident management and regular emergency management structures. These two subsystems had an uneasy co-existence, and the differences between their command hierarchy and practices made cooperation difficult.

Ignoring the long shadow of crisis

The most complex leadership challenges of crisis management often arise after the operational demands of the incident response have been addressed. When exhausted policy makers are ready to return to the "normal" issues of running the government, they discover that emergencies cast a long shadow.

They must continue to engage in the politics of crisis management. This involves designing and implementing recovery programs in a climate of trauma and, often, recrimination. Typically, policy makers focus on the calls for inquiries, and the public discussion concentrates on blame and liability. The goal of these debates is to determine what changes should occur as a result of the crisis experience, but these changes bring both gains and losses to various stakeholders. Time and again, the activities of victim groups, journalists, lawyers, parliamentary oppositions, and inquiry bodies have demonstrated the crucial importance of this post-emergency phase. Underestimating the potential for a "crisis after the emergency" is yet another avoidable failure of crisis planning.

This report does not go into detail on the challenges of the post-emergency period. Still, it is evident to even the most casual observer of post-Fukushima politics that the crisis had devastating political consequences for the prime minister, the government, and the nuclear establishment. What remains to be seen is whether Japan will move away from blame assignment and emotion-laden policy reflexes and move toward more reflective forms of learning from disaster.

What is to be done?

What best practices should Japan aspire to? There is considerable consensus among crisis management scholars on this point. Let me end this essay by briefly mentioning the key elements of that consensus.

First, effective crisis planning combines a generic, all-hazards approach with a suite of specific contingency plans for priority crisis scenarios. Good preparation is based on the realization that every disaster is unique and may take on unknown (and unknowable) proportions. At the same time, it aims to build organizational capacities to deal with known dangers that can be expected to occur.

Second, emergency planners should focus not on outputs (such as reams of paper produced to meet legal obligations) but on the development of carefully aligned activity clusters. The necessary activities include ongoing risk monitoring, continuous education of personnel, and training through rigorous field exercises. Throughout, planners must endeavor to build trust and productive working relationships among the organizations that may become involved in crisis response.

Third, planners must manage public expectations by acknowledging the inherent trade-offs of crisis management. For example, the public should be educated on both the cost of optimal safety measures and key stakeholders' willingness to pay that cost. Education campaigns can also draw attention to the inherent limitations of official crisis responses; for example, it's unrealistic to expect crisis managers to gain quick control over a catastrophe like a tsunami or mega-fire.

Fourth, effective disaster planning entails proactive communication with relevant citizen communities. Crisis planning shouldn't just ensure that the government knows what to do in the event of a disaster. By discussing how government, business, and community sectors will respond during a crisis, planners can enhance community resilience. This type of planning requires a government that values and actively solicits community participation, and that takes pains to extend the dialogue beyond the well-educated middle class to the most vulnerable people.

Fifth, high-level policy makers must be trained to deal with crises. They must learn to recognize and intervene in the recurrent patterns of emergency management. They must be ready for the political issues that will emerge, the faltering information flows, the complex dilemmas and the impossible choices, the toll of stress, and the search for scapegoats. When they're called upon to make challenging decisions they must balance short- and long-term effects and make sure they're hearing contrarian views. They must know when not to meddle, must leave operational decisions to the professionals, and should facilitate emerging coordination rather than imposing preexisting plans. Leaders must also be able to engage with the media in a proactive manner.

Sixth, frequent and rigorous crisis simulations make for better emergency management performance. The former mayor of New York City, Rudolph Giuliani, credited the series of crisis management exercises held before the 9/11 terrorist attacks for the effective actions of the New York City response teams on the day of the attack. Experts agree that regular simulation exercises nurture awareness of crisis management complexities, hone decision-making skills, and allow members of the response network to get to know each other.

Seventh, effective crisis preparedness includes the forging of relationships among response agencies, as well as with media representatives, external stakeholders, and a variety of experts. Once an emergency has arisen, there is usually no time to look for the right people. Successful crisis response relies strongly on preexisting cooperative networks built and maintained painstakingly during the preceding years. Leaders in both the public and private sectors should do everything they can to foster the growth of such networks and should not tolerate the persistence of silo mentalities or turf wars.

Eighth, crisis management systems should be audited on a regular basis by independent experts. Critical examination by outsiders is an essential quality assurance system. Such inspections can introduce accountability in a normally obscure area, which typically only comes under scrutiny after disasters have occurred. Audits force the crisis management team to explain why the system looks the way it does, and to reflect on the strengths and weaknesses of current arrangements. This discussion can generate new insights that can make the system more effective. Effective emergency preparedness involves the continuous updating of plans in light of experiences gained in training exercises, the coming and going of stakeholders, waves of policy and organizational reforms, and analysis of experiences in other jurisdictions.

Finally, system-wide crisis preparedness will not happen without the active, continuous involvement and visible commitment of political-administrative elites. These leaders must nurture a culture of inquiry, in which everyone is invited to consider vulnerabilities and propose better ways to organize a resilient system. In words and deeds, these leaders must signal that crisis management is considered a crucial activity by those at the very top—and all the time, not just in the wake of terrible tragedies like Fukushima.

References

Abe, S. (2011) Some points regarding severe accidents and safety goals. In: *2010 Fall Planning Session of the Atomic Energy Safety Subcommittee of the Atomic Energy Society of Japan*, September 17. Available at: www.soc.nii.ac.jp/aesj/division/safety/H221021siryou2.pdf (accessed February 28, 2012). (In Japanese.)

Acton, J.M. and Hibbs, M. (2012) *Why Fukushima was Preventable*. Report. Washington, DC: Carnegie Endowment for International Peace.

Acton, J.M. and Hibbs, M. (2012) Fukushima could have been prevented. *International Herald Tribune*, March 9.

Advisory Committee for Natural Resources and Energy (2001) *On securing the nuclear energy safety base*. Report, Nuclear and Industrial Safety Subcommittee. Tokyo: METI.

Alexey, Y., Vassily, N., and Alexey, N. (2009) *Chernobyl: Consequences of The Catastrophe for People and the Environment*. (Annals of the New York Academy of Science Vol. 1181, December.) Boston: Blackwell Publishing, the New York Academy of Sciences.

Alvarez, R., Beyea, J., Janberg, K., Kang, J., Lyman, E., Macfarlane, A., Thompson, G. and von Hippel, F.N. (2003) Reducing the hazards from stored spent power-reactor fuel in the United States. *Science and Global Security* 11: 1. Available at: irss-usa.org/pages/documents/11_1Alvarez.pdf

Arima, T. (2008) *Nuclear Power Plants, Shoriki, and the CIA—the background history of the Showa Era discovered through confidential documents*. Tokyo: Shincho Shinsho. (In Japanese.)

Asahi Shimbun (2011) Nuclear Safety Commission: Full transcript of Chief Cabinet Secretary Edano's press conference. April 4 and April 5. (In Japanese.)

Asahi Shimbun (2011) US military prepares comprehensive aid list, anticipates large-scale dispersion in response to the nuclear accident. Available at: www.asahi.com/international/update/0521/TKY201105210528.html (accessed February 28, 2012).

Asahi Shimbun (2011) Untitled. Morning Edition, September 15. (In Japanese.)

Asahi Shimbun (2011) Untitled. Morning Edition, October 8. (In Japanese.)

Asahi Shimbun (2011) Untitled. Morning Edition, October 23. (In Japanese.)

Asahi Shimbun (2012) Government probe: Many Fukushima hospital patients died during botched rescue operation. July 24. Available at: http://ajw.asahi.com/article/0311disaster/life_and_death/AJ201207240092

Asahi Shimbun (2013) Experts urge Japan to break away from "failed" nuclear reprocessing program. August 28. Available at: http://ajw.asahi.com/article/0311 disaster/fukushima/AJ201308010057

Atomic Energy Council (2001) *The station blackout incident of the Maanshan NPP unit 1*. Report, April 18. Available at: www.aec.gov.tw/webpage/UploadFiles/report_file/1032313985318Eng.pdf (accessed February 28, 2012).

Atomic Energy Society of Japan (2010) Meeting notes from Special Investigatory Committee on Earthquake Safety. In: *The logic of seismic safety in the design and evaluation of nuclear power plants*, Section 7: Summary and future work. Tokyo: AESJ.

ATOMICA (1998) Japan's response to the Three Mile Island accident. Available at: www.rist.or.jp/atomica/data/dat_detail.php?Title_No=02-07-04-06 (accessed August 31, 2013). (In Japanese.)

Bingham, A. (2011) NRC chair: "No water in the spent fuel pool" at unit 4. In: ABC News, *The Note*, March 16. Available at: http://abcnews.go.com/blogs/politics/2011/03/nrc-chair-no-water-in-the-spent-fuel-pool-at-unit-4/ (accessed August 31, 2013).

Brenner, D. (2011) We don't know enough about low-dose radiation risk. *Nature*, April 5. DOI: 10.1038/news.2011.206. Available at: www.nature.com/news/2011/110405/full/news.2011.206.html (accessed June 21, 2013).

Cabinet Office, Government of Japan (2001) *Nuclear power safety white paper*. Report. Tokyo: Cabinet. (In Japanese.)

Cabinet Office, Government of Japan (2011) *Chronology report on Fukushima accident*. Report. Tokyo: Cabinet.

Cabinet Office, Government of Japan (2011) *Report of the working group on risk management of low-dose radiation exposure*. Report, December 22. Available at: www.cas.go.jp/jp/genpatsujiko/info/twg/111222a.pdf (accessed June 21, 2013). (In Japanese.)

Cabinet Office, Government of Japan (2011) *Interim report of the Government Investigation Committee on the Accident at TEPCO's Fukushima Nuclear Power Stations*. Report, Government Investigation Committee on the Accident at the Fukushima Nuclear Power Stations, December 26. Tokyo: Cabinet. Available at: http://jolisfukyu.tokai-sc.jaea.go.jp/ird/english/sanko/hokokusyo-jp-en.html

California Institute of Technology (2011) Caltech researchers release first large observational study of 9.0 Tohoku-Oki earthquake. Available at: www.caltech.edu/content/caltech-researchers-release-first-large-observational-study-90-tohoku-oki-earthquake (accessed August 31, 2013).

Chernobyl Forum (2003–2005) *Chernobyl's legacy: health, environmental and socio-economic impacts: Recommendations to the governments of Belarus, the Russian Federation and Ukraine (second revised version)*. Vienna: IAEA. Available at: www.iaea.org/Publications/Booklets/Chernobyl/chernobyl.pdf

Chugoku Shimbun (2011) Dissenting voices heard on Fukushima response. July 11. (In Japanese.)

Clarke, L. and Chess, C. (2008) Elites and panic: More to fear than fear itself. *Social Forces* 87(2): 993–1014. Available at: www.bupedu.com/lms/admin/uploded_article/eA.491.pdf

Council for Nuclear Fuel Cycle (2003) Nuclear power plants represent a fateful partnership: An interview with Futaba town Mayor Tadao Iwamoto. *Plutonium* 42. (In Japanese.)

Cox, A., Ericson, M. and Tse, A. (2011) The evacuation zones around the Fukushima Daiichi Nuclear Plant. *The New York Times*, March 25. Available at: www.nytimes.com/interactive/2011/03/16/world/asia/japan-nuclear-evacuation-zone.html?_r=0.

Criticality.org (2013) Kei Sugaoka, the GE/Tepco Whistleblower. Available at: www.youtube.com/watch?v=fBjiLaVOsI4&feature=youtu.be (accessed August 31, 2013). (In Japanese).

Diet, Government of Japan (2011) Diet deliberations. House of Councilors Budget Committee. April 18.

Edano, Y. (2011) Press conference, Office of the Prime Minister. 3:30 pm, March 13. Available at: www.youtube.com/watch?v=3orovQa2K3w

Epstein, P.R., Buonocore, J.J., Eckerle, K., Hentryx, M., Stout III, B.M., Heinberg, R., Clapp, R.W., May, B., Reinhart, N.L., Ahern, M.M., Doschi, S.K. and Glustrom, L. (2011) Full cost accounting for the life cycle of coal. *Annals of the New York Academy of Sciences* 1219 (February).

European Committee on Radiation Risk (2010) *2010 recommendations of the European Committee on Radiation Risk: The health effects of exposure to low doses of ionizing radiation*. Report, regulators' edition. Brussels: ECRR. Available at: www.euradcom.org/2011/ecrr2010.pdf (accessed June 21, 2013).

Frommer, F. (2011) Nuke crisis reignites debate on protective pills. *Associated Press*, March 30.

Fukushima Medical University (2013) *Survey results: Proceedings of the 11th prefectural oversight committee meeting for Fukushima Health Management Survey*. Report, June 5. Available at: www.fmu.ac.jp/radiationhealth/results/20130605.html (accessed August 30, 2013).

Fukushima Prefecture (2011) Health management survey for prefecture residents. Available at: www.cms.pref.fukushima.jp/pcp_portal/PortalServlet?DISPLAY_ID=DIRECT&NEXT_DISPLAY_ID=U000004&CONTENTS_ID=24287 (accessed June 21, 2013.)

Fukushima Prefecture (2011) *An overview of the basic tests (total exposure to external radiation) and thyroid tests*. Report, Health Management Survey for Prefecture Residents Survey Committee, December 13. Available at: www.pref.fukushima.jp/imu/kenkoukanri/231213gaiyo.pdf (accessed June 21, 2013).

Funabashi, Y. and Kitazawa, K. (2012) Fukushima in review: A complex disaster, a disastrous response. *Bulletin of the Atomic Scientists* 68(2): 9–21. Available at: http://bos.sagepub.com/content/68/2/9.full.

Ganjoho (2011) Cancer statistics in Japan. Available at: http://ganjoho.jp/public/statistics/backnumber/2011_en.html. (In Japanese.)

Global Security (2013) Nuclear ship *Mutsu*—1974 incident. Available at: www.globalsecurity.org/military/world/japan/ns-mutsu-1974.htm (accessed August 31, 2013).

Goffman, J. (1991) *Ningen to Hoshasen*. Tokyo: Shakai Shisosha. (Translated to Japanese.) Original text is: Goffman, J. (1981) *Radiation and Human Health*. San Francisco: Sierra Club Books.

Grodzinsky, D. (translated by Sato, M.) (1966) *Houshasen Seibutsugaku Nyuumon (Introduction to Radiobiology)*. Tokyo: Kagaku fukyu shinsho, Tokyo Tosho.

Health Physics Society (1996) *Radiation risk in perspective*. Report, a position statement of the Health Physics Society XXIV(3). Available at: www.hps.org/documents/radiationrisk.pdf (accessed June 21, 2013).

Heine, M. (2011) Tokio in todesangst. *Die Welt*, March 16. Available at: www.welt. de/print/die_welt/politik/article12841234/Tokio-in-Todesangst.html (accessed August 31, 2013). (In German.)

Hirano, M. (2011) The history of countermeasures against severe accident and "residual risk." *Journal of the Atomic Energy Society of Japan (ATOMOΣ)* 53(11): 22–28. (In Japanese.)

Iitatemura, Fukushima (2011) The decontamination plan of Iitatemura. Available at: www.vill.iitate.fukushima.jp/saigai/wp-content/uploads/2011/10/b2eb22467554edc1286c0f22672344be (accessed September 4, 2013). (In Japanese.)

Institute of Nuclear Power Operations (2011) *Special report on the nuclear accident at the Fukushima Daiichi Nuclear Power Station*. Report. Atlanta: INPO.

International Atomic Energy Agency (1986) *Convention on early notification of a nuclear accident*. Report. Vienna: IAEA. Available at: www.iaea.org/Publications/Documents/Infcircs/Others/infcirc335.shtml (accessed August 31, 2013).

International Atomic Energy Agency (1991) *Basic safety principles for nuclear power plants*. Report, International Nuclear Safety Advisory Group. Vienna: IAEA. Available at: www-pub.iaea.org/MTCD/publications/PDF/P082_scr.pdf.

International Atomic Energy Agency (2000) *Safety of nuclear power plants: Design*. Report, IAEA Safety Standard Series, NS-R-1. Vienna: IAEA. Available at: www-pub.iaea.org/MTCD/publications/PDF/Pub1099_scr.pdf.

International Atomic Energy Agency (2007) Integrated Regulatory Review Service to Japan: Report to the Japanese government. Report. Vienna: IAEA.

International Atomic Energy Agency (2009) *Safety assessment for facilities and activities*. Report, IAEA Safety Standard Series, GSR-Part 4. Vienna: IAEA.

International Atomic Energy Agency (2013) OSART: Operational safety review teams. Brochure. Available at: www-ns.iaea.org/downloads/ni/s-reviews/osart/OSART_Brochure.pdf (accessed August 31, 2013).

International Atomic Energy Agency (2013) The international nuclear event scale. Available at: www-ns.iaea.org/tech-areas/emergency/ines.asp (accessed August 31, 2013).

Japan Act on Compensation for Nuclear Damage, Article 16.

Japan Atomic Energy Agency (2012) *Report of the results of the decontamination model projects—decontamination technologies*. Report, Cabinet Office's Team in Charge of Assisting the Lives of Disaster Victims, March 26. Available at: www.jaea.go.jp/fukushima/decon04/english/2-2-2%20Decontamination%20Technologies.pdf.

Japan Atomic Energy Commission (1977) Item 9: Design requirements to protect against power outages. In: *Safe design inspection guidelines*.

Japan Atomic Energy Commission (2003) *All about nuclear energy: Wisdom of coexisting with the Earth*. Report, All About Nuclear Energy Editorial Committee Edition. Tokyo: JAEC. (In Japanese.)

Japan Atomic Energy Commission (2012) The current situation of plutonium management in Japan. Available at: www.aec.go.jp/jicst/NC/iinkai/teirei/plutonium_management.htm.

Japan Atomic Power Company (2011) *On the status of the Tokai Daini Power Plant and its safety precautions after the Tohoku Pacific earthquake*. Report (current as of the end of June 2011).

Japan Coastal Ocean Predictability Experiment (2012) Fukushima radionuclide dispersion simulation in the ocean using JCOPE. Available at: www.jamstec.go.jp/frcgc/jcope/htdocs/e/fukushima.html (Accessed August 30, 2013).

Japan Electric Association (2009) *Quality assurance regulation for nuclear power plant safety*. Report, JEAC-4111. Tokyo: JEA. (In Japanese.)

Japan Nuclear Energy Safety Organization (2012) *Third-party investigative committee on inspections and other operations*. Report. Tokyo: JNES.

Japan Nuclear Energy Safety Organization (2013) Seawater flooding of pump building triggers shutdown of Kalpakkam unit 2 for safety purposes. Available at: www.atomdb.jnes.go.jp/content/000023842.pdf (accessed August 31, 2013). (In Japanese.)

Kainuma, H. (2011) *A Theory of Fukushima: How Did the Nuclear Power Village Arise?* Tokyo: Seidosha. (In Japanese.)

Kan, N. (2012) Panel discussion. In: Annual Meeting of the World Economic Forum, January 6. Davos: World Economic Forum.

Kikkawa, T. (2011) *TEPCO: The True Nature of Failure*. Tokyo: Toyo Keizai Shimbunsha. (In Japanese.)

Kondo, S. (1990) *Genshiryoku no Anzensei (Nuclear Power Safety)*. Tokyo: Doubun Shoin.

Kondo, S. (2011) *Outline of the Worst-Case Scenario for the Fukushima Daiichi Nuclear Power Station*. Report, Japanese government, March 25.

Kondo, S. (2011) *Where Japan is and where Japan will go: Update of the Fukushima accident and the deliberation of post-Fukushima nuclear energy policy in Japan*. Presentation, Japan Atomic Energy Commission, December 2. Available at: www.aec.go.jp/jicst/NC/about/kettei/111202b.pdf (accessed August 31, 2013).

Lamarsh, J. (2003) *Introduction to Nuclear Engineering Vol. 2*. Tokyo: Pearson Education.

Legal Defense Counsel (1979) *Atomic energy and social dispute on safety: criticism of Ikata nuclear plant court decision*. Report, Nuclear Energy Technology Research Edition. Tokyo: Technology and Humanity. (In Japanese.)

Lochbaum, D. (2004) *US Nuclear Plants in the 21st Century: The Risk of a Lifetime*. Cambridge, MA: Union of Concerned Scientists.

Maher, K. (2011) *The Japan that Can't Decide*. Tokyo: Bunshun Shinsho. (In Japanese.)

Mainichi Shimbun (2002) Untitled. Morning Edition, December 27. (In Japanese.)

Mainichi Shimbun (2011) IAEA report on the Fukushima Daiichi Nuclear Power Station: Japan slow to make decisions during the crisis. June 18. (In Japanese.)

Miller, C., Cubbage, A., Dorman, D., et al. (2011) *Recommendations for enhancing reactor safety in the 21st century: The Near Term Task Force review of insights from the Fukushima Dai-ichi accident*. Report, Nuclear Regulatory Commission, July 12. Available at: http://pbadupws.nrc.gov/docs/ML1118/ML111861807.pdf

Ministry of Economy, Trade, and Industry (1992) The status of accident management measures within the safety regulatory system. Press release, July.

Ministry of Economy, Trade, and Industry (1994) Report on the review of accident management measures to be developed at light water nuclear power reactor facilities. Press release, October.

Ministry of Education, Culture, Sports, Science, and Technology (2011) Plan to strengthen environmental monitoring. Press release, April 22, Nuclear Accident Response Center, April 22.

Ministry of Education, Culture, Sports, Science, and Technology (2011) A forum to discuss the creation of maps illustrating the geographical distribution of radiation

References

levels and other issues. Press release, Nuclear Accident Response Center, May 16. Available at: www.mext.go.jp/b_menu/shingi/chousa/gijyutu/017/gaiyo/1307559.htm (accessed June 21, 2013).

Ministry of Education, Culture, Sports, Science, and Technology (2011) Short-term policy initiatives to assist victims of the nuclear accident. Press release, Nuclear Accident Response Center, May 17.

Ministry of Education, Culture, Sports, Science, and Technology (2011) Monitoring information of environmental radioactivity level. Available at: http://radioactivity.mext.go.jp/ja/1910/2011/10/17485.pdf (accessed February 28, 2012). (In Japanese.)

Ministry of Education, Culture, Sports, Science, and Technology (2011) Monitoring information of environmental radioactivity level: Results for various isotopes of plutonium and strontium. Available at: http://radioactivity.mext.go.jp/ja/distribution_map_around_FukushimaNPP/0002/5600_0930.pdf (accessed February 28, 2012). (In Japanese.)

Ministry of Internal Affairs and Communications (2011) White paper information and communication in Japan. Available at: www.soumu.go.jp/johotsusintokei/statistics/statistics05a.html (accessed September 4, 2013). (In Japanese.)

Morokuzu, M. (2011) Now is the time for a deeper understanding of defense in depth. *Journal of the Atomic Energy Society of Japan (ATOMOΣ)* 53(12): 794–795. (In Japanese.)

MSN Economics News (2011) Press conference with Chief Cabinet Secretary Edano regarding radiation leakage: "I never said everything was fine." March 25. Available at: http://sankei.jp.msn.com/politics/news/110325/plc11032518580033-n1.htm (accessed February 28, 2012). (In Japanese.)

Murray, R. (1965) *Genshikaku Kougaku Nyuumon (Introduction to Nuclear Engineering)*. Tokyo: Pearson Education.

Nagashima, A. (2011) Proposal to reconvene US-Japan coordination capabilities in response to nuclear power generation. Memo, Ministry of Defense, March.

Nagashima, A. (2011) Panel discussion. In: *Sasakawa Peace Foundation Third US–Japan Public Policy Forum*, November 2011. Tokyo: Sasakawa Peace Foundation.

Nakagawa, Y. (1991) *A History of Radiation Exposure*. Tokyo: Akashi Shoten.

Nanbara, S. (1947) Nuclear power and the second Industrial Revolution. *The Mainchi*, October 27. (In Japanese.)

Nishiwaki, Y. (2007) Issues and observations regarding regulations of nuclear power plants. *Transactions of the Atomic Energy Society of Japan* 6(3): 239–252. (In Japanese.)

Nishiwaki, Y. (2011) Changes in Japan's protections against severe accidents: Where did regulations go wrong? Genshiryoku Eye, September/October. (In Japanese.)

Nishiwaki, Y. (2011) *Rethinking the nature of the nuclear safety agencies and the regulatory system for nuclear reactors*. Report, JNES Technical Information Seminar Materials. Tokyo: JNES.

Nöggerath, J., Geller, R.J. and Gusiakov, V.K. (2011) Fukushima: The myth of safety, the reality of geoscience. *Bulletin of the Atomic Scientists* 67(37). Available at: http://thebulletin.org/2011/september/fukushima-myth-safety-reality-geoscience.

Nuclear and Industrial Safety Agency (2008) Press release, March 14. Available at: http://warp.ndl.go.jp/info:ndljp/pid/286890/www.meti.go.jp/press/20080314007/irrs.pdf (accessed August 31, 2013). (In Japanese.)

Nuclear and Industrial Safety Agency (2010) Nuclear and Industrial Safety Agency's mission and action plans. Available at: www.nisa.meti.go.jp/oshirase/2010/files/220617-6-1.pdf (accessed February 28, 2012). (In Japanese.)

Nuclear Emergency Response Headquarters, Government of Japan (2011) *Additional Report of the Japanese Government to the IAEA: The Accident at TEPCO's Fukushima Nuclear Power Stations*. Report, September 15. Available at: www.iaea.org/newscenter/focus/fukushima/japan-report2/

Nuclear Energy Legal System Research Group (2009) *Science and legal structure study group report*. Report, Department of Nuclear Engineering and Management. Tokyo: School of Engineering, University of Tokyo.

Nuclear Safety Commission (1990) *Inspection guidelines for safety design of light-water type nuclear reactor for power generation*. Report, Ministry of Education, Culture, Sports, Science, and Technology. Tokyo: NSC.

Nuclear Safety Commission (1990) *Interim report of the common issues discussion group of Nuclear Safety Commission's special committee on safety standards and guides*. Report, February 19. Tokyo: NSC.

Nuclear Safety Commission (1992) *Accident management for the severe accidents at light water nuclear power reactor facilities*. Report, May 28.

Nuclear Safety Commission (1993) *The event of total loss of AC power at a nuclear power plant*. Report, Total AC Power Loss Event Working Group of the Deliberation Committee on Analysis and Evaluation of Accidents and Failures in Nuclear Installations.

Nuclear Safety Commission (2006) Decision No. 60, item 18. In: *Regulatory guide for reviewing seismic design of nuclear power reactor facilities*. Report, September 19. Tokyo: NSC. Available at: pbadupws.nrc.gov/docs/ML0803/ML080310851.pdf

Nuclear Safety Commission (2006) *Regulatory guide for reviewing seismic design of nuclear power reactor facilities*. Report, September 19. Tokyo: NSC. Available at: pbadupws.nrc.gov/docs/ML0803/ML080310851.pdf

Nuclear Safety Commission (2006) Section 1: Basic policies (commentary). In: *Regulatory guide for reviewing seismic design of nuclear power reactor facilities*. Report, September 19. Tokyo: NSC. Available at: pbadupws.nrc.gov/docs/ML0803/ML080310851.pdf

Nuclear Safety Commission (2008) Chairman's comment regarding the results of the IAEA/IRRS assessment. Press release, March 17.

Nuclear Safety Commission (2011) Item 27: Design considerations for protection against power outages. In: *Report on Investigations of Total AC Power Loss* (draft). Report, subcommittee item 4-1-1. Available at: www.nsc.go.jp/senmon/shidai/anzen_sekkei/anzen_sekkei4/siryo4-1-1.pdf (accessed February 28, 2012). (In Japanese.)

Nuclear Safety Commission (2011) *The 7th Nuclear Safety Commission Subcommittee for Investigating Safety Design regulatory guide*. Report, November 16. Available at: www.nsc.go.jp/senmon/shidai/anzen_sekkei/anzen_sekkei8/siryo1-2.pdf (accessed February 28, 2012). (In Japanese.)

Nucleonics Week (2002) IAEA aims for thaw with Japan on OSART, safety cooperation. *Nucleonics Week* 43(39), September 26.

Nucleonics Week (2002) TEPCO says it has cooperated with IAEA on OSART findings. *Nucleonics Week* 43(41), October 10.

Okamoto, K. (2012) On the possibility of total power loss. *Journal of the Atomic Energy Society of Japan (ATOMOΣ)* 54(1): 27–31. (In Japanese.)

Omoto, A., Juraku, K. and Tanaka, S. (2011) Why was the accident not prevented? In: *University of Tokyo global COE program GoNERI symposium 2011: Rethinking nuclear power education and research after the accident at TEPCO's Fukushima Daiichi Nuclear Power Station.* Tokyo, Japan. (In Japanese.)

Permanent Mission of Japan to IAEA, Government of Japan (2011) *Report of Japanese government to the IAEA Ministerial Conference on Nuclear Safety: The accident at TEPCO's Fukushima Nuclear Power Stations.* Report, June 7. Available at: www.iaea.org/newscenter/focus/fukushima/japan-report/

President's Commission on the Accident at Three Mile Island (1979) *Report of the President's Commission on the Accident at Three Mile Island: The need for change, the legacy of Three Mile Island.* Report, October. US Government Printing Office: Washington, DC. Available at: www.threemileisland.org/downloads/188.pdf

President's Commission on the Accident at Three Mile Island (1979) *Report on the Emergency Preparedness and Response Task Force.* Staff Report to the President's Commission on the Accident at Three Mile Island. US Government Printing Office: Washington, DC.

Prime Minister's Office, Government of Japan (2011) Great East Japan earthquake. Global communication activities of Prime Minister's Office. Press release, September 23.

Radiation Safety Research Center (2013) Aiming at further understanding of the biological effects of low-dose radiation. Central Research Institute of Electric Power Industry, Nuclear Technology Research Laboratory, Radiation Safety Research Center. Available at: http://criepi.denken.or.jp/jp/ldrc/index.html (accessed August 30, 2013).

Rasmussen, N. (1979) *Reactor safety study (WASH-1400).* Report, US Nuclear Regulatory Commission.

Roisman, A.Z., Honaker, E. and Spaner, E. (2009) Regulating nuclear power in the new millennium (the role of the public). *Pace Environmental Law Review* 26(317).

Sankei Shimbun (2011) Japanese government pressured to explain ... dumping of contaminated water with no advance warning. April 6. (In Japanese.)

Sankei Shimbun (2011) Statement from the Ministry of Education, Culture, Sports, and Technology: "On the May 27 statement that 'We target a level of under 1 mSv per year for schools,' which reflected 'our thinking at that time.'" July 20.

Sasaki, Y. (2012) Sociotechnology and the Fukushima Daiichi Nuclear Power Station accident. In: *8th annual sociotechnology research symposium,* January 28.

Sato, K. (2011) *The Logic of Nuclear Safety* (revised edition). Tokyo: Nikkan Kogyo Shimbunsha. (In Japanese.)

Shimbun Kenkyu (2011) Untitled. September. (In Japanese.)

Stewart, P. (2011) US readies to fly military families out of Japan. *Reuters,* March 17. Available at: www.reuters.com/article/2011/03/17/japan-quake-usa-military-id USN1716166220110317 (accessed August 31, 2013).

Takemori, S. (2011) *The National Policy-Private Operation Trap: The Hidden Nuclear Policy Struggle.* Tokyo: Nikkei Publishing. (In Japanese.)

Tasaka, H. (2012) *An Inside Look at the Reality of the Nuclear Accident from the Prime Minister's Office.* S.l.: Kobunsha. (In Japanese.)

Tirone, J. (2011) Nuclear watchdog says Japan falls short supplying information. *Bloomberg*, March 16. Available at: www.bloomberg.com/news/2011-03-16/nuclear-watch dog-says-japan-falls-short-supplying-information.html (accessed August 31, 2013).

Tokyo Electric Power Company, Inc. (2002) *Investigation report on the items pointed out by GE on the periodic inspections and repair for our nuclear power plants*. Report. Tokyo: TEPCO.

Tokyo Electric Power Company, Inc. (2011) *Fukushima nuclear accidents analysis report (interim report) supplement*. Report, December 2. Tokyo: TEPCO. Available at: www.tepco.co.jp/en/press/corp-com/release/betu11_e/images/111202e14.pdf (accessed August 31, 2013).

Tokyo Electric Power Company, Inc. (2011) *On hypotheses regarding the status of damage to units 1–3 reactor cores*. Report. Tokyo: TEPCO.

Tokyo Electric Power Company, Inc. (2011) *Report of the TEPCO Management and Finance Investigation Committee*. Report, Management and Finance Investigation Committee. Tokyo: TEPCO. Available at: www.cas.go.jp/jp/seisaku/keieizaimutyousa/dai10/siryou1.pdf (accessed August 30, 2013). (In Japanese.)

Tokyo Electric Power Company, Inc. *Manual of Operational Procedures in Case of an Accident*. Tokyo: TEPCO.

Tsuchiya, T. (2011) Issues on earthquake and tsunami risk assessments and seismic design—interim report of the expert hearing. In: *International symposium on joint fact finding: The possibility of earthquake risk analysis for nuclear power plants*. Tokyo: Tokyo University Policy Alternatives Research Institute. (In Japanese.)

United Nations Scientific Committee on the Effects of Atomic Radiation (2008) *Sources and Effects of Ionizing Radiation*. Report, Vol. II, Annex D: Health effects due to radiation from the Chernobyl accident.

US Code of Federal Regulations: Title 10, Part 50.63.

US Environmental Protection Agency (1992) *Manual of protective action guides and protective actions for nuclear incidents*. Report, Office of Radiation Programs, 400-R-92-001. Washington, DC: EPA. Available at: www.epa.gov/radiation/docs/er/400-r-92-001.pdf.

US Environmental Protection Agency (2013) Radon. Available at: www.epa.gov/radiation/radionuclides/radon.html (accessed June 21, 2013).

US Food and Drug Administration (2001) *Guidance: potassium iodide as a thyroid blocking agent in radiation emergencies*. Report.

US National Research Council (2006) *Health Risks from Exposure to Low Levels of Ionizing Radiation: BEIR VII—Phase 2*. Washington, DC: The National Academies Press.

US Nuclear Regulatory Commission (1975) *Reactor Safety Study: An Assessment of Accident Risks in U.S. Commercial Nuclear Power Plants*. Report, Executive Summary.

US Nuclear Regulatory Commission (1979) *NRC statement on risk assessment and the reactor safety study report (WASH-1400) in light of the risk assessment review group report*. Report, January 18.

US Nuclear Regulatory Commission (1986) *Safety goals for the operation of nuclear power plants*. Report, 51-FR-28044, August 4. Available at: www.nrc.gov/reading-rm/doc-collections/commission/policy/51fr30028.pdf

US Nuclear Regulatory Commission (1995) *Final policy statement on the use of probabilistic risk assessment methods in nuclear regulatory activities*. Report, SECY-95-126, August

16. Available at: www.nrc.gov/reading-rm/doc-collections/commission/policy/60fr42622.pdf

US Nuclear Regulatory Commission (2002) Excessive Japanese requirements led to cover-ups, managers say. *Inside NRC* 24(22): 348–349.

US Nuclear Regulatory Commission (2002) *NRC's regulation of Davis-Besse regarding damage to the reactor vessel head*. Report, Inspector General's Office, case no. 02-03S, December 30. Available at: www.nrc.gov/reading-rm/doc-collections/insp-gen/2003/02-03s.pdf (accessed January 4, 2013).

US Nuclear Regulatory Commission (2004) Changes to adjudicatory process. *Federal Register* 69(9): 2182.

US Nuclear Regulatory Commission (2011) Document ML1128A114. Obtained by Freedom of Information Act request. Available at: www.nuclearfreeplanet.org/nrc-foia-documents.html

US Nuclear Regulatory Commission (2011) Document ML11257A101. Obtained by Freedom of Information Act request. Available at: www.nuclearfreeplanet.org/nrc-foia-documents.html

US Nuclear Regulatory Commission (2011) Document ML11269A172. Obtained by Freedom of Information Act request. Available at: www.nuclearfreeplanet.org/nrc-foia-documents.html

US Nuclear Regulatory Commission (2011) NRC sees no radiation at harmful levels reaching US from damaged Japanese nuclear power plants. Press release, 11-046, March 13, 2011.

US Nuclear Regulatory Commission (2011) NRC analysis continues to support Japan's protective actions. Press release, 11-049, March 15.

US Nuclear Regulatory Commission (2011) NRC provides protective action recommendations based on US guidelines. Press release, 11-050, March 16.

US Nuclear Regulatory Commission (2011) Briefing on the progress of the task force: Review of NRC processes and regulations following the events in Japan. Transcript, March 21. Washington, DC: NRC. Available at: www.nrc.gov/reading-rm/doc-collections/commission/tr/2011/20110321.pdf

US Nuclear Regulatory Commission (2011) Letter from Chairman Gregory B. Jaczko to Senator Jim Webb providing information on assumptions used in recommending a 50-mile evacuation for U.S. citizens following the Fukushima Daiichi nuclear facility events. Letter, June 17. Available at: www.nrc.gov/reading-rm/doc-collections/congress-docs/correspondence/2011/webb-06-17-2011.pdf

US Nuclear Regulatory Commission (2012) Press conference. Available at: www.nsr.go.jp/kaiken/data/20120919sokkiroku.pdf (In Japanese.)

Yomiuri Shimbun (1953) Let us put hydrogen to peaceful use. January 1. (In Japanese.)

Yomiuri Shimbun (2002) TEPCO: Conceals nuclear plant damage; 29 incidents, 11 document falsifications still uncorrected. August 30. (In Japanese.)

Yomiuri Shimbun (2011) Prime Minister Kan interview. (In Japanese.)

Yoshioka, H. (1999) *A Social History of Nuclear Power: Its Development in Japan*. Tokyo: Asahi Shimbun. (In Japanese.)

Index

Page numbers in *italic* refer to illustrations.

accident management, Fukushima Daiichi: chain reactions stopped 4, 98; cooling systems *xxxiii, xxxiv*, 4–5, 9–11, 24, 26, 98; emergency generators 5, 6, 109; fire engines *xxxv*, 17–18, 19, 26, 31, 33; generator trucks 8–9, 17, 19; helicopter water drops 31, 32–3, 169; seawater cooling *xxxv*, 18, 22–4, 25, 26, 33, 188–9; spent fuel pools cooling 31–3; US assistance 171–2; venting *xxxv*, 11, 12–13, 14–15, 16–17, 18–19, 25, 27; *see also* Cabinet response; evacuations
accident-management measures 99, 115–18
accident preparedness 41, 67–8, 71, 87, 95–123; defense-in-depth 97–100, 115, 187; design-basis events 95–6; drills 2; earthquakes 104, 105, 111; floods 106–7; and nuclear security 186–7; off-site centers 122–3, 179; probabilistic safety assessments 96–7, 103, 104, 113, 114, 115–18, 119–20, 210–11; severe accidents 96–7, 111–20, 178–80; SPEEDI 120–2, 179–80; station blackouts 107–11; tsunamis 2, 78, 79–81, 101–6, 107, 181–2
accidents, nuclear: and antinuclear movement 56; Chernobyl 13, 51, 113, 114, 126, 141, 207, 208, 212; drills 2; Monju 46–7; *Mutsu* 43, 44; and safety myth 51; severe accidents 96–7, 111–20, 178–80; Three Mile Island 46, 51, 56, 112–13, 208, 210, 211, 217; Tokai-mura 47, 51, 56, 122; Windscale 41
Act on Compensation for Nuclear Damage 39–40, 41–2, 135
acute radiation syndrome 37
advertising nuclear power 57
AEC *see* Atomic Energy Commission
aftershocks 9, 35
Agency for Natural Resources and Energy (ANRE) *xxviii*, 48, 117–18
agriculture *xxxix*, 126–7, 129–33
aid, foreign 158–9, 160–1; liability exemptions 172–3
amakudari 72
Amano, Yukiya 92
antinuclear movement 43–4, 45–6, 55, 56, 60
Araki, Hiroshi 76, 81
Asahi City 129–31
Atomic Energy Basic Act 1, 38, 45
Atomic Energy Commission (AEC) *xxviii*, 1, 38, 39, 40, 45
Atomic Energy Society of Japan 104
Atoms for Peace program 36, 39

batteries 20–1, 24–5, 26
beef *xxxix*, 131–3
BEIR (Committee on the Biological Effects of Ionizing Radiations) 139, 140, 142
Belarus 207

boric acid 22, 24
Brenner, David 139
budget, nuclear energy 36
business community 36, 54; *see also* nuclear village

Cabinet response 6–9, 11–13, 14–17, 20–3, 189–94, 216–17; deciding evacuation limits 120–2, 180, 189; evacuation of workers 27–30, 185, 190; global information sharing 151–3; press conferences 14–15, 21–2, 126, 141, 146–51, 225; and seawater cooling 22–3, 25; Twitter account 156–7; worst-case scenario 34–5, 192–3; *see also* communications
cancer 100, 139, 212
car batteries 24–5, 26
Casto, Charles 169, 170
cattle *xxxix*, 131–3
cesium *xxxvi*, 99, 125, 131, 132
Chernobyl 13, 51, 113, 114, 126, 141, 207, 208, 212
Chiba Prefecture *xxxix*, 127, 129–31
children, protection of 133, 134–5
China 92
China Syndrome meltdown 28
Chubu Electric Power 88
Churchill, Winston 199
Citizens' Nuclear Information Center 44
communications 145–57; Cabinet and TEPCO 29–30, 189–90, 224; global information sharing 151–3; of health risks 141–2, 150–1, 211–21; lack of disclosure 145–6, 201–2, 222; press conferences 14–15, 21–2, 126, 141, 146–51, 225; social media 146, 151, 153–7, 194; with United States 159, 161–6; US–Japan coordination conferences 170–2
community meetings 48
compensation strategies 39–40, 41–2, 135
concealment, culture of 81–3
concrete-pumping vehicles 33
containment ventilation systems 114
contamination *see* radioactive material releases
control rods 4, 98

Convention on Early Notification of a Nuclear Accident 92, 93
Convention on Nuclear Safety 87, 90
cooling: fire engines *xxxv*, 17–18, 19, 26, 31, 33; helicopter water drops 31, 32–3, 169; with seawater *xxxv*, 18, 22–4, 25, 26, 33, 188–9; of spent fuel pools 31–3
cooling systems *xxxiii*, *xxxiv*, 4–5, 9–11, 24, 26, 98
crisis management 188–94, 195, 199–200, 203, 216–17; *see also* Cabinet response

Daigo Fukuryu Maru (fishing boat) 37
decontamination work 34, 99, 135–6
defense-in-depth 97–100, 115, 187
Democratic Party of Japan 55, 56, 60
Denryoku Soren (Federation of Electric Power Related Industry Worker's Unions of Japan) 54–5, 56
deregulation, electricity industry 75–6
design-basis events 95–6
disaster management *see* accident management, Fukushima Daiichi
disaster preparedness *see* accident preparedness
Don't Pay One More Yen movement 74
drinking water contamination *xxxvii*, 127–8, 133–4, 157
dry cask storage, spent fuel 210
dry ventings 15

early notification convention 92, 93
earthquake-resistant building 111, 194
earthquakes: accident preparedness 104, 105, 111; aftershocks 9, 35; historical events 101–2; nuclear accident drill for 2; Seismic Design Regulatory Guide 102–3; Tohoku earthquake 4–5, 109
ECRR *see* European Committee on Radiation Risk
Edano, Yukio 7, 8, 28, 29–30, 34, 121, 152, 163, 168, 190; press conferences 14–15, 21, 126, 141, 146, 148, 149, 150–1, 225
Electricity Business Act 76
electricity supplies 3; accident preparedness 107–11; emergency

Index

generators 5, 6, 40, 109, 110; generator trucks 8–9, 17, 19; reestablishment 33–4; station blackouts 6, 107–11, 179

electric power companies 36; advertising nuclear power 57; and compensation 39–40, 41–2; creation 73; deregulation 76–7; and international peer reviews 88, 91; primary responsibility for safety 63, 64; and safety myth 53–7; severe accident preparedness 115–18, 178–9; state-planned privately operated approach 39–40, 51, 78–9, 184–6; tsunami preparedness 2, 78, 79–81, 101–2, 105–6, 181–2; *see also* TEPCO

Electric Technology Research Institute 38–9

Emergency Control Center 6, 7, 192

emergency generators 5, 6, 40, 109, 110

emergency planning 223, 226–8; *see also* accident preparedness

emergency preparedness *see* accident preparedness

Emergency Response Special Review Board 67

Emergency Response Support System (ERSS) 121

Emergency Technology Advisory Committee 67

employment 58

Endo, Tetsuya (committee member) 196–7

English-language communications 151–3

Enomoto, Toshiaki 85

Europe 114, 119, 215–16; flood preparedness 106–7

European Committee on Radiation Risk (ECRR) 141, 143

evacuations: deciding limits 120–2, 180, 189; foreign visitors 152–3; of public 13–14, 15, 16–17, 21–2, 30–1, 208–9; US citizens 159, 165–9; of workers 27–30, 185, 190; worst-case scenario 35, 176–7, 209

explosions 10, 16; and spent fuel pools 32; Unit 1 reactor *xxx*, 19, 21; Unit 2 reactor 27, 30; Unit 3 reactor 25–6; Unit 4 reactor 27; worst-case scenario 35

Federation of Electric Power Companies 54, 91

Federation of Electric Power Related Industry Worker's Unions of Japan (*Denryoku Soren*) 54–5, 56

financial incentives 44, 58, 62, 75–6

Finland 119

fire engines *xxxv*, 17–18, 19, 26, 31, 33

firefighters 31, 33

fishery contamination *xxxviii*

fixed asset tax 58, 75

flood preparedness 106–7

food contamination *xxxix*, 126–7, 128, 129–33, 150

France 107, 114, 158

fuel rods 4; dry cask storage 210; meltdown 11, 19; *see also* spent fuel pools

Fujii, Hirohisa 29

Fujii, Mariko (committee member) 201–2

Fukano, Hiroyuki 48, 187

Fukushima Daiichi Nuclear Power Station: international peer review 91; location *xxvii, xxxii*; Tohoku earthquake and tsunami 3–6, 106, 109; *see also* accident management, Fukushima Daiichi

Fukushima Daini Nuclear Power Station *xxvii*, 87, 88, 91

Fukuyama, Tetsuro 7, 8, 21, 122, 152, 170, 191

Furukawa, Motohisa 151

Futaba 1, 13–14

generators, emergency 5, 6, 40, 109, 110

generator trucks 8–9, 17, 19

Germany 114

government *see* Cabinet response; crisis management; safety governance

Gozan no Okuribi bonfire festival 138

greenhouse gas reductions 75

Hamaoka Nuclear Power Station *xxvii*, 88

Hart, Paul 't (review of report) 218–28
Hattori, Takuya 70
Hayano, Ryugo 154
Health Physics Society, US 140–1
health risks 100, 138–42, 150–1, 211–21
helicopter water drops 31, 32–3, 169
high-pressure coolant injection (HPCI) system *xxxiii*, 24
Hiraiwa, Gaishi 81
Hirose, Kenkichi 64–5
Hitachi 39, 77, 83
Hokuriku Electric Power Company 102
Hosono, Goshi 8, 20, 23, 28, 29–30, 31, 34, 152, 170, 171
hydrogen *xxx*, 11, 19, 25, 27; *see also* explosions

IAEA *see* International Atomic Energy Agency
Ibaraki Prefecture *xxxix*, 51, 105–6, 126–7
Ichimiya, Ryo 154
ICRP *see* International Commission on Radiological Protection
Ikata Nuclear Power Station *xxvii*, 45
imported nuclear technology 39, 40, 42
incinerator ash 137
indoor-confinement zones 31, 99, 176–7, 189
infants, protection of 133, 134
inspections: agencies responsible for 68–9, 72; falsification of reports 46, 47, 48, 69–70, 81–2; increased paperwork 46, 70–1, 84; international 86, 88; lawsuit over validity 45–6
Integrated Regulatory Review Service (IRRS) 87, 88–91, 182–3
international aid *see* aid, foreign
International Atomic Energy Agency (IAEA) 63–4, 86–91, 92, 95, 113–14, 128; levels of defense 98–9; Nuclear Safety Standards 48, 64, 86
International Commission on Radiological Protection (ICRP) 128, 140, 142
International Conference on the Peaceful Uses of Atomic Energy, Geneva 37
International Nuclear and Radiological Event Scale 126

international peer reviews 86–91
International Regulatory Review Team (IRRT) 86
Internet 146, 151, 153–7
iodine 99, 125, 126–7, 129, 133, 134, 157
iodine tablets 168, 172, 206–8
IRRS *see* Integrated Regulatory Review Service
Ishikawa, Hiroshi 154–5
isolation condensers *xxxiii*, *xxxiv*, 5, 9–11, 98, 178
Iwamoto, Tadao 59

Jaczko, Gregory 32, 162, 164–5, 167–8
Japan Atomic Energy Commission (JAEC) 1, 43, 71, 72, 162
Japan Atomic Industrial Forum (JAIF) 39
Japan Atomic Power Company (JAPC) 39, 80–1, 105–6
Japan Business Federation (*Keidanren*) 39, 54, 73–4
Japan–China–Korea Nuclear Safety Initiative 92
Japan Nuclear Energy Safety Organization (JNES) 49, 68–9, 72, 105, 107, 162–3
Japan Power Engineering and Inspection Corporation 72
Japan Society of Radiation Safety Management 136
Joint Emergency Response Headquarters 30–1, 190, 192

Kaieda, Banri 8, 14, 15, 16, 22, 23, 27–8, 29–30, 121, 122, 190
Kan, Naoto 6, 7, 13, 20–3, 29–31, 34, 193; early-notification protocols 92; evacuation of workers 29–30, 185; and foreign media 152, 153; leadership style 199–200, 203; media interaction 146–8, 225; micromanagement 20–1, 190–1, 224–5; and safety myth 56–7, 60; SPEEDI 121; US communications 163, 168; visit to Fukushima Daiichi 15–16
Kano, Tokio 74
Kashiwazaki-Kariwa Nuclear Power Station *xxvii*, 111

Katsumata, Tsunehisa 30
Keidanren (Japan Business Federation) 39, 54, 73–4
Kikkawa, Takeo 78–9
Kitamura, Toshiro 80–1
Kitazawa, Toshimi 163, 169
kokusaku minei see state-planned privately operated approach
Kondo, Shunsuke 34–5, 104, 162, 170, 176–7, 209
Kono, Ichiro 39
Kosako, Toshiso 142
Kuboyama, Aikichi 37
Kukita, Yutaka 22–3, 35, 67
Kyodo News 55
Kyoto Protocol 75

labor unions 54–5, 56
Lee Myung-bak 92
legislation 1, 38, 45, 72; compensation 39–40, 41–2, 135; deregulation 76; Nuclear Emergency Act 6, 7, 8, 9, 122; power-siting laws 44, 58, 62
lettuce (*sanchu*) 129–31
liability exemptions 172–3
Liberal Democratic Party (LDP) 54–5, 74
local government: and safety myth 52, 57–9, 61, 62; siting of nuclear plants 43, 44, 57–9, 62, 75–6
local response centers 12, 122–3, 179
low-level radiation exposure 138–42, 150–1
Lucky Dragon 5 (fishing boat) 37

Maanshan Nuclear Power Station, Taiwan 110
Madarame, Haruki 12–13, 15–16, 21, 22–3, 28, 29, 34, 67
manual, nuclear emergency-response 2–3, 12, 193, 194
manufacturers, nuclear plant 39, 77, 83
marketing nuclear power 57
Matsumoto, Ryu 7, 29
Matsumoto, Takeaki 93, 168
Matthews, Jessica Tuchman (review of report) 213–17
media: advertising nuclear power 57; foreign 152–3; press conferences 14–15, 21–2, 126, 141, 146–51, 225; and safety myth 53; social 146, 151, 153–7, 194; support of nuclear power 55–6
meltdowns 5, 11, 19, 28, 148–9, 176–7
METI *see* Ministry of Economy, Trade, and Industry
MEXT *see* Ministry of Education, Culture, Sports, Science, and Technology
MHLW *see* Ministry of Health, Labor, and Welfare
micromanagement 20–1, 190–1, 224–5
Minami, Nobuya 76, 77, 81, 82–3
Ministry of Agriculture, Forestry, and Fisheries (MAFF) 132
Ministry of Economy, Trade, and Industry (METI) xxviii, xxix, 48, 68–9, 70, 72, 90, 117; *see also* Kaieda, Banri
Ministry of Education, Culture, Sports, Science, and Technology (MEXT) xxviii, xxix, 48, 60, 71–2, 124–5, 127, 134–5, 141, 142
Ministry of Environment xxix, 49, 135, 136, 137; *see also* Nuclear Regulatory Authority
Ministry of Health, Labor, and Welfare (MHLW) 126, 127, 128
Ministry of International Trade and Industry (MITI) 37, 53–4, 101, 114–16, 117
Ministry of Land, Infrastructure, Transport, and Tourism 48
MITI *see* Ministry of International Trade and Industry
Mitsubishi 39
Miyagi Prefecture xxxix, 102
Mizuta, Mikio 42
Monju Nuclear Power Station 46–7
Moriyama, Kinji 43
Muto, Sakae 16, 23
Mutsu (nuclear powered ship) 43, 44

Nagashima, Akihisa 169–70
Nakagawa, Keichi 154
Nakamura, Koichiro 148–9
Nakasone, Yasuhiro 36, 37

Nasu, Shou 81
Nei, Hisanori 149
NHK 156
Niigata Prefecture 2, 75, 111
NISA *see* Nuclear and Industrial Safety Agency
Nishiwaki, Yoshihiro 69, 117
Nishiyama, Hidehiko 149
Noguchi, Tetsuo 149
NRC *see* Nuclear Regulatory Commission, US
NSC *see* Nuclear Safety Commission
nuclear accidents *see* accidents, nuclear
Nuclear and Industrial Safety Agency (NISA) *xxviii*, *xxix*, 2, 68–71; accident preparedness 68, 71, 105, 107, 110, 116, 119, 179; ambiguous role 48, 49, 64, 182–4; disaster response 12, 14–15, 20, 21, 28, 29, 49; and international peer reviews 88–9, 90; nuclear security 187; press conferences 148–9; TEPCO scandal 48, 69–70, 71; US assistance 161, 162, 163–4; *see also* Terasaka, Nobuaki
Nuclear and Industrial Safety Subcommittee 47
Nuclear Emergency Act 6, 7, 8, 9, 122
nuclear emergency-response manual 2–3, 12, 193, 194
nuclear energy development, Japan: in 1950s 36–41; in 1970s 42–6; *see also* safety governance
nuclear fission 4
Nuclear Power Engineering Corporation 72, 117–18
Nuclear Power Research Council 39
nuclear power stations: construction rates 44; location map *xxvii*; siting of 43, 44, 57–9, 62, 75–6; *see also* individual *power stations*
nuclear reactors *xxxi*; normal operation 4–5; *see also* individual *Units 1 to 6*
Nuclear Regulation Act 72
Nuclear Regulatory Authority (NRA) *xxix*, 49, 61
Nuclear Regulatory Commission (NRC), US 158, 160, 163, 165–7, 170, 171, 205–6, 207, 214; nuclear security guidelines 186–7; probabilistic safety assessments 113, 119, 210–11; station blackout regulations 108–9; Tokyo Task Force 161, 162, 165, 167
Nuclear Safety Commission (NSC) *xxviii*, *xxix*, 45, 46, 65–8, 170; accident preparedness 67–8, 110, 113, 115, 116, 119–20, 179; ambiguous role 48, 49, 64, 182–3; food contamination guidelines 128; and international peer reviews 88, 90; safety guidelines 3, 48, 64, 66; Seismic Design Regulatory Guide 102–3; SPEEDI 121; station blackout guidelines 108–9, 179; *see also* Madarame, Haruki
Nuclear Safety Network (NS Net) 91
nuclear security 186–7
nuclear technology: imported 39, 40, 42; safeguarding 38, 72
nuclear tests 37
nuclear village 50–1, 52–3, 181, 205; central branch 53–7, 60–1; regional branch 57–9, 61

Obama, Barack 160, 168
Office of Global Communications 151–2, 153
off-site centers 12, 122–3, 179
Okuma 1, 13–14, 122
Omoto, Akira 71, 83, 84, 85
Onagawa Nuclear Power Station *xxvii*, 181–2
Operational Safety Review Team (OSART) 86, 87–8
Operation Castle Bravo 37

peer-review system 86–91
Periodic Safety Management Examination system 70
Perrow, Charles 220, 221
Persbo, Andreas 92
personnel rotation system 45, 71, 90
Plan for Nuclear Energy Research and Development (1955) 37
plutonium 125
Poland 207
police 8–9, 31

political contributions 54–5, 74
Poneman, Dan 162
potassium iodide 206–8
power-generator trucks 8–9, 17, 19
power-siting laws 44, 58, 62
power supplies *see* electricity supplies
pregnant women 133
Preparatory Council on Peaceful Uses of Atomic Energy 37
press conferences 14–15, 21–2, 126, 141, 146–51, 225
Prime Minister's Office: role in nuclear development 38, 40, 47; *see also* Cabinet response
privately administered government policy *see* state-planned privately operated approach
probabilistic safety assessments 96–7, 103, 104, 113, 114, 115–18, 119–20, 210–11
public disclosure 46, 47, 48, 69, 145–6, 201–2, 222; *see also* communications
public opinion 43–4, 46, 47, 48; and breakdowns of trust 145–6, 148–9, 203; and safety myth 50, 59–61

radiation: dangers to workers 18–19; data given on social media 154–5; doses received 143–4; health risks 100, 138–42, 150–1, 211–21; levels in food 129, 131–3; levels in water 125, 133–4; low-level exposure 138–42, 150–1; monitoring 72, 121, 124–6; safe dose estimates 125, 128, 134–5, 142, 166–7; SPEEDI 72, 120–2, 145, 179–80
radioactive material releases *xxxvi*, 124–44; Chernobyl 113, 114, 126; of contaminated water 34, 93–4; drinking water contamination *xxxvii*, 127–8, 133–4, 157; fears of major 28, 29, 34–5; fishery contamination *xxxviii*; food contamination *xxxix*, 126–7, 128, 129–33, 150; low-level radiation exposure 138–42, 150–1; from meltdowns 19, 99; mitigation strategies 99; monitoring 121, 124–6; and schools 134–5; soil contamination 125, 126; from spent fuel pools 31, 32, 35; Three Mile Island 112; Tokai-mura accident 47; from venting 11, 12–13, 15; worst-case scenario 35, 176–7; *see also* evacuations
radioactive waste treatment 137–8
radon 143
RASCAL (Radiological Assessment System for Consequence Analysis) program 167
Rasmussen, Norman 96–7
RaSSIA (Radiation Safety and Security Infrastructure Appraisal) 86–7
reactor core isolation cooling systems 5, 11, 24, 26, 98
recriticality 22–3
regional government *see* local government
regulation *see* safety governance
risk denial *see* safety myth
risk inflation 219–21
Roos, John 163, 168, 169, 170
Russia 93, 158, 207

safety: concerns about 42–4, 45–6, 47; lack of guidance 38; lack of research 40–1; *see also* accident preparedness; inspections
Safety 21 program 114–15, 117
safety governance *xxviii*, *xxix*, 63–72, 197–9; and accidents 46–7; ambiguous responsibilities 40–1, 44–6, 47–9, 64, 72, 182–6; compensation 39–40, 41–2; international 86–94; and safety myth 53–7, 60–1; state-planned privately operated approach 39–40, 51, 78–9, 184–6; and TEPCO scandal 46, 48, 69–70, 71, 81–2, 83–4, 87–8; *see also* NISA (Nuclear and Industrial Safety Agency); Nuclear Safety Commission
safety guidelines 3, 48, 53, 64, 66; food contamination 128; Seismic Design Regulatory Guide 102–3; station blackouts 108–9, 179; US nuclear security 186–7; US severe-accident management 114, 215
safety myth 2, 42, 49, 50–62, 181–2, 204–6, 218–19; after Fukushima

Daiichi disaster 60–1; central branch of nuclear village 53–7, 60–1; public opinion 50, 59–61; regional branch of nuclear village 57–9, 61; United States 113, 205
safety standards 42, 64, 66; international 48, 64, 86; structural strength 118–19
sanchu (lettuce) 129–31
schools 134–5, 136
Science and Technology Agency (STA) 38, 40, 42, 45, 47, 53–4, 69, 71, 118
scramming 4, 98
seawater cooling *xxxv*, 18, 22–4, 25, 26, 33, 188–9
security, nuclear 186–7
Seismic Design Regulatory Guide 102–3
Self Defense Forces 9, 18, 26, 31, 32–3, 160–1, 169, 186
Sengoku, Yoshito 170
severe accidents 96–7, 111–20, 178–80
Shika Nuclear Power Station *xxvii*, 102
Shikata, Noriyuki 151
Shimizu, Masataka 27–8, 29–30, 31, 190
Shoriki, Matsutaro 39
Socialist Party 43, 56
social media 146, 151, 153–7, 194
societal anxiety 141–2, 211–21
societal failures 51–2
Sohyo (General Council of Trade Unions of Japan) 56
soil contamination 125, 126; *see also* food contamination
South Korea 92, 93
special event declarations 6, 7, 9
SPEEDI (the System for Prediction of Environment Emergency Dose Information) 72, 120–2, 145, 179–80
spent fuel pools *xxxi*, 4, 31–3, 35, 164–5, 169, 176–7, 185, 209–10
sports stadiums 75–6
STA *see* Science and Technology Agency
state of nuclear emergency 8
state-planned privately operated approach (*kokusaku minei*) 39–40, 51, 78–9, 184–6
station blackouts 6, 107–11, 179
stress tests 60–1
strontium 125

structural strength 118–19
students, foreign 152–3
subsidies 44, 58, 62, 75–6
suicide squad 16, 185–6
supermarkets 130
sustainability standards 81
Sweden 119

Tadaki, Keiichi (committee member) 197–9
Taiwan 110, 215
Tajima, Eizo 42–3
Takagi, Jinzaburo 44
Takekuro, Ichiro 8, 12, 13, 15, 22–3
Tanaka, Kakuei 44
Tasaka, Hiroshi 146
taxation 58, 62, 75
TEPCO (Tokyo Electric Power Company) 1, 73–85, 183, 200; awareness of safety risks 84–5; Cabinet distrust 9, 13, 20, 203, 224; communication problems 29–30, 163, 189–90, 224; contaminated water release 93–4; cost reductions 77–8; culture of concealment 81–3; evacuation of workers 27–30, 185, 190; falsification scandal 46, 48, 69–70, 71, 81–2, 83–4, 87–8; financial incentives for siting plants 75–6; and international peer reviews 88, 91; local trust 59; severe accident preparedness 178–9; tsunami preparedness 2, 78, 79–81, 101–2, 105, 181–2; Twitter account 157; US assistance 163–4, 165, 170, 172; *see also* accident management, Fukushima Daiichi; Takekuro, Ichiro
TEPCO labor union 55
Terada, Manabu 8, 20, 29–30
Terasaka, Nobuaki 8, 149
terrorism 186–7
Three Mile Island 46, 51, 56, 112–13, 208, 210, 211, 217
thyroid cancer 100, 207
thyroid protection 206–8
Tohoku earthquake and tsunami *xxxii*, 3–6, 106, 109
Tohoku Electric Power Company 181–2

Index 247

Tokai Daini Nuclear Power Station xxvii, 39, 81, 105–6
Tokai-mura nuclear accident 47, 51, 56, 122
Tokyo: water restrictions 128, 133–4; in worst-case scenario 35
Tokyo Electric Power Company *see* TEPCO
Toshiba 39, 77, 83
tourists 153
TranSAS (Transport Safety Appraisal Services) 87
trust, breakdowns of 188, 202–3; Cabinet and TEPCO 9, 13, 20, 203, 224; public 145–6, 148–9, 203
tsunamis: accident preparedness 2, 78, 79–81, 101–6, 107, 181–2; height predictions *xxxii*, 2, 5, 79–81, 101–2, 105–6, 181–2; historical events 101–2; protection works 80–1, 105–6; regulatory standards 2, 5, 101; simulations 78, 79–81, 85, 104, 182; Tohoku tsunami *xxxii*, 3, 5–6, 106, 109
turnkey agreements 40
Twitter 151, 154, 156–7

Ukraine 207
Union of Concerned Scientists, US 205–6
unions, labor 54–5, 56
Unit 1 reactor: actions to save *xxxv*, 4, 11, 14–15, 16–19, 22–4, 32, 33; construction 40; cooling system *xxxiii*, 5, 9–11; electricity supply 6, 9, 33; explosion *xxx*, 19, 21; meltdown *xxxiv*, 11, 19
Unit 2 reactor: actions to save 4, 26–7, 33; cooling system 11, 26; electricity supply 6, 17, 33; suspected explosion 27, 30
Unit 3 reactor 4, 24–5, 33
Unit 4 reactor 27, 32–3, 35, 164–5, 176–7
Unit 5 reactor 4, 6, 33–4
Unit 6 reactor 4, 6
United Kingdom 39, 41
United Nations 37

United Nations Scientific Committee on the Effects of Atomic Radiation (UNSCEAR) 140, 142
United States: Atoms for Peace program 36, 39; communication problems 159, 161–6; critique of nuclear regulation 214–16; disaster assistance 158–9, 160–1, 171–2; Emergency Planning Zones 43; evacuation of citizens 159, 165–9; future of alliance with Japan 173–4; nuclear security guidelines 186–7; nuclear tests 37; potassium iodide 207; probabilistic safety assessments 96–7, 113, 114, 119, 210–11; safety myth 113, 205; severe-accident guidelines 114, 215; spent fuel pools 209; station blackout regulations 108–9; sustainability standards 81; Three Mile Island 46, 51, 56, 112–13, 208, 210, 211, 217; US–Japan coordination conferences 170–2
UNSCEAR *see* United Nations Scientific Committee on the Effects of Atomic Radiation
uranium 4; *see also* fuel rods

venting *xxxv*, 11, 12–13, 14–15, 16–17, 18–19, 25, 27
VERTIC 92
von Hippel, Frank (review of report) 204–12

WANO *see* World Association of Nuclear Operators
WASH-1400 report 96–7
waste treatment 137–8
Watanabe, Toshitsuna 14
water cannons 33
water contamination: drinking water *xxxvii*, 127–8, 133–4, 157; leaks at plant 34; ocean 93–4, 125
Weightman, Michael 91
Wen Jiabao 92
wet venting 11, 13
Windscale, UK 41
workers: dangers from radiation 18–19; evacuation of 27–30, 185, 190;

injuries 19, 25–6; suicide squad 16, 185–6; *see also* accident management, Fukushima Daiichi
World Association of Nuclear Operators (WANO) 87, 91
worst-case scenarios: of disaster 34–5, 176–7, 192–3, 209; for tsunamis 2, 5

Yagi, Makoto 57
Yamaji, Kenji (committee member) 202–3

Yanase, Tadao 22–3
Yasui, Masaya 28
Yomiuri Shimbun 55–6
Yoshida, Masao 9, 10, 13, 28, 177, 200, 222; Kan's visit 15, 16; seawater cooling 18, 23, 25, 188–9
Yoshida, Shigeru 36
Yoshioka, Hitoshi 53

zirconium 4, 11, 19, 25, 32